Springer Complexity

Springer Complexity is an interdisciplinary program publishing the best research and academic-level teaching on both fundamental and applied aspects of complex systems—cutting across all traditional disciplines of the natural and life sciences, engineering, economics, medicine, neuroscience, social and computer science.

Complex Systems are systems that comprise many interacting parts with the ability to generate a new quality of macroscopic collective behavior the manifestations of which are the spontaneous formation of distinctive temporal, spatial or functional structures. Models of such systems can be successfully mapped onto quite diverse "real-life" situations like the climate, the coherent emission of light from lasers, chemical reaction-diffusion systems, biological cellular networks, the dynamics of stock markets and of the Internet, earthquake statistics and prediction, freeway traffic, the human brain, or the formation of opinions in social systems, to name just some of the popular applications.

Although their scope and methodologies overlap somewhat, one can distinguish the following main concepts and tools: self-organization, nonlinear dynamics, synergetics, turbulence, dynamical systems, catastrophes, instabilities, stochastic processes, chaos, graphs and networks, cellular automata, adaptive systems, genetic algorithms and computational intelligence.

The three major book publication platforms of the Springer Complexity program are the monograph series "Understanding Complex Systems" focusing on the various applications of complexity, the "Springer Series in Synergetics", which is devoted to the quantitative theoretical and methodological foundations, and the "Springer Briefs in Complexity" which are concise and topical working reports, case studies, surveys, essays and lecture notes of relevance to the field. In addition to the books in these two core series, the program also incorporates individual titles ranging from textbooks to major reference works.

Editorial and Programme Advisory Board

Understanding Complex Systems

Founding Editor: S. Kelso

Future scientific and technological developments in many fields will necessarily depend upon coming to grips with complex systems. Such systems are complex in both their composition–typically many different kinds of components interacting simultaneously and nonlinearly with each other and their environments on multiple levels–and in the rich diversity of behavior of which they are capable.

The Springer Series in Understanding Complex Systems series (UCS) promotes new strategies and paradigms for understanding and realizing applications of complex systems research in a wide variety of fields and endeavors. UCS is explicitly transdisciplinary. It has three main goals: First, to elaborate the concepts, methods and tools of complex systems at all levels of description and in all scientific fields, especially newly emerging areas within the life, social, behavioral, economic, neuro- and cognitive sciences (and derivatives thereof); second, to encourage novel applications of these ideas in various fields of engineering and computation such as robotics, nano-technology, and informatics; third, to provide a single forum within which commonalities and differences in the workings of complex systems may be discerned, hence leading to deeper insight and understanding.

UCS will publish monographs, lecture notes, and selected edited contributions aimed at communicating new findings to a large multidisciplinary audience.

More information about this series at http://www.springer.com/series/5394

Themis Matsoukas

Generalized Statistical Thermodynamics

Thermodynamics of Probability Distributions and Stochastic Processes

Springer

Themis Matsoukas
Chemical Engineering
Pennsylvania State University
University Park
PA, USA

ISSN 1860-0832 ISSN 1860-0840 (electronic)
Understanding Complex Systems
ISBN 978-3-030-04148-9 ISBN 978-3-030-04149-6 (eBook)
https://doi.org/10.1007/978-3-030-04149-6

Library of Congress Control Number: 2018964605

This Springer imprint is published by the registered company Springer Nature Switzerland AG.
The registered company address is: Gewerbestrasse 11, 6330 Cham, Switzerland

To Kristen and Melina

Preface

The problem that motivates this undertaking is evolution and change in generic populations. We imagine individuals who come together and form groups or clusters. Groups gain or lose members; they merge or split; or they may undergo any combination of these transformations simultaneously. These are the primitive processes that shape the distribution of individuals among clusters. The "individuals" (units) may be physical particles coupled via a bath, people waiting in a queue, nodes on a graph, or geographical areas that become infected with a disease. The underlying physics[1] that gives rise to the distributed nature of the population is unimportant. It is the *distribution* that is the object of interest. The physical problems that motivate our pursuit are many and diverse: atomic clusters, colloidal particles, ecosystems, and galaxies. They are all examples of populations that are distributed with respect to some attribute that we may call *size*. The goal is to present a *thermodynamic calculus* for generic populations, in fact, a thermodynamic calculus for *distributions*.

Why *Thermodynamics*?

We typically associate thermodynamics with the equilibrium state of matter. Ultimately, however, thermodynamics is about a *probability distribution*: the distribution of microstates. The main thesis of this work is that the web of mathematical relationships we recognize as thermodynamics is a probabilistic calculus that exists independently of physics. Our goal is to present a theory of thermodynamics that

[1] We use *physics* in a broader sense to refer to the rules for assigning probabilities to the clusters in the population. In a system of interacting particles these rules are set by the physics of interactions. In more abstract cases, such as random networks, they are set by the *model* we adopt for the process.

stands on its own feet and is completely independent of the particular problem to which it may be applied. We call this theory *generalized statistical thermodynamics*.

Gibbs introduced the notion of the ensemble as a collection of identical systems of material particles governed by identical and fully deterministic laws, whose initial conditions are neither known nor knowable (Gibbs 1902). Every feasible microstate i is present with probability p_i. Gibbs obtained the probability distribution at equilibrium and discovered a remarkable property: It maximizes the *entropy functional*,

$$S[p] = -\sum_i p_i \log p_i,$$

with respect to all probability distributions that are consistent with the observable state. Shannon arrived at the same functional from quite a different perspective in pursuit of an unrelated problem, the transmission of signals (Shannon 1948). He obtained entropy as a measure of the *uncertainty* associated with probability distribution p_i. The more information we have about a random variable, the less the uncertainty and the higher the chance of predicting its behavior. Conversely, if we are seeking to determine p_i, we are obligated, once all available evidence has been factored in, to choose the distribution with the *maximum* entropy. Picking any other distribution amounts to *assuming* information that is not present in the data. This is the principle of *maximum entropy* as articulated by Jaynes (1957).

Here is how this principle is put to work. Suppose we have a discrete random variable x_i and seek its probability distribution p_i but all we know is its mean $\bar{x}$. The unknown distribution satisfies the conditions

$$\sum_i p_i = 1, \quad \sum_i p_i x_i = \bar{x},$$

which express everything we know about p_i: it is a normalized distribution and its mean is $\bar{x}$. The number of distributions that satisfy these conditions is infinite, and the problem of determining p_i is intractable. This indeterminacy is resolved by requiring that the distribution maximize the entropy functional. The mathematical problem is the maximization of the entropy functional subject to the two constraints on the probability distribution p_i. The maximum entropy distribution under these constraints is

$$p_i = \frac{e^{-\beta x_i}}{q},$$

where β and q are Lagrange multipliers that correspond to the two constraints on p_i. We calculate them by back-substituting p_i into the constraints:

$$q = \sum_i e^{-\beta x_i}, \quad \bar{x} = \frac{1}{q}\sum_i x_i e^{-\beta x_i} = -\frac{\partial q}{\partial \beta}.$$

These results, obtained with no reference to material particles, are identical to those we encounter in statistical thermodynamics. Jaynes saw a larger principle hiding in this result: If we adopt the constrained maximization of entropy as the fundamental postulate, under constraints that serve to narrow the search based on what we know about the unknown distribution, we have a generalized method of inference under incomplete information. Statistical mechanics is then obtained as a special application. Combining different constraints appropriately, we obtain various other distributions. Practically any distribution can be derived as a maximum entropy distribution under suitable constraints.

The maximum entropy method has a combinatorial interpretation that brings us back to Gibbs's notion of ensemble. Suppose we have a large but finite population of size M, which we divide into N groups ("clusters") such that $M/N = \bar{x}$ is fixed. We construct the ensemble of all possible ways that M can be subdivided into N ordered groups and *assert that any partition is as likely as any other.* As an immediate consequence, certain *distributions* are more likely than others. If $\mathbf{n} = (n_1, n_2, \cdots)$ is the distribution of clusters, where n_i is the number of clusters that contain i members, this distribution comes with a multiplicity that counts the number of permutations in the order in which cluster sizes appear, as if they were drawn randomly from an urn that contains every cluster of that distribution. This multiplicity is given by the multinomial factor

$$\mathbf{n}! = \frac{N!}{n_1!n_2!\cdots}.$$

Distributions with large diversity of values have a higher multinomial coefficient and higher probability compared to distributions of lower diversity. The most probable distribution in this ensemble is the one that maximizes the multinomial coefficient under the constraints

$$\sum_i n_i = N; \quad \sum_i i n_i = M; \quad k = 1, 2, \cdots .$$

This constrained maximization leads to the same exponential distribution as the one we obtained for p_i—not surprisingly, since the log of the multinomial coefficient *is* the entropy functional,

$$\log \mathbf{n}! = -N \sum_i \frac{n_i}{N} \log \frac{n_i}{N} \equiv S(\mathbf{n}).$$

Gibbs, Shannon, and Jaynes are working with a common calculus, and the nexus is entropy.

Jaynes developed maximum entropy as a tool for statistical inference, but this principle has an appeal with a far-reaching implication: It suggests that the formalism of statistical thermodynamics can be extended to *any* problem that involves unknown distributions, whether this is the probability distribution of a sparsely sampled random variable or the size distribution of colloidal clusters undergoing aggregation under some rate law. If we could generalize Jaynes's method, we could make the toolbox of thermodynamics—a very powerful toolbox—available to a host of new problems. Of particular interest are problems that involve the formation of a *giant component*, a large-scale coherent structure that is often seen to emerge from finite-size populations: polymer gelation, forest fires, epidemics, network connectivity, and monopolies are examples of systems that exhibit this dynamic behavior. The common feature in all of these is the emergence of a percolated state that coexists with the un-percolated population from which it emerged, much like a large infectious outbreak in the midst of smaller pockets of infected individuals. This process has the hallmarks of a thermodynamic *phase transition*, at least at a qualitative level. The possibility that such connection could be made rigorously is intriguing.

But there is a problem: Maximum entropy offers no organized methodology to construct constraints that represent our knowledge for these types of problems. Here is an example: Suppose that a population of clusters grows by adding one member at a time with probability $P_{i \to i+1} \propto i^\alpha$, where i is the size of the cluster that receives a new member. With $\alpha = 0$ (size-independent growth), the long-time distribution is Gaussian. With $\alpha = 1$ (size-proportional growth), the distribution is exponential.[2] Both results can be derived via maximum entropy arguments by constraining the mean size (exponential distribution) or the variance (Gaussian distribution), but there is no obvious connection between these constraints and the rate laws that govern the physical problem. We lack a principle that would allow us to convert a *physical model* (here, a rate law) into a set of "constraints" of the type needed in the maximum entropy formalism. We need a different approach.

For guidance we turn to statistical mechanics, which we regard, as did Jaynes, as a successful but narrow demonstration of a more powerful yet unclear principle. How does *physics* enter into the development of the thermodynamic ensembles? The textbook derivation of the canonical ensemble goes like this. We form an ensemble of microstates i, each with energy E_i, volume V_i, and number of particles N_i such that the total energy E in the ensemble, total volume V, and total number of particles N are fixed. From this ensemble we randomly draw a large number of systems with distribution of microstates $\mathbf{n} = (n_1, n_2, \cdots)$, where n_i is the number of systems in microstate i. We assert that the probability of distribution $\mathbf{n}$ is proportional to the multinomial factor $\mathbf{n}!$ and proceed to identify the most probable distribution by maximizing multiplicity under the constraints that define the ensemble:

$$\sum_i n_i E_i = E, \quad \sum_i n_i V_i = V, \quad \sum_i n_i N_i = E.$$

The result is the generalized canonical distribution, an exponential distribution in the energy E_i, volume V_i, and number of particles N_i of microstate i:

$$\tilde{n}_i \propto e^{-\beta E_i - \alpha V_i - \gamma N_i},$$

[2]See Matsoukas and Lin (2006).

with β, α, and γ identified as measurable quantities related to temperature, pressure, and chemical potential:

$$\beta = 1/kT, \quad \alpha = p/kT, \quad \gamma = -\mu/kT.$$

The generalized canonical distribution generates all relationships of statistical thermodynamics, and yet it is remarkable how little input from physics is required to derive it. None of the physical details of molecular interactions matter. These are of course very important when we seek to calculate macroscopic properties from molecular models of matter but completely irrelevant in deriving the methodology that is used for their calculation. Where is physics in this derivation?

Physics has entered via the postulate of equal a priori probabilities, a *model assumption* that renders all microstates equally probable. The postulate allows us to take the probability of distribution $\mathbf{n}$ to be proportional to the multiplicity factor $\mathbf{n}!$ and thus identify the most probable distribution as the distribution with the maximum multiplicity. Under this more careful reading of the derivation, we recognize that our knowledge and assumptions about the system—the *physical model*—are encoded not in the constraints but in the rule that prescribes the *assignment of probabilities*. Suppose we adopt a different physical model by introducing a probability that is nonuniform but biased somehow toward certain microstates. This model would produce a different distribution of microstates and would lead to new predictions for the thermodynamic behavior of the system. Whether such revised model would be appropriate for material systems is not the point. The point is that, if our assumptions about the system change but our observations do not, we must reassign probabilities but leave the constraints alone.

The Cluster Ensemble

We are ready now to state the basic premise of our theory. The system of interest is defined by a small number of external extensive properties (*observables*) that are consistent with a very large number of internal variables (*configurations*) that are distributed according to any possible distribution $\mathbf{n}$. We form the ensemble of all possible distributions that are consistent with fixed values of the extensive variables and assign probabilities so that the probability of distribution $\mathbf{n}$ is proportional not only to the multiplicity $\mathbf{n}!$ but also to a *selection functional* $W(\mathbf{n})$. The case $W(\mathbf{n}) = 1$ corresponds to the postulate of equal a priori probabilities. By allowing freedom in the selection of W, we have the flexibility to apply a *bias* in the sampling of distributions from the ensemble and obtain a most probable distribution other than the exponential. To state the problem mathematically, we consider the simplest possible ensemble that is formed when M identical units are distributed among N groups. The extensive variables that define the state are M and N; the internal degrees of freedom are the various ways in which we can satisfy the external state by rearranging the members of the populations between the groups. All distributions in the ensemble satisfy the two fundamental constraints:

$$\sum_i n_i = N, \quad \sum_i i n_i = M.$$

For the probability of distribution $\mathbf{n}$, we write

$$P(\mathbf{n}) = \mathbf{n}! \frac{W(\mathbf{n})}{\Omega_{MN}},$$

where Ω_{MN} is the partition function. Thermodynamics arises naturally upon letting M and N become large. By *thermodynamics* we refer to the *calculus* that governs the most probable distribution in this ensemble. This calculus appears as a set of relationships between the primary quantities of the ensemble, M, N, and W, the most probable cluster distribution $\tilde{\mathbf{n}} = (\tilde{n}_1, \tilde{n}_2, \cdots)$, and the *partition function* Ω. We may construct a Markov process that transitions reversibly between distributions of the cluster ensemble with probability such that its stationary distribution is the same as the most probable distribution of the cluster ensemble. We may view then the most probable distribution of the static ensemble as the *equilibrium* distribution of the associated kinetic Markov ensemble. This mapping between a biased static ensemble of distributions and a kinetic ensemble in which distributions are converted into each other via reversible reactions allows us to pass the entire calculus of equilibrium thermodynamics to the cluster ensemble independently of any details about the assignment of probabilities. It is so even if the population undergoes changes that are unidirectional in real time, i.e., irreversible in the *physical* sense. In Chaps. 8, 9, and 10, we obtain results for precisely such irreversible processes. When we talk of *equilibrium* or *reversibility* in the cluster ensemble, it is *not* with respect to real time; it is with respect to the sampling of the phase space via a reducible Markov process that has converged to the most probable distribution.

How is this different from the standard maximum entropy method? In two important ways. First, the ensemble is always subject to the same *two* constraints that define the observable state of the system. These constraints, and these only, are always present, regardless of the physical details that govern the system. All other information is represented by W, a functional that shapes the probability space of distributions. Second, this approach takes the focus away from entropy and puts it on the partition function Ω_{MN}, which expresses the composite effect of entropy and of the selection functional. This is the central quantity of the ensemble, a function whose variational properties define the "equilibrium" state of the population. The partition function and its partial derivatives together produce a network of relationships that we will recognize as "thermodynamic relationships," as well as criteria for coexistence between two distinct populations—a *phase* transition that we will demonstrate. These ideas suggest that thermodynamics is a stochastic calculus that arises naturally in stochastic systems.

Our goal with this book is twofold: first, to derive a thermodynamic calculus independently of physical interactions using as a basis the mathematical construct of the cluster ensemble and second, to show how this formalism may be applied to systems stochastic processes other than the microstate of matter. The specific stochastic problem that motivates this work is population balances and is the problem that provides the case studies examined in the book. The theory however is far more general and applicable to *any* stochastic process that exhibits extensive behavior with respect to the scale of the system. The hope is to open new doors and offer fresh insights to these problems.

Here is how the material is organized:

- Chapter 1 sets the background. Here we review the central concept of familiar thermodynamics, *entropy*, and discuss the work of Gibbs, Shannon, and Jaynes as they relate to our development.
- In Chap. 2 we formulate the core of the theory in the discrete phase space of the cluster ensemble. Like Bernoulli's urn, the cluster ensemble is a mathematical construct that allows us to visualize the phase space as a box that contains every conceivable distribution—a mother of all distributions, of sorts.
- In Chap. 3 we introduce the thermodynamic limit (ThL), define the most probable distribution (MPD), and derive the thermodynamic relationships of the ensemble.
- In Chap. 4 we discuss the most probable distribution in greater detail and consider the inverse problem, namely, how much information about the selection functional we may recover from the knowledge of the distribution itself.
- In Chap. 5 we formulate stability criteria. We define the giant cluster (analogous to the giant component in networks and the gel phase in polymerization) and show that it represents a formal thermodynamic phase. Specifically, we construct the tie line of a two-phase population using the familiar tools of vapor-liquid equilibrium.
- In Chap. 6 we extend the theory to bicomponent populations. We define the selection functional for random mixing and express special cases of nonrandom mixing through departure functions analogous to excess functions in solution thermodynamics.
- In Chap. 7 we reformulate the theory in the language of calculus of variations in a way that makes no reference to the cluster ensemble. The core of the theory is a thermodynamic functional that generates the entire network of thermodynamic relationships for any probability distribution.

The remaining three chapters address applications of the theory to stochastic processes, two problems in particular, binary aggregation and binary fragmentation, both irreversible in real time, both of which exhibit a phase transition.

- In Chap. 8 we formulate the problem of irreversible binary aggregation in the language of the cluster ensemble. We construct the discrete phase space, obtain the partition function for arbitrary aggregation kernels, establish the connection to the classical Smoluchowski theory of aggregation, and solve two standard problems in this field, the constant and the sum kernel.
- In Chap. 9 we consider the problem of kinetic gelation for the product kernel. This problem has attracted a large body of literature by virtue of its strong analogies to familiar phase transitions. We show gelation is a true phase transition and use the tools of Chap. 5 to construct the phase diagram in the pre-gel and post-gel regions.
- In Chap. 10 we formulate the problem of binary fragmentation. This irreversible process advances on the same phase space as aggregation, only in the reverse direction, and exhibits a phase transition, *shattering*, analogous to gelation in aggregation. We obtain the partition function of this system and show how to study shattering using stability analysis.

E. T. Jaynes: A Personal Tribute

The intellectual trajectory from which this work branches out begins with Gibbs, who had the insight to move past dynamics to bring the focus to the ensemble of all internal states that are possible under the specified observables. Gibbs removed *time* from the picture.[3] By doing so all noise subsides and out of chaos emerges an orderly and elegant construct that we call statistical mechanics. Shannon stumbled on Gibbs' main piece of machinery, entropy, while in pursuit of seemingly unrelated goals. By doing so he brought entropy outside the realm of the material world. Jaynes saw these dots to be connected and declared entropy maximization to be the universal and unifying principle from which statistical mechanics and information theory both arise as special cases. The evolution of ideas from Gibbs to Shannon and Jaynes is nothing less than fascinating; but it is Jaynes in particular to whom this work owes a special intellectual debt. Jaynes took the ideas of statistical thermodynamics and used them to build a bridge to an unrelated universe, statistical inference. I view the main contribution of my work to be the bridging two worlds that have existed in separation—thermodynamics and stochastic processes. In his paper on information theory and statistical mechanics, Jaynes writes (Jaynes 1957, the emphasis is mine):

> ...we have now reached a state where *statistical mechanics* is no longer dependent on physical hypotheses, but may become simply an example of *statistical inference*.

[3]Boltzmann, the other giant in this field and the first to point out the connection between probability distributions and entropy, chose to pursue the connection between thermodynamics and *time*, a direction that had little influence in our project.

I would like to paraphrase Jaynes:

> We have now reached a state where *statistical thermodynamics* is no longer dependent on physical hypotheses, but may be shown to be a *probabilistic calculus* that applies to stochastic processes in general.

University Park, PA, USA
August 2018

Themis Matsoukas

Acknowledgments

When I walked into my first undergraduate thermodynamics class as an assistant professor at Penn State, 27 years ago almost to the day these lines are written, I could not have imagined it would be the beginning of the long, winding, and rewarding trajectory that today brings me to this book. My first thanks go to my numerous students, undergraduate and graduate, whose expected and unexpected questions, misconceptions, and sudden insights taught me to think hard before answering their questions and to keep digging deeper into this material until it all felt so clear. It is because of them that I learned thermodynamics. As a graduate student at the University of Michigan, I became interested in population balances, coagulation, gelation, and the Smoluchowski equation, seeds from which this book has sprouted. I had the good fortune to learn from Bob Ziff, a teacher and mentor, who sustained my interest in these problems and helped keep the physicist inside me alive. During the preparation of the manuscript, I often had to rely on two technical communities for answers to questions of immediate urgency: the LaTeX community and the community of *Mathematica*. I am thankful to both. The team at Springer has been great to work with. Chris Coughlin, my editor, offered continuous support and flexibility throughout this project. During the year of writing, on sabbatical from Penn State, I had the opportunity to visit the Chemical Engineering Department at the Pontificia Universidad Católica de Valparaíso where I presented a series of short workshops on thermodynamics at the invitation of Prof. Cristian Antonnuci. I am grateful to Cristian for making this an unforgettable visit. Parts of the book were written in Athens and Arta, thanks to the warm hospitality of Christos, Angelos, and Diana. Sean Cotner graciously initiated me into the mysteries of σ-algebra and other mathematical esoterica. My wife Kristen and daughter Melina indulged me with patience during my year of sabbatical. Our trip to Sydney together provided a memorable setting for some inspired writing. Throughout the year it took to complete the manuscript, in near solitary confinement in my armchair with my laptop, two friends never left my side. Lulu, my little dog, sat to my left and probably read most of the book. Beau, my cat, perched on my right, walked on the keyboard enough times that I will gladly credit her with coauthoring the typos in the book.

Contents

1 Evolution of Ideas on Entropy ... 1
1.1 Classical Entropy ... 1
1.2 Gibbs Entropy ... 4
1.3 Shannon Entropy ... 7
1.4 Jaynes Entropy ... 11
1.5 Entropy and Probability Distributions ... 13

2 The Cluster Ensemble ... 23
2.1 Microcanonical Ensemble ... 23
2.2 Entropy and Multiplicity ... 26
2.3 Selection Functional ... 31
2.4 Connection to the Multinomial Distribution ... 35
2.5 Entropies in the Cluster Ensemble ... 38
2.6 The Binary Exchange Reaction ... 39
2.7 Linear Ensembles ... 43
2.8 The Unbiased Ensemble ... 46
2.9 Canonical Ensemble ... 49
2.10 M'-Canonical Ensemble ... 61

3 Thermodynamic Limit (ThL) ... 65
3.1 Convergence in the ThL ... 65
3.2 Most Probable Distribution (MPD) ... 68
3.3 The Microcanonical Equations ... 72
3.4 Canonical Probability in the Thermodynamic Limit ... 74
3.5 M-Canonical Probability in the ThL ... 80
3.6 Convergence of Ensembles ... 86
3.7 Microcanonical Surface and the Quasistatic Process ... 87
3.8 Generalized Cluster Ensemble ... 88

4 The Most Probable Distribution in the Continuous Limit ... 99
4.1 Properties in the Continuous Limit ... 100
4.2 Unbiased Ensemble ... 102
4.3 Linear Ensemble with Power-Law Bias ... 106
4.4 Nonlinear Ensembles ... 110
4.5 The Inverse Problem ... 116
4.6 Canonical Representation of Some Common Distributions ... 119

5 Phase Transitions: The Giant Cluster ... 125
5.1 Variational Properties of the Partition Function ... 125
5.2 The Equilibrium Phase ... 128
5.3 The Giant Cluster ... 129
5.4 Sol-Gel Equilibrium ... 133
5.5 A Case Study: Linear Ensemble with $w_i = i^{-3}$ ... 135
5.6 Nucleation, Fluctuations, and the Order Parameter ... 148
5.7 An Abrupt Phase Transition: The i^2 Model ... 151

6 The Bicomponent Ensemble ... 163
6.1 Bicomponent Cluster Ensemble ... 163
6.2 Microcanonical Probability ... 170
6.3 Selection Functional ... 175
6.4 The Bicomponent MPD ... 176
6.5 The Sieve-Cut Ensemble ... 179
6.6 Random Mixing ... 183
6.7 Nonrandom Mixing ... 188

7 Generalized Thermodynamics ... 197
7.1 Generating Distributions by Random Sampling ... 197
7.2 Biased Sampling ... 199
7.3 Canonical Phase Space ... 202
7.4 Microcanonical Phase Space ... 203
7.5 Entropy ... 205
7.6 Curvature ... 208
7.7 The Linearized Selection Functional ... 213
7.8 What *Is* Thermodynamics? ... 218
7.9 Contact with Statistical Mechanics ... 220
7.10 Contact with the Maximum Entropy Principle ... 221
7.11 What Is W? ... 227

8 Irreversible Clustering ... 241
8.1 The Binary Clustering Graph ... 242
8.2 Transition Probabilities ... 244
8.3 Parent–Offspring Relationships ... 246
8.4 The Master Equation ... 250
8.5 Partition Function and Canonical Parameters ... 252

8.6 Exact Results: Linear Ensembles ... 254
8.7 Mean Distribution ... 268
8.8 Connection to the Smoluchowski Theory ... 270
8.9 Continuous Domain ... 274
8.10 Contact with the Literature ... 278

9 Kinetic Gelation ... 289
9.1 Product Kernel ... 290
9.2 Stability and Phase Diagram ... 293
9.3 Gel Branch and the Sol-Gel Tie Line ... 297
9.4 Monte Carlo Simulation of Kinetic Gelation ... 299
9.5 A Closely Related Linear Ensemble ... 304
9.6 Flory and Stockmayer (But Mostly Stockmayer) ... 308
9.7 Power-Law Kernels ... 313
9.8 Contact with the Literature ... 319

10 Fragmentation and Shattering ... 325
10.1 The Process ... 325
10.2 The Partition Function of Fragmentation ... 330
10.3 An Exact Solution: Unbiased Ensemble ... 333
10.4 Mean Distribution ... 335
10.5 Homogeneity in the Thermodynamic Limit ... 336
10.6 Shattering: A Phase Transition ... 341

List of Symbols ... 349

References ... 351

Index ... 355

Chapter 1
Evolution of Ideas on Entropy

Our stated goal is to develop a general theory of thermodynamics that we may apply to stochastic processes, but what *is* thermodynamics? At the mathematical level thermodynamics is the calculus of entropy. The inequality of the second law forms the starting point for a large number of mathematical relationships that are written among the set of primary variables, entropy, energy, volume and number of particles, and a number of defined functions based on the primary set. The mathematical framework of classical thermodynamics is established as soon as the second law is formulated in mathematical form. The subsequent development of statistical mechanics left this framework intact, while making the connection to the microscopic structure of matter. Even as the basic mathematical framework did not change, the revolution lead by Gibbs introduced a statistical view of entropy that opened an entirely new viewpoint. Starting with Shannon's work on information theory and Jaynes's formulation of maximum entropy has now escaped from the realm of physics to invade other fields. In this chapter we review the evolution of ideas on entropy, from the classical view, to Gibbs, Shannon, and Jaynes, and cover the background that gives rise to our development in the chapters that follow.

1.1 Classical Entropy

Classical thermodynamics is a self-contained subset of thermodynamics that can be formulated with no reference to constitutive theories of matter. And even though classical thermodynamics is superseded by statistical mechanics, there is an advantage in starting with the classical formulation of entropy, in that this approach minimizes the need to talk about physical matter in any detail and this allows us to focus on the attributes of thermodynamics independently of the particulars of

T. Matsoukas, *Generalized Statistical Thermodynamics*,
Understanding Complex Systems, https://doi.org/10.1007/978-3-030-04149-6_1

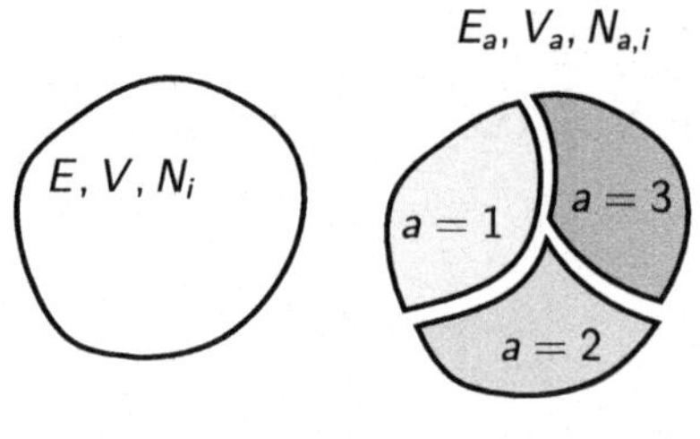

Fig. 1.1 Schematic partitioning of system with fixed energy E, volume V, and number of particles N_k of type k into two noninteracting subsystems

the physical system to which it is applied. The classical state of a mixture of K components is defined by a set of extensive variables, energy E, volume V, and number of particles N_k of component $k = 1, 2 \cdots K$. The *equilibrium* state is given by entropy, a homogeneous function of E, V, and N_k with degree 1, such that[1] $S(E, V, N_k)$ is at a maximum with respect to any partitioning of energy, volume, and number of particles into $\mathcal{N}$ noninteracting parts, as shown schematically in Fig. 1.1. That is,

$$S(E, V, N_k) = \max_{\{E_a, V_a, N_{k,a}\}} \sum_a S(E_a, V_a, N_{k,a}), \tag{1.1}$$

under the constraints

$$\sum_a E_a = E, \quad \sum_a V_a = V, \quad \sum_a N_{k,a} = N_k. \tag{1.2}$$

A set $\{E_a, V_a, N_{k,a}\}$ that satisfies the constraints in Eq. (1.2) represents a feasible partitioning of the extensive set $\{E, V, N_k\}$.[2] At equilibrium the total entropy of the parts is a function of the total E, V, and N_k, rather than of the particular distribution of these totals over the parts. By virtue of the homogeneity condition $S(E, V, N_k)$ satisfies Euler's theorem (see Appendix),

$$S(E, V, N_k) = E\left(\frac{\partial S}{\partial E}\right)_{V,N_k} + V\left(\frac{\partial S}{\partial V}\right)_{E,N_k} + \sum_k N_k\left(\frac{\partial S}{\partial N_k}\right)_{V,N_{j\neq k}}, \tag{1.3}$$

which constitutes the fundamental equation of thermodynamics. We express this in condensed form as

$$S(E, V, N_k) = \beta E + \alpha V + \sum_k \gamma_k N_k, \tag{1.4}$$

---

[1]We write $S(E, V, N_k)$ as a shorthand for $S(E, V, N_1, \cdots)$. Summations over k are understood to go over the total number of components.

[2]The index $a = 1, 2, \cdots \mathcal{N}$ counts the number of subsystems to which the larger system is divided. The number $\mathcal{N}$ of partitions may be arbitrarily large as long as each subsystem contains enough molecules to be treated as a continuum.

where β, α, and γ_k stand for the corresponding partial derivatives, all of which are homogeneous functions of E, V, and N_k with degree 0. The differential of entropy is

$$dS(E, V, N_k) = \beta dE + \alpha dV + \sum_k \gamma_k dN_k. \tag{1.5}$$

Equation (1.1) expresses the second law: it is the foundational postulate of classical thermodynamics that establishes the variational properties of entropy and defines the equilibrium state; Eqs. (1.4) and (1.5) taken together[3] form the basis for all thermodynamic relationships, by which we mean any relationships that can be written among the group of variables

$$S,\ E,\ V,\ N_k,\ \beta,\ \alpha,\ \gamma_k.$$

and any other functions that may be defined in terms of these variables.

Note 1.1 (Contact with Physics)

The mathematical formulation of thermodynamics given here has required input from the physical world that was not explicitly acknowledged. The first point of contact with physics has been made via two assumptions about the nature of the state of matter, both based on empirical observation: First, that the macroscopic state is described in terms of energy, volume, and number of particles of each type present; and second, that the intensive state of matter in equilibrium remains unchanged when all extensive variables are multiplied by the same factor. The former condition establishes entropy as a function of (E, V, N_k) and the latter establishes its *homogeneity* with respect to these variables with degree 1. Another point of contact is needed in order to establish the physical meaning of β, α, and γ_k. We do this by considering a closed system of nonreactive particles. In this case $dN_k = 0$ and Eq. (1.5) solved for dE gives

$$dE = \frac{dS}{\beta} - \frac{\alpha}{\beta} dV. \tag{1.6}$$

Since $-pdV$ represents mechanical work, we interpret this relationship as a statement of energy conservation for a quasistatic process (dE, dV) on a closed system: we identify α/β as *pressure*, dS/β as *heat*, and β as inverse temperature $1/k_B T$ where k_B is the Boltzmann constant. Finally, based on the standard definition of the chemical potential of component k,

$$\mu_k = -k_B T \left(\frac{\partial S}{\partial n_k}\right)_{E,V,N_{j\neq k}},$$

[3] Equations (1.4) and (1.5) taken together are equivalent to the condition that entropy is homogeneous in E, V, and N_k with degree 1.

we identify $\gamma_k = -\mu_k/k_B T$. Returning to the process (dE, dV) described by Eq. (1.6), *quasistatic* refers to the fact that we are applying the equilibrium differential in Eq. (1.5) to an evolving state. This restricts its applicability to paths that trace a sequence of equilibrium states.

★

1.2 Gibbs Entropy

Gibbs introduced the probabilistic notion of *microstate*. The microstate is a detailed specification of the internal microscopic state of matter, and an intrinsically unknowable quantity. Every microstate i that is feasible under the macroscopic specifications is assigned a probability p_i, and the problem of determining the equilibrium state is reduced to the problem of determining the equilibrium probability distribution p_i^* among all distributions that are feasible under the macroscopic specification of the state of the system. The two key results that allow the solution to this seemingly intractable problem were both obtained by Gibbs. The first of them establishes the connection between entropy and probability of microstate[4]:

$$S(\mathbf{p}) = -\sum_i p_i \log p_i, \tag{1.7}$$

where $\mathbf{p} = (p_1, p_2 \cdots)$ is the vector representation of the probability distribution and the summation goes over all feasible microstates. The second one establishes the fact that the equilibrium probability p_i^* maximizes the entropy functional in Eq. (1.7) among all feasible probability distributions p_i. Gibbs's derivation is based on a particular microscopic model of matter, interacting particles that obey Newtonian mechanics[5]; the significance of the result extends, however, well beyond the scope of its original derivation.

The maximization of the entropy functional has a combinatorial interpretation that brings it closer to the interpretation of classical entropy. We partition a system of fixed energy E, volume V, and number of particles N_k of type k into $\mathscr{N}$ subsystems and count the number of systems n_i in microstate k. The vector

$$\mathbf{n} = (n_1, n_2, \cdots) \tag{1.8}$$

[4]The entropy in Eq. (1.7) refers to an ensemble of microstates under fixed macroscopic state and is expressed in intensive form per microstate.

[5]Gibbs derives the variational properties of entropy in a series of theorems in Chap. XI of his book (Gibbs 1902).

with

$$\sum_i n_i = \mathcal{N} \tag{1.9}$$

is a feasible partitioning of microstates and $p_i = n_i/\mathcal{N}$ is a feasible probability distribution of microstates. A partitioning **n** has a multiplicity that is given by the binomial coefficient,

$$\mathbf{n}! = \frac{\mathcal{N}!}{n_1!n_2!\cdots} \tag{1.10}$$

and corresponds to the number of ways to assign the distribution of microstates to $\mathcal{N}$ distinguishable subsystems. The partitioning is similar as in Eq. (1.1) but it is the distribution of microstates that is now the object of interest. The macroscopic constraints in Eq. (1.2) are now written in terms of **n**:

$$\sum_i n_i = \mathcal{N}, \quad \sum_i n_i E_i = E, \quad \sum_i n_i V_i = V, \quad \sum_i n_i N_{k,i} = N_k, \tag{1.11}$$

where E_i, V_i, and $N_{k,i}$ respectively are the energy, volume, and number of particles of each type in microstate i. We seek the distribution $\mathbf{n}^*$ with the maximum multiplicity. Equivalently we maximize the logarithm of multiplicity, which in the Stirling approximation is

$$\log \mathbf{n}! = -\sum_i n_i \log \frac{n_i}{\mathcal{N}} = -\mathcal{N} \sum_i p_i \log p_i = \mathcal{N} S(\mathbf{p}). \tag{1.12}$$

Therefore, the distribution with the maximum multiplicity is also the distribution with the maximum entropy among all feasible distributions, i.e., among all distributions that satisfy the conditions in Eq. (1.11). We obtain the distribution that maximizes entropy by the method of Lagrange multipliers. We construct the objective function

$$\begin{aligned}\mathcal{F} = &-\sum_i n_i \log \frac{n_i}{\mathcal{N}} - \lambda \left(\sum_i n_i - \mathcal{N} \right) \\ &- \beta \left(\sum_i n_i E_i - E \right) - \alpha \left(\sum_i n_i V_i - V \right) - \sum_k \gamma_k \left(\sum_i n_i N_{k,i} - N_i \right)\end{aligned}$$

where λ, β, α, and γ_k are Lagrange multipliers corresponding to the constraints in Eq. (1.11). We obtain the equilibrium distribution n_i^* by maximizing $\mathcal{F}$ with respect to n_i at fixed $\mathcal{N}$, treating all n_i as independent variables. The result is

$$\frac{n_i^*}{\mathcal{N}} = \frac{e^{-\beta E_i - \alpha V_i - \gamma_1 N_{1,i} - \gamma_2 N_{2,i} \cdots}}{\mathcal{Q}^*} \equiv p_i^*, \tag{1.13}$$

where $\mathscr{A} = e^{1+\lambda}$. We insert this result into Eq. (1.12) to calculate the entropy of this distribution. The result is

$$\begin{aligned}\mathscr{N} S(\mathbf{p}^*) &= \mathscr{N} \sum_i p_i \left(\beta E_i + \alpha V_i + \sum_k \gamma_k N_{k,i} \right) - \log \mathscr{A} \\ &= \mathscr{N} \left(\beta \bar{E} + \alpha \bar{V} + \sum_k \gamma_k \bar{N}_k + \frac{\log \mathscr{A}}{\mathscr{N}} \right), \qquad (1.14)\end{aligned}$$

where $\bar{E} = E/\mathscr{N}$, $\bar{V} = V/\mathscr{N}$, and $\bar{N}_k = N_k/\mathscr{N}$ are the mean values of energy, volume, and number of particles of each type per part of the larger system. For distribution p_i to represent the probability of microstate in common material systems, it must be invariant if E, V, N_k, and $\mathscr{N}$ increase simultaneously by the same factor; or, equivalently, if $\mathscr{N}$ is increased at fixed $\bar{E}$, $\bar{V}$, and $\bar{N}_k$. Then from Eq. (1.14) we find that β, α, and γ_k must be invariant by this change of scale and $(\log \mathscr{A})/\mathscr{N} \to 0$.[6] The Gibbs entropy then becomes

$$S(\mathbf{p}^*) = \beta \bar{E} + \alpha \bar{V} + \sum_k \gamma_k \bar{N}_k \equiv S(\bar{E}, \bar{V}, \bar{N}_k). \qquad (1.15)$$

With this result we have made contact between Gibbs's entropy functional and entropy in the classical development. Here we are recycling the same symbol S to indicate entropy as a *function* of (E, V, N_k), and entropy as a *functional* of the equilibrium probability distribution $\mathbf{p}^*$.[7]

Note 1.2 (Probabilistic Interpretation of the Equilibrium Distribution)
If we take the view that all $\mathbf{n}!$ permutations of distribution $\mathbf{n}$ are equally probable and form the ensemble of all distributions $\mathbf{n}$ that can be formed given E, V, and N_k, then the probability of distribution $\mathbf{n}$ is proportional to its multiplicity $\mathbf{n}!$,

$$P(\mathbf{n}) = \frac{\mathbf{n}!}{\Omega},$$

where Ω is the normalization factor. In this view, the distribution $\mathbf{n}^*$ that maximizes the multinomial coefficient is *the most probable distribution* among all distributions that have the same total energy, volume, and number of particles of each type. The ensemble of distributions with fixed (E, V, N_k) is known in statistical mechanics

[6] This derivation is discussed in more detail in Hill (1987). Hill calls this the *generalized* ensemble and uses it to obtain the familiar ensembles (microcanonical, canonical, and grand canonical) as special cases.

[7] A brief review of functionals and variational calculus is given in the appendix of Chap. 7.

as *microcanonical* and consists of equiprobable partitions. Since distribution $\mathbf{n}$ is represented by $\mathbf{n}!$ partitions, the most probable distribution is that which maximizes multiplicity, or equivalently, the entropy functional. That is, the equilibrium distribution is the maximum entropy distribution as well as the most probable distribution among all distributions of the ensemble.

★

1.3 Shannon Entropy

Shannon (1948) arrived at the entropy functional through a very different route than that of Gibbs. Following Shannon we pose the question: *what is a suitable measure of uncertainty associated with a probability distribution?* Intuition suggest very convincingly that uncertainty indeed varies. The uncertainty associated with a coin toss (possible outcomes "heads" or "tails") is smaller than that in the toss of a die (outcomes 1, 2, $\cdots$ 6) but still higher than the uncertainty in the average number of "heads" in 1000 tosses of a coin. We propose that a measure $S(\mathbf{p})$ of the uncertainty[8] associated with discrete distribution $\mathbf{p} = (p_1, p_2, \cdots)$ should satisfy the following conditions:

1. *(Continuity)* $S(\mathbf{p})$ must be a continuous function of the p_i.
2. *(Monotonicity)* If all p_i are equal, i.e.,

$$p_i = \frac{1}{\mathscr{N}}, \quad i = 1, \cdots \mathscr{N},$$

   then $S(\mathbf{p})$ must be a monotonically increasing function of $\mathscr{N}$.
3. *(Additivity)* Suppose we partition the phase space into K nonoverlapping groups. Let $\mathbf{q} = (q_1, q_2, \cdots)$ be the probability distribution of the partitions such that

$$q_k = \sum_{i \in [k]} p_i$$

   is the probability of the kth partition, with the summation going over all elements of phase space in partition k; and let $\mathbf{w}_k = (w_{1|k}, w_{2|k}, \cdots)$ be the conditional probability distribution within partition k such that $w_{i|k}$ is the probability of element i in partition k. Additivity requires the uncertainties of $\mathbf{p}$, $\mathbf{q}$, and $\mathbf{w}_k$ to satisfy

$$S(\mathbf{p}) = S(\mathbf{q}) + \sum_k q_k S(\mathbf{w}_k). \tag{1.16}$$

[8] Shannon used the symbol H to denote the uncertainty of probability distribution. We use S to indicate that we are discussing the same functional as in Gibbs's treatment, though from a fairly different point of view.

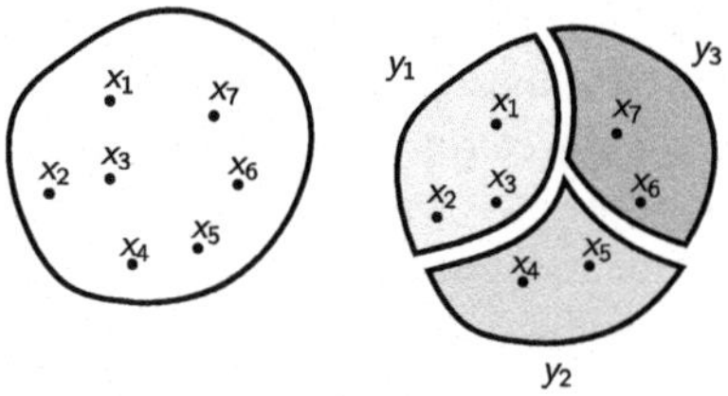

Fig. 1.2 Partitioning of a phase space $\{x_i\}$, $(i = 1 \cdots, 7)$ into three nonoverlapping partitions, $Y_1 = \{x_1, x_2, x_3\}$, $Y_2 = \{x_4, x_5, x_6\}$, and $Y_3 = \{x_6, x_7\}$. The condition of additivity equates the entropies of the partitioned system to that of the whole. The probabilities that appear in Eq. (1.16) are: $p_i = \text{Prob}(x = x_i)$, $q_k = \text{Prob}(y = y_k)$, and $w_{i|k} = \text{Prob}(x = x_i | y = y_k)$

The condition of additivity expresses the uncertainty of the partitioned system as a sum of two contributions, one that arises from the probability of partition k, and one that arises from the conditional distribution within the partition. The latter component contributes to the total uncertainty in proportion to the probability of the partition. Equation (1.16) essentially states that regardless of how the phase space is partitioned the uncertainty should remain the same because knowing both q_k and $w_{k,i}$ is equivalent to knowing p_i. To use a thermodynamic analogy, the partitioning of phase space is a reversible process because it preserves the state of knowledge: accordingly, uncertainty is conserved.

Note 1.3 (Additivity of the Uncertainty Measure)

The additivity condition is demonstrated graphically in Fig. 1.2. In this example a phase space with 7 elements,

$$\{x_1, x_2, x_3, x_4, x_5, x_6, x_7\},$$

and probability distribution $p_i = \text{Prob}(x = x_i)$ is partitioned into three nonoverlapping subsets,

$$y_1 = \{x_1, x_2, x_3\}, \quad y_2 = \{x_4, x_5\}, \quad y_3 = \{x_6, x_7\}.$$

The probability distribution of y_i is given by the additive property of probabilities,

$$q_1 = p_1 + p_2 + p_3, \quad q_2 = p_3 + p_5, \quad q_3 = p_6 + p_7,$$

and is properly normalized. The conditional probabilities $w_{i|k}$ to find element x_i in partition k is

$$w_{1|1} = p_1/q_1, \; w_{2|1} = p_2/q_1 \; w_{3|1} = p_3/q_1,$$
$$w_{4|2} = p_4/q_2, \; w_{5|2} = p_5/q_2,$$
$$w_{6|3} = p_6/q_3, \; w_{7|3} = p_7/q_3,$$

with all other $p_{i|k} = 0$. The uncertainty of the partition is

$$S(q_1, q_2 \cdot q_3) + q_1 S(w_{1|1}, w_{2|1}, w_{3|1}) + q_2 S(w_{4|2}, w_{5|2}) + q_3 S(w_{6|3}, w_{7|3})$$

and according to the property of additivity this must be equal to the uncertainty of the original probability distribution, $S(p_1, p_2, p_3, p_4, p_5)$. Notice that zero elements do not change the value of uncertainty. For example, the complete form of the conditional distribution $\mathbf{w}_2$ is

$$\mathbf{w}_2 = (w_{1|2}, w_{2|2}, w_{3|2}, w_{4|2}, w_{5|2}, w_{5|2}, w_{6|2}, w_{7|2}) = (0, 0, 0, p_4/q_2, p_5/q_2, 0, 0)$$

and its uncertainty is

$$S(0, 0, 0, p_4/q_2, p_5/q_2, 0, 0) = S(p_4/q_2, p_5/q_2).$$

This equality reflects the fact that adding new events with zero probability does not alter our state of knowledge.

1.3.1 The Uncertainty Functional

We now set out to identify the functional that satisfies the conditions we placed on the measure of uncertainty. We consider a phase space with N equiprobable events $x_i, i = 1, 2, \cdots N$. By the monotonicity property, $S(1/N, 1/N, \cdots)$ must be a monotonically increasing function of N. Let $A(N)$ be that function:

$$S(\mathbf{p}) = A(N).$$

Accordingly, $A(N)$ is the entropy of any equiprobable space with N possible events. Next we partition the phase space in some fashion to produce a probability distribution $q_k = n_k/N$ of partitions, where n_k is the number of elements in partition k. For large enough N the partition may be constructed such that q_k may represent any possible distribution. The uncertainty of the partition is

$$S(\mathbf{q}) + \sum_k q_k S(\mathbf{w}_k) = A(N). \tag{1.17}$$

Since all nonzero elements in any partition are equally probable, the uncertainty of the conditional probabilities is $S(\mathbf{w}_k) = A(n_k)$, where n_k is the number of elements in partition k. The uncertainty of the partitioning now becomes

$$S(\mathbf{q}) + \sum_k q_k A(n_k) = A(N). \tag{1.18}$$

This must be true for any partitioning of the phase space. We choose to partition into K parts of equal size M. Then $N = KM, n_k = M$ for all k, $p_k = 1/K$. Substituting these results into Eq. (1.18) we obtain

$$A(K) + A(M) = A(KM). \tag{1.19}$$

This is a condition on function A that must be satisfied for any K and M. We solve it by setting $A(z) = c \log z$ where c is an arbitrary constant. This constant must be positive in order to satisfy the second condition on uncertainty. We set $c = 1$. In the final step we substitute the expression for $A(x)$ into Eq. (1.18) and solve for $S(\mathbf{q})$:

$$S(\mathbf{q}) = -\sum_k q_k \log n_k + \log N = -\sum_k q_k \log \frac{n_k}{N}, \tag{1.20}$$

and since $q_k = n_k/N$, the final result is

$$S(\mathbf{q}) = -\sum_k q_k \log q_k. \tag{1.21}$$

This is the uncertainty of the probability distribution of partitions. Since we can partition a large equiprobable phase space such as to obtain any possible probability distribution q_k, the above functional is a general measure of uncertainty for any such probability distribution q_k.

Note 1.4 (Shannon's Functional and Thermodynamic Entropy)

As lore of metascience has it, Shannon was unsure what to call his newly discovered uncertainty functional and asked John von Neumann for his opinion. Von Neumann quipped[9]:

> You should call it entropy for two reasons: first because that is what the formula is in statistical mechanics but second and more important, as nobody knows what entropy is, whenever you use the term you will always be at an advantage!

Von Neumann was right to suggest *entropy* as the proper name for Shannon's functional. Not only does it have the same mathematical form as the Gibbs entropy, they both refer to the same fundamental concept. To derive Shannon's functional we construct a reversible partition of an equiprobable phase space. This is entirely analogous—in fact *equivalent*—to the partitioning of the microcanonical phase space of statistical mechanics. In the context of information theory *reversibility* refers to a process that preserves the state of knowledge (the probability of the whole can be reproduced from knowledge of the probabilities of the parts); in thermodynamics it refers to the equilibrium state; in both cases entropy is preserved.

[9]As quoted in Tribus and McIrvine (1971).

Shannon and Gibbs are working with the *same calculus*. It is only the *problems* they work on that are different. Our goal is to bring this calculus into full view.

⋆

1.4 Jaynes Entropy

A connection between the Gibbs and Shannon entropies was first drawn by E. T. Jaynes. Jaynes was motivated by a quite different problem: *inference* under insufficient evidence. Suppose we are presented with a die whose possible outcomes are $x_i = i$, $i = 1, 2 \cdots 6$: what is the probability distribution of its outcomes if the die were to be tossed? Nothing else is known about the die, its exact shape, material of construction, distribution of weight, and so on. The problem as stated is indeterminate: *any* nonnegative sequence $(p_1, p_2, p_3, p_4, p_5, p_6)$ that satisfies the normalization condition

$$\sum_i p_i = 1, \tag{1.22}$$

is a possible answer. Nonetheless, intuition suggests that of this infinite set of feasible distributions the one that "makes sense" is the uniform distribution $p_i = 1/6$. Jaynes proposed a mathematical procedure that leads to the same answer. If we adopt the entropy functional as a measure of uncertainty we ought to pick the distribution with the *maximum uncertainty* under the current state of knowledge. Picking any other distribution amounts to *assuming* information that is not present in the available data. This is known as the *maximum entropy method* (MEM) and works as follows: construct mathematical constraints on the unknown probability distribution that represent what is known about it, then maximize the entropy functional with respect to the unknown distribution under the given constraints. In the case of the die the constraint is Eq. (1.22) and represents the only thing we know: p_i is a probability distribution over a phase space of six outcomes. To maximize the entropy of p_i under the constraint in Eq. (1.22) we use the Lagrange multiplier λ to construct the objective function

$$-\sum_i p_i \log p_i - \lambda \left(\sum_i p_i - 1 \right),$$

which we maximize with respect to p_i treating all p_i as independent. We set the derivative with respect to p_i equal to zero,

$$-1 - \log p_i - \lambda = 0,$$

and solve for p_i to obtain

$$p_i = e^{-1-\lambda}.$$

This states that all p_i have the same value. Using the normalization condition to eliminate λ_0 we obtain

$$p_i = 1/6, \quad i = 1, 2, \cdots 6. \tag{1.23}$$

We have recovered the result suggested by intuition, but this time using a methodology which we may now apply to more complicated problems.

Suppose that a new piece of information becomes available: the mean of the unknown distribution $\bar{x}$ is given. To simplify the math we allow this die to have an infinite number of possible outcomes (the methodology still applies to a regular six sided die). This piece of information is represented by the condition

$$\sum_i x_i p_i = \bar{x}, \tag{1.24}$$

and this, along with the normalization condition in Eq. (1.22), forms the constraints imposed on the unknown distribution. The Lagrange function now is

$$-\sum_i p_i \log p_i - \lambda_0 \left(\sum_i p_i - 1 \right) - \lambda_1 \left(\sum_i x_i p_i - \bar{x} \right),$$

with two Lagrange multipliers, λ_0 and λ_1, that correspond to the two constraints. The probability distribution that maximizes this function is

$$p_i = e^{-1-\lambda_0-\lambda_1 i}. \tag{1.25}$$

Setting $Q = e^{1+\lambda_0}$ this becomes

$$p_i = \frac{e^{-\lambda_1 x_i}}{Q}. \tag{1.26}$$

If we interpret i as the microstate of a material system with fixed volume and number of particles, x_i as the energy of the microstate, λ_1 as the inverse temperature $1/k_B T$, and Q as the canonical partition function, Eq. (1.26) is the probability distribution in the canonical ensemble of statistical mechanics.

One might argue that we have done nothing new and are merely repeating the steps of the standard textbook derivation. To which we answer, *precisely!*—except that we now interpret our procedure in a more abstract and potentially far more powerful way. Our latest derivation of the canonical probability of a material system makes absolutely no reference to the properties of matter beyond the fact that system possesses "energy" and its "volume" and "mass" are fixed. By contrast, Gibbs's derivation is intimately tied to *Newtonian mechanics*. In Jaynes's view the maximization of entropy is not dictated by *physics* but by *logic*: by maximizing entropy under the constraints that represent our state of knowledge, the logical choice is to pick the most general—most *uncertain*—distribution possible. The canonical probability distribution is exponential because this is the most general probability distribution among all distributions with fixed mean.

Note 1.5 (Entropy, Logic, and Intuition)
The assignment of equal a priori probabilities to all outcomes of the die is so natural that Laplace and Bernoulli accepted it as obvious.[10] This is known as principle of *indifference*, *insufficient reasoning*, or *uninformed prior*; in statistical mechanics it is known as the *postulate of equal a priori probabilities*. In Jaynes's view, probability is degree of belief in a hypothesis, and entropy maximization is a logical quantifiable principle for assigning truth values to such beliefs that vary continuously from 0 (false) to 1 (true). The principle of entropy maximization provides a formalized mathematical process that involves no "guessing" and yet produces the same answer as intuition does in simple cases. It is difficult to escape the philosophical implications of this result. To Jaynes the implications were so profound that one of his early manuscripts on this topic was titled *How does the brain do plausible reasoning*. (It got rejected.[11]) Jaynes has been criticized for his subjective view of probability. His response was that his expanded interpretation allowed him to solve more complex problems without ever contradicting any results of the conventional frequency approach. The debate between Bayesian and frequency views of probability is a long running one and we do not intend to join it. We suggest, however, that one does not have to abandon the frequency view of probability to appreciate the elegance of the Bayesian view.

★

1.5 Entropy and Probability Distributions

The most attractive element of Jaynes's approach is that any probability distribution may be obtained as a maximum entropy distribution under suitable constraints. We have already seen that the normalization condition alone produces the uniform distribution and that adding a constraint for the mean produces the exponential distribution. If the constraint on the mean is replaced by a constraint on the variance, the maximum entropy distribution is Gaussian. By careful construction of other constraints, typically in the form of known moments or other quantities

[10]By *a priori* we mean probabilities that are assigned before we even toss the die to study its outcomes, based on our total knowledge up to that point.

[11]This title suggest that entropy maximization is perhaps encoded in the human brain. This is a hyperbole, but Jaynes was mischievously fond of making strong pronouncements. The manuscript, its review, and the author's response in Jaynes's inimitable style can be found at https://bayes.wustl.edu/etj/node1.html.

expressed as summations of known functions over the probability distribution,[12] any distribution may be interpreted as a maximum entropy distribution. Problems abound in the physical sciences where the unknown is a probability distribution. In all stochastic processes this is the generic question: given a set of rules that govern a stochastic process (the *model*), determine the probability distribution of its outcomes. If the unknown distribution can be obtained by maximizing entropy, then we have introduced *thermodynamics* to the stochastic process in an unforced manner. *How?* Suppose the model of the stochastic process reduces into two constraints, the normalization in Eq. (1.22), and the known mean in Eq. (1.24). The maximum entropy solution is the exponential distribution in Eq. (1.26). We return this distribution into the two constraints, which we express in the form

$$Q = \sum_i e^{-\lambda_1 x_i}, \tag{1.27}$$

and

$$\bar{x} = \frac{1}{Q} \sum_i x_i e^{-\lambda_1 x_i} = -\frac{d \log Q}{d\lambda}. \tag{1.28}$$

The last result relates the mean $\bar{x}$ to the derivative of the partition function, and has an exact counterpart (in fact several[13]) in statistical mechanics. Next we insert the probability distribution into the entropy functional:

$$S(\mathbf{p}) = -\sum_i p_i \log p_i = \sum_i p_i \left(\lambda_1 x_i + \log Q\right) = \lambda_1 \bar{x} + \log Q. \tag{1.29}$$

This too has its counterpart in statistical thermodynamics.[14] The entire calculus of thermodynamics then becomes available to the stochastic process under consideration.

Jaynes (1957) obtained these results as corollaries of the maximum entropy principle and his conclusion was that the constrained maximization of entropy represents a general principle of inference, of which statistical mechanics is a special case. Our

[12] See Kapur (1989) for a discussion of many common distributions, discrete and continuous, and how they can be obtained by the maximum entropy method under suitable constraints.

[13] Recall the following results from statistical mechanics:

$$\bar{E} = -\left(\frac{\partial \log Q}{\partial \beta}\right)_{V,N_k}, \quad \bar{N}_k = -\left(\frac{\partial \log \Xi}{\partial \gamma_k}\right)_{V,N_{j\neq k}},$$

where $Q(\beta, V, N_k)$ is the canonical partition function, $\Xi(\beta, V, \gamma_k)$ is the grand canonical partition function, $\beta = 1/k_B T$, and $\gamma_k = -\mu_k/k_B T$. Similar expressions can be written $\bar{E}$ in the grand canonical ensemble, $\bar{V}$ in the (E, p, N_k) ensemble, and so on.

[14] With $\bar{x} = \bar{E}, \beta = 1/k_B T$, this reverts to $S = \beta\bar{E} + \log Q$, a familiar result from thermodynamics.

reading is different: a door has opened that allows thermodynamics to move outside the realm of *physics* into the realm of stochastic process—of which material systems is but one example. The intuitive urge to apply thermodynamics to problems outside the original scope (systems of physical particles exerting interactions) is difficult to resist. Some problems invite analogies to thermodynamics that are simply too strong to discount. In percolation, the formation of a system-spanning giant cluster has all the hallmarks of a phase transition, yet it is unclear how to apply thermodynamics to a system that lacks the thermal coupling of molecular systems. The maximum entropy principle suggests that an answer to this question may in fact be possible, but provides no organized methodology to accomplish this task. It tells us what to do once we have distilled our knowledge about the process into a set of algebraic constraints on the unknown distribution p_i, but it does not tell us *how* to convert our knowledge into such constraints. Nor does it tell us whether thermodynamics is still possible under constraints different from those that lead to the exponential distribution.

Note 1.6 (Other Entropies)

Following Shannon's groundbreaking work a number of other measures of uncertainty have been developed and are also known under the generic name *entropy*. Renyi entropy of order α is defined as

$$S_\alpha(\mathbf{p}) = \frac{1}{1-\alpha} \log \left(\sum_i p_i^\alpha \right), \tag{1.30}$$

and reduces to the Gibbs entropy functional for $\alpha = 1$. The Tsallis entropy,

$$S_\alpha(\mathbf{p}) = \frac{1}{\alpha - 1} \left(1 - \sum_i p_i^\alpha \right) \tag{1.31}$$

also reduces to the familiar functional for $\alpha = 1$ and has the unusual property that it is not extensive, i.e., it is not additive under reversible partitioning. These and other generalized measures of uncertainty, useful as they may be for their intended applications, have no relevance in thermodynamics except in the trivial limit that they reproduce the standard entropy functional.

★

Appendix: The Mathematical Calculus of Entropy

The mathematical relationships we recognize as thermodynamics are based on three key concepts: curvature (concavity/convexity), homogeneity, and Legendre transformations. These are briefly reviewed here, then are applied to obtain certain key results in thermodynamics.

Curvature

A function $f(x_1, x_2 \cdots)$ is concave with respect to x_1 if

$$f(\lambda_1 x_1 + \lambda_1' x_1', x_2 \cdots) \geq \lambda_1 f(x_1, x_2 \cdots) + \lambda_1' f(x_1', x_2 \cdots)$$

for all positive $\lambda_1, \lambda_1' = 1 - \lambda_1$ at fixed $x_2, \cdots$. The second derivative of a concave function is negative:

$$\frac{\partial^2 f}{\partial x_1^2} \leq 0. \tag{1.32}$$

For a convex function these inequalities are inverted. If f is concave with respect to several independent variables, the concave inequality applies to each variable at a time, keeping all other variables constant. A multivariate function may have different curvatures with respect to different variables. For example, if $f(x_1, x_2)$ is concave with respect to x_1 and convex with respect to x_2, then

$$f(\lambda x_1 + (1-\lambda)x_1', x_2) \geq af(x_1, x_2) + (1-\lambda) f(x_1', x_2). \tag{1.33}$$

and

$$f(x_1, \lambda x_2 + (1-\lambda)x_2') \leq af(x_1, x_2') + (1-\lambda) f(x_1, x_2'). \tag{1.34}$$

It is an elementary property that f and $(-f)$ have opposite curvatures: if f is concave, then $(-f)$ is convex, and vice versa.

Homogeneity

A multivariate[15] function $f(x_1, x_2)$ is homogeneous in x_1 and x_2 with degree ν if

$$f(\lambda x_1, \lambda x_2) = \lambda^\nu f(x_1, y_1) \tag{1.35}$$

for all λ. We differentiate with respect to λ to obtain

$$x_1 \frac{\partial f}{\partial(\lambda x_1)} + x_2 \frac{\partial f}{\partial(\lambda x_2)} = \nu \lambda^{\nu-1} f, \tag{1.36}$$

[15]We use a bivariate function to demonstrate homogeneity and Euler's theorem. The extension to any number of variables is straightforward.

then setting $\lambda = 1$,

$$x_1 f_1 + x_2 f_2 = \nu f, \tag{1.37}$$

where f_i is a shortcut notation for the partial derivative with respect to x_i. This is Euler's theorem for homogeneous functions of degree ν. For degree $\nu = 1$ we obtain

$$f = x_1 f_1 + x_2 f_2. \tag{1.38}$$

The derivatives f_1 and f_2 are homogeneous in x_1 and x_2 with degree 0, i.e.,[16]

$$f_i(\lambda x_1, \lambda x_2) = f_i(x_1, x_2). \tag{1.39}$$

Taking the differential of f in Eq. (1.38) with respect to all x_i and f_i we have

$$df = f_1 dx_1 + f_2 dx_2 + f_1 dx_1 + x_2 df_2, \tag{1.40}$$

and since $df = f_1 dx_1 + x_2 df_2$,

$$f_1 dx_1 + x_2 df_2 = 0. \tag{1.41}$$

This is the Gibbs-Duhem equation associated with the Euler form in Eq. (1.38).[17] It expresses the fact that homogeneity, by virtue of being a special condition on f, imposes the constraint that the variations of the partial derivatives are not all independent.

If $f(x_1, \cdots x_m)$ is homogeneous with degree 1 only with respect to $x_1, \cdots x_k$, then the Euler and the Gibbs-Duhem equation apply to these variables,

$$f = \sum_{i=1}^{k} x_i f_i, \tag{1.42}$$

$$\sum_{i=1}^{k} f_i dx_i = 0, \tag{1.43}$$

with the understanding that the remaining variables $x_{k+1}, \cdots x_m$ are held constant.

---

[16]In general, if f is homogeneous in x with degree ν, its nth derivative with respect to n is homogeneous with degree $\nu - k$. The proof is left as an exercise.

[17]In solution thermodynamic the name *Gibbs-Duhem* is specifically associated with Eq. (1.55), which is a special case of the above result.

Legendre Transformations

Given a monotonic function $f(x_1, x_2, \cdots)$ with partial derivatives $f_i, i = 1, 2\cdots,$ its Legendre transformation with respect to variable x_1 is[18]

$$F^{(1)}(f_1, x_2) = f - x_1 f_1, \tag{1.44}$$

The Legendre transformation takes a function of (x_1, x_2) and turns it into a function of (f_1, x_2), where f_1 is the derivative of the original function with respect to the transformed variable. The usefulness of the Legendre may be appreciated better if we write the differentials of the original and of the transformed functions (the derivations are straightforward and can be found in the standard literature):

$$df = +f_1 dx_1 + f_2 dx_2$$
$$df F^{(1)} = -x_1 df_1 + f_2 dx_2.$$

The Legendre transformation with respect to x_1 changes the independent variable from x_1 to f_1, and the corresponding partial derivative from f_1 to $-x_1$, i.e.,

$$\left(\frac{\partial F^{(1)}}{\partial f_1}\right)_{x_2,\cdots} = -\left(\frac{\partial f}{\partial x_1}\right)_{x_2\cdots}. \tag{1.45}$$

These derivatives will usually be notated more simply as $\partial F^{(1)}/\partial f_1$ and $\partial f/\partial x_1$ with the understanding that they are taken with respect to the proper set of independent variables of each function. If we Legendre-transform $F^{(1)}$ with respect to f_1 we obtain the original function f. This makes the Legendre transformation an involution—its inverters transformation is itself. The Legendre transformation can be applied to any subset of variables. For example, transforming f with respect to x_1 and x_2 we obtain

$$F^{(1,2)} = f - x_1 f_1 - x_2 f_2, \tag{1.46}$$

whose differential is

$$dF^{(1,2)} = -x_1 df_1 - x_2 df_2 + f_3 dx_3 \cdots . \tag{1.47}$$

[18]We are using the notation $F^{(i,j,\cdots)}$ to indicate the Legendre transformation of $F(x_1, x_2\cdots)$ with respect to $x_i, x_j, \cdots$.

The Legendre transformation inverts curvature with respect to the transformed variable and preserves it for all untransformed variables. If $f(x_1, x_2 \cdots)$ is concave in all x_i, then $F^{(1)}(f_1, x_2, \cdots)$ is convex in f_1 and concave in $x_2, x_3 \cdots$.

Thermodynamic Relationships

We assert that function $S = S(E, V, N_k)$, defined in the positive quadrant of the (E, V, N_k) plane, has the following properties: it is positive, concave, and homogeneous with degree 1 with respect to all of its independent variables. These conditions reproduce all relationships of classical thermodynamics, as we demonstrate below.

Second Law

We apply the concave inequality to entropy. With $\lambda + \lambda' = 1$ $(\lambda, \lambda' \geq 0)$, we have

$$\begin{aligned} S(\lambda E + \lambda' E', V, N_k) \geq \\ \lambda S(E, V, N_k) + \lambda' S(E', V, N_k) = \\ S(\lambda E, \lambda V, \lambda N_k) + S(\lambda' E', \lambda' V, \lambda' N_k). \end{aligned}$$

Setting

$$\begin{aligned} E_1 &= \lambda E, & V_1 &= \lambda V, & N_{k,1} &= \lambda N_k, \\ E_2 &= \lambda' E', & V_2 &= \lambda' V, & N_{k,2} &= \lambda' N_k, \end{aligned}$$

the concave inequality becomes

$$S(E_1 + E_2, V_1 + V_2, N_{k,1} + N_{k,2}) \geq S(E_1, V_1, N_{k,1}) + S(E_2, V_2, N_{k,2}), \tag{1.48}$$

which is the inequality of the second law. It is always possible to obtain an infinite set of positive $V_1, V_2, N_{k,1}, N_{k,2}$, that satisfy these equations for any positive V, N_k, and any $\lambda, \lambda' = 1 - \lambda$, such that $1 \geq \lambda \geq 0$. The inequality in Eq. (1.48) therefore applies for any positive $U_1, U_2, V_1, V_2, N_{k,1}$, and $N_{k,2}$.

Entropy Equation

By Euler's theorem,[19]

$$\frac{S(E,V,N_k)}{k_B} = E\left(\frac{\partial S}{\partial E}\right)_{V,N_k} + V\left(\frac{\partial S}{\partial V}\right)_{E,N_k} + \sum_k \gamma_k \left(\frac{\partial S}{\partial N_k}\right)_{E,V,N_{j\neq k}}$$
$$= \beta E + \alpha V + \sum_k \gamma_k N_k, \tag{1.49}$$

where β, α, and γ_k are the derivatives of entropy:

$$\beta = \frac{1}{k_B T}, \quad \alpha = \frac{p}{k_B T}, \quad \gamma_k = -\frac{\mu_k}{k_B T}. \tag{1.50}$$

These derivatives are homogeneous with degree 0, i.e., they are intensive functions of the state.

Thermodynamic Potentials

A family of thermodynamic potentials is obtained by Legendre transforming entropy with respect to various combinations of its independent variables. We give one example: the Gibbs function G, which is commonly defined as

$$G = E - TS - pV = E - \frac{S/k_B}{\beta} - \frac{\alpha V}{\beta}. \tag{1.51}$$

We express this in the equivalent form

$$-\beta G = \frac{S^{(1,2)}(E,V,N_k)}{k_B} \tag{1.52}$$

which identifies the product $-\beta G$ as the Legendre transformation of $S(E,V,N_k)/k_B$ with respect to E and V. The product $-\beta G$ is a function of β, α, and N_k (as is G), and its differential is written immediately by application of Eq. (1.47)

$$-d(\beta G) = -E d\beta - V d\alpha + \sum_k \gamma_k dN_k. \tag{1.53}$$

[19]Boltzmann's constant on the left-hand side of Eq. (1.49) gives entropy the dimensions of energy over temperature. It is a historical accident that temperature (and heat) was given its own units rather than the same units as energy, which would have set $k_B = 1$ and would have made entropy dimensionless. While the dimensions of *heat* were eventually corrected to match those of energy, the same was never done for temperature.

Substituting Eq. (1.50) for β, α, and γ_k we obtain the more familiar result

$$dG = -SdT + Vdp + \sum_k \mu_k dN_k. \tag{1.54}$$

At fixed T and p, the Gibbs energy is homogeneous in N_k with degree 1. From Euler's theorem for a mixed set of extensive and intensive variables, given in Eq. (1.42), we obtain

$$G = \sum_k \mu_k N_k, \quad \text{(const. } p, T). \tag{1.55}$$

Applying the Gibbs-Duhem equation with respect to all extensive variables while keeping the intensive variables constant we have

$$0 = \sum_k N_k d\mu_k \quad \text{(const. } p, T). \tag{1.56}$$

In thermodynamics the Gibbs-Duhem equation is specifically associated with this result. We use *Gibbs-Duhem* in a more general sense to refer to a condition on the simultaneous variation of intensive properties as a companion of the Euler equation, which governs the variation of the extensive variables.

Stability Criteria

The concave condition on S implies that the second derivatives of entropy are negative:

$$\left(\frac{\partial^2 S}{\partial E^2}\right)_{V,N_k} \le 0, \quad \left(\frac{\partial^2 S}{\partial V^2}\right)_{E,N_k} \le 0, \quad \left(\frac{\partial^2 S}{\partial N_k^2}\right)_{E,V,N_{j\neq k}} \le 0.$$

All stability criteria are obtained by manipulation of these derivatives. To develop stability criteria for the Gibbs energy, we first note that the product $-\beta G$ (see Eq. (1.52)) is concave with respect to the untransformed variables N_k, as is entropy. Accordingly, at fixed β and α the Gibbs energy is convex in N_k (the negative sign flips the curvature). Thus the equilibrium Gibbs energy is at a minimum with respect to partitioning of components at fixed β and α, or equivalently, at fixed T and p:

$$G(T, p, N_k) \le G(T, p, \lambda_k N_k) + G(T, p, (1 - \lambda_k)N_k). \tag{1.57}$$

Similar expressions can be written for the other thermodynamic potentials (energy, free energy, and enthalpy). We will not write them down as our intention is not to reproduce the complete set of thermodynamic relationships but rather to demonstrate how to obtain these results by combining curvature, homogeneity, and Legendre transformations.

Chapter 2
The Cluster Ensemble

The generic population that is the subject of our study consists of M indistinguishable members assembled into N groups, or clusters, such that no group is empty. The "member" is the fundamental unit of the population and plays the same role as "monomer" in a polymeric system, or "primary particle" in granular materials. The *cluster* is characterized by the number of members it contains. We will refer to the number of members in the cluster as the *size* or *mass* of the cluster and will use the terms interchangeably. The goal in this chapter is to define a sample space of distributions of clusters and assign a probability measure over it. This probability space of distributions will be called *cluster ensemble* and forms the basis for the development of generalized thermodynamics. In Chap. 7 we will reformulate the theory on the basis of a more abstract space of distributions.

2.1 Microcanonical Ensemble

The cluster is a collection of indistinguishable units. Clusters are distinguishable only with respect to size[1]; clusters of the same size are indistinguishable from each other, but distinguishable with respect to clusters of other sizes. The cluster *configuration* is an ordered sequence of N clusters with total mass M. In the language of number theory this is an ordered partition of integer M into N parts and is called a composition. We represent configurations in vector form as

$$\mathbf{m} = (m_1, m_2, \cdots m_N),$$

[1]We are assuming that the population consists of a single component. We will later generalize to clusters made of any number of distinguishable *components*.

T. Matsoukas, *Generalized Statistical Thermodynamics*,
Understanding Complex Systems, https://doi.org/10.1007/978-3-030-04149-6_2

where m_k is the size of the cluster in position k of the ordered list. By definition,

$$\sum_{k=1}^{N} m_k = M. \tag{2.1}$$

We define the *cluster ensemble* as the set of all configurations that can be constructed under fixed M and N. We call this *microcanonical cluster ensemble* and will refer to it by the number of clusters it contains and its total mass.

A configuration $\mathbf{m}$ is characterized by its distribution $\mathbf{n} = (n_1, n_2, \cdots)$ that gives the number n_i of clusters in the configuration with mass i. All distributions of the (M, N) ensemble satisfy the conditions

$$\sum_i n_i = N, \tag{2.2}$$

$$\sum_i i n_i = M. \tag{2.3}$$

In these summations i runs over all cluster masses i that are present in the distribution. The smallest possible mass in a configuration is 1; the maximum possible mass is

$$i_{\max} = M - N + 1, \tag{2.4}$$

and corresponds to a configuration that contains $N-1$ particles with unit mass, plus one particle with the rest of the mass, $M - N + 1$. Since zero elements in $\mathbf{n}$ do not contribute to summations in Eqs. (2.2) and (2.3), we adopt the convention $n_i = 0$ for all $i > M - N + 1$ and allow the limits of summations in n_i to run from $i = 1$ to ∞. For simplicity in the notation, these limits will not be shown explicitly.

The number of permutations of a configuration with distribution $\mathbf{n}$ is given by the multinomial coefficient $\mathbf{n}!$,

$$\mathbf{n}! = \frac{N!}{n_1! n_2! \cdots}. \tag{2.5}$$

All $\mathbf{n}!$ permutations of a configuration have the same distribution. Conversely, a distribution $\mathbf{n}$ is represented by $\mathbf{n}!$ configurations. In this sense the multinomial coefficient represents the *natural* multiplicity of the distribution in the microcanonical ensemble.

If the configuration is given, the distribution is easily calculated, and similarly, given the distribution we can construct all $\mathbf{n}!$ configurations associated with it. We may view the microcanonical ensemble either as a collection of configurations or of distributions. We will take the distribution as the primary mathematical object of interest and view configuration as a "primitive" object that attaches physical meaning to the distribution and its multiplicity but is of secondary importance relative to the distribution itself.

The total number of configurations in the (M, N) ensemble is the *volume* of the ensemble. Since each configuration is present $\mathbf{n}!$ times, the volume is the sum of the multinomial factors,

$$\mathcal{V}_{M,N} = \sum_{\mathbf{n}} \mathbf{n}!, \tag{2.6}$$

with the summation going over all distributions with N clusters and M total mass. The result of this summation is remarkably simple and is given by

$$\mathcal{V}_{M,N} = \binom{M-1}{N-1}. \tag{2.7}$$

The proof is straightforward[2]: to divide M objects into N groups, line up the objects, and draw $N-1$ bars between them. There are $M-1$ spaces where we can draw the $N-1$ bars. The answer then is equal to the number of ways to choose $N-1$ out of a total of $M-1$ objects, and is given by the binomial term on the right-hand side of Eq. (2.7).

Example 2.1 (Microcanonical Volume)

Consider the microcanonical ensemble $M = 8$, $N = 5$, whose total number of configurations is

$$\mathcal{V}_{8,5} = \binom{7}{4} = 35.$$

These configurations are enumerated in Table 2.1. They are represented by three distributions: distribution $\mathbf{n}_A$ contains 4 monomers and one tetramer (rows 1 through 5); distribution $\mathbf{n}_B$ contains three monomers, one dimer, and one trimer (rows 6–25); distribution $\mathbf{n}_C$ contains two monomers and three dimers (rows 26–35). Using vector notation, the distributions and their multiplicities are:

$$\mathbf{n}_A = (4, 0, 0, 1), \quad \mathbf{n}_A! = \frac{5!}{4!\,1!} = 5$$

[2] See Bóna (2006), p. 90.

Table 2.1 Microcanonical table for $M = 8$, $N = 5$ in Example 2.1. The ensemble consists of 35 configurations represented by three distributions

#	m_1	m_2	m_3	m_4	m_5
1	4	1	1	1	1
2	1	4	1	1	1
3	1	1	4	1	1
4	1	1	1	4	1
5	1	1	1	1	4

#	m_1	m_2	m_3	m_4	m_5
6	3	2	1	1	1
7	3	1	2	1	1
8	3	1	1	2	1
9	3	1	1	1	2
10	2	3	1	1	1
11	2	1	3	1	1
12	2	1	1	3	1
13	2	1	1	1	3
14	1	3	2	1	1
15	1	3	1	2	1
16	1	3	1	1	2
17	1	2	3	1	1
18	1	2	1	3	1
19	1	2	1	1	3
20	1	1	3	2	1
21	1	1	3	1	2
22	1	1	2	3	1
23	1	1	2	1	3
24	1	1	1	3	2
25	1	1	1	2	3

#	m_1	m_2	m_3	m_4	m_5
26	2	2	2	1	1
27	2	2	1	2	1
28	2	2	1	1	2
29	2	1	2	2	1
30	2	1	2	1	2
31	2	1	1	2	2
32	1	2	2	2	1
33	1	2	2	1	2
34	1	2	1	2	2
35	1	1	2	2	2

$$\mathbf{n}_B = (3, 1, 1, 0), \quad \mathbf{n}_A! = \frac{5!}{3!\,1!\,1!} = 20$$

$$\mathbf{n}_C = (2, 3, 0, 0), \quad \mathbf{n}_A! = \frac{5!}{2!\,3!} = 10$$

The sum of multiplicities is

$$\mathbf{n}_A! + \mathbf{n}_B! + \mathbf{n}_C! = 5 + 2 + 10 = 35,$$

and is equal to the volume of the ensemble.

★

2.2 Entropy and Multiplicity

We define the entropy of distribution **n** as the logarithm of its multiplicity:

$$S(\mathbf{n}) = \log \mathbf{n}! \tag{2.8}$$

Using the Stirling formula $x! \to \sqrt{2\pi x}e^x/x$, the multinomial factor takes the form

$$\mathbf{n}! = \frac{1}{(2\pi N)^{(K-1)/2}} \prod_{i=1}^{K} \left(\frac{n_i}{N}\right)^{-n_i-1/2}, \tag{2.9}$$

where K is the number of nonzero elements in $\mathbf{n}$. The logarithm of the multiplicity factor is

$$\log \mathbf{n}! = -N \sum_i \frac{n_i}{N} \log \frac{n_i}{N} - \frac{1}{2} \sum_i \log \frac{n_i}{N} - \frac{K-1}{2} \log(2\pi N). \tag{2.10}$$

For large n_i the terms of order $\log n_i$ make negligible contribution relative to the terms of order $n_i \log n_i$ and may be dropped. Then we have

$$S(\mathbf{n}) = \log \mathbf{n}! \to -N \sum_i \frac{n_i}{N} \log \frac{n_i}{N}. \tag{2.11}$$

Equations (2.8) and (2.11) differ only at small n_i. For large n_i we may as well consider Eq. (2.11) as an equivalent definition of entropy. Where the two definitions converge we obtain complementary interpretations of entropy as the log of multiplicity, or as the functional of distribution defined in Eq. (2.11). If we multiply all elements of distribution $\mathbf{n}$ by the same factor λ, the entropy in Eq. (2.15) is multiplied by the same factor:

$$S(\lambda \mathbf{n}) = \lambda S(\mathbf{n}). \tag{2.12}$$

This makes entropy a homogeneous function of $\mathbf{n}$ with degree 1. The partial derivative of $S(\mathbf{n})$ with respect to n_i under all $n_{j \neq i}$ constant is

$$\left(\frac{\partial S(\mathbf{n})}{\partial n_i}\right)_{n_{j\neq i}} = -\log \frac{n_i}{N}. \tag{2.13}$$

Then Eq. (2.11) can be written in the equivalent form,

$$S(\mathbf{n}) = \sum_i n_i \left(\frac{\partial S(\mathbf{n})}{\partial n_i}\right)_{n_{j\neq i}}, \tag{2.14}$$

which is a statement of Euler's theorem for homogeneous functions with degree 1.

Equation (2.8) defines entropy as a functional of extensive distribution $\mathbf{n}$ with integer elements. This definition may be extended to any discrete list with elements that are not necessarily integers. First we write Eq. (2.11) in the form

$$S(\mathbf{n}) = -\sum_i n_i \log \frac{n_i}{J_0(\mathbf{n})}, \tag{2.15}$$

where $J_0(\mathbf{n})$ is the zeroth order moment of $\mathbf{n}$:

$$J_0(\mathbf{n}) = \sum_i n_i = 1. \tag{2.16}$$

In this form of Eq. (2.15) entropy can be applied to any vector whose elements are nonnegative numbers. We will apply it to the normalized cluster distribution $p_i = n_i/N$. This may be interpreted as the probability distribution of cluster mass i in a configuration with distribution $\mathbf{n}$ and satisfies the normalization

$$J_0(\mathbf{p}) = \sum_i p_i = 1.$$

With $n_i = Np_i$ in Eq. (2.11) we obtain

$$S(\mathbf{n}) = S(N\mathbf{p})] = -N \sum_i p_i \log p_i = NS(\mathbf{p}),$$

where $S(\mathbf{p})$ is the entropy functional in the form of Eq. (2.15) applied to $\mathbf{p}$,

$$S(\mathbf{p}) = -\sum_i p_i \log p_i. \tag{2.17}$$

We recognize the entropy functional of Gibbs and Shannon as the intensive entropy of the distribution, namely, as the entropy of the normalized distribution $\mathbf{p}$.

2.2.1 Entropy Inequality

Suppose we have two populations of clusters, A and B, with corresponding distributions $\mathbf{n}_A$ and $\mathbf{n}_B$ and combine them to produce the composite distribution

$$\mathbf{n}_{A+B} = \mathbf{n}_A + \mathbf{n}_B.$$

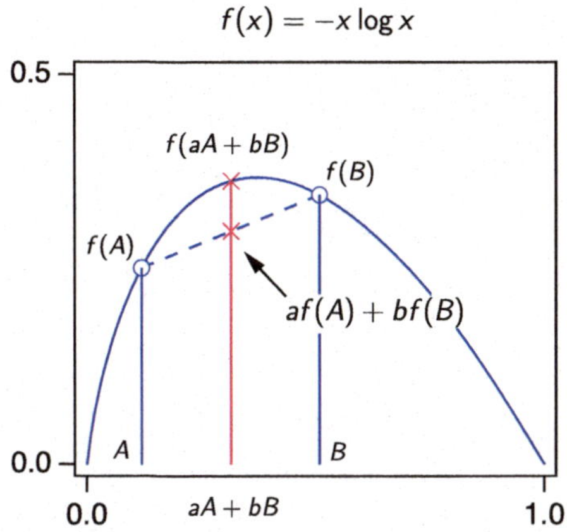

Fig. 2.1 Function $f(x) = -x \log x$ is concave and this implies the inequality $S(\mathbf{n}_A) + S(\mathbf{n}_B) \le S(\mathbf{n}_A + \mathbf{n}_B)$

We will compare the entropy of the composite system to that of its constituent distribution. Using $p_{iA} = n_{iA}/N_A$ and $p_{iB} = n_{iB}/N_B$ to denote the normalized populations, and $a = N_A/(N_A + N_B)$ and $b = N_B/(N_A + N_B)$ to indicate their relative amounts, we have:

$$\begin{aligned} S(\mathbf{n}_A) + S(\mathbf{n}_B) &= -(N_A + N_B) \sum_i \left(a p_{iA} \log p_{iA} + b p_{iB} \log p_{iA}\right) \\ &\le -(N_A + N_B) \sum_i (a p_{iA} + b p_{iB}) \log(a p_{iA} + b p_{iB}) \\ &= S(\mathbf{n}_A + \mathbf{n}_B). \end{aligned}$$

The last inequality follows from the concave property of the function $f(x) = -x \log x$,

$$a f(x) + b f(y) \le f(ax + bx); \quad a + b = 1,$$

and is demonstrated graphically in Fig. 2.1. Therefore we have

$$S(\mathbf{n}_A + \mathbf{n}_B) \ge S(\mathbf{n}_A) + S(\mathbf{n}_B). \tag{2.18}$$

In the special case $\mathbf{n}_A = \lambda \mathbf{n}_B$, i.e., the two distributions are homogeneous copies of each other, the inequality reduces to an exact equality:

$$S(\mathbf{n}_A) + S(\mathbf{n}_B) = S(\lambda \mathbf{n}_B) + S(\mathbf{n}_B) = (\lambda + 1) S(\mathbf{n}_B) = S((\lambda + 1)\mathbf{n}_B) = S(\mathbf{n}_A + \mathbf{n}_B).$$

We can express the concave property of entropy as follows:

> The entropy of the whole is greater than the entropy of the parts unless the parts are homogeneous copies of each other, in which case entropy becomes an additive property of the parts.

Note 2.2 (Homogeneity, Additivity, and Extensivity)

The entropy inequality in Eq. (2.18) is a general property of concave homogeneous functionals with degree 1. If $F(\mathbf{n})$ is such functional, then

$$F(\mathbf{n}_A + \mathbf{n}_B) \geq F(\mathbf{n}_A) + F(\mathbf{n}_B). \tag{2.19}$$

The homogeneous–concave inequality implies *additivity in isolation*: the right-hand side is the total property F of n_A and n_B in isolation from each other, and is given by the sum of property F for each distribution. The inequality states that F is not always conserved upon combining the isolated parts into a single distribution.

The equal sign applies in the special case that the distributions are homogeneous copies of each other, i.e., they both have the same normalized distribution:

$$\frac{\mathbf{n}_A}{N_A} = \frac{\mathbf{n}_B}{N_B} = \mathbf{p},$$

where N_A and N_B are the total number of clusters in each distribution. This is true for any concave homogeneous functional with degree 1. For convex functionals the inequality is reversed but the conditions for the equality to hold remain the same. Therefore, with homogeneous functionals regardless of curvature we obtain exact additivity when the parts are homogeneous copies of the whole. This implies extensivity. In the remainder of this book we will use the term "extensive" as synonymous to "homogeneous with degree 1" and the term "intensive" as synonymous to "homogeneous with degree 0."

To complete the discussion of curvature and additivity, we consider the special case that F is a linear functional:

$$F(\mathbf{n}) = \sum_i a_i n_i,$$

where n_i depends on i but not on n_i. In this case the homogeneous–concave inequality is an exact equality for all $\mathbf{n}$ and property F is conserved when the parts are combined into one whole.

★

2.3 Selection Functional

We have defined the microcanonical phase space of distributions $\mathbf{n}$, and we will now apply a probability measure over that space. We define the probability of distribution $\mathbf{n}$ to be proportional to its multiplicity $\mathbf{n}!$ and to a functional $W(\mathbf{n})$, which we call *selection functional*, or *selection bias*. By this definition the probability of $\mathbf{n}$ is

$$P(\mathbf{n}) = \frac{\mathbf{n}!W(\mathbf{n})}{\Omega_{M,N}}. \tag{2.20}$$

The product $\mathbf{n}!W(\mathbf{n})$ in the numerator is the *microcanonical weight* of distribution $\mathbf{n}$; the normalization constant in the denominator is the *microcanonical partition function* and is given by

$$\Omega_{M,N} = \sum_{\mathbf{n}} \mathbf{n}!W(\mathbf{n}), \tag{2.21}$$

with the summation running over all distributions of the (M, N) ensemble.

The selection functional fixes the probability of distribution and in this sense W embodies the stochastic model that gives rise to the population. We will leave W unspecified[3] except for certain mathematical properties. Specifically, we require W to be such that the log of the microcanonical weight is concave in $\mathbf{n}$ and homogeneous with degree 1. The purpose of these conditions will become clear in Chap. 3; until then we will simply point out that we require of the microcanonical weight the same curvature and homogeneous properties of entropy. Since the log of the microcanonical weight is

$$\log \mathbf{n}!W(\mathbf{n}) = S(\mathbf{n}) + \log W(\mathbf{n})$$

and $S(\mathbf{n})$ is homogeneous in $\mathbf{n}$ with degree 1, $\log W(\mathbf{n})$ must be as well:

$$\log W(\lambda\mathbf{n}) = \lambda W(\mathbf{n}). \tag{2.22}$$

[3]Specific examples will be given in Chaps. 8, 9, and 10 for the case of aggregation and fragmentation processes. Until then, W will be treated as a mathematical functional.

With respect to concavity, since entropy is concave, it is sufficient that $\log W$ be linear or concave in $\mathbf{n}$. The homogeneous property of $\log W$ allows us to write it in the Euler form

$$\log W(\mathbf{n}) = \sum_i n_i \left(\frac{\partial \log W(\mathbf{n})}{\partial n_i}\right)_{n_{j\neq i}} \equiv \sum_i n_i \log w_{i|\mathbf{n}}, \tag{2.23}$$

where $\log w_{i|\mathbf{n}}$ is the partial derivative of $\log W$ with respect to n_i:

$$\log w_{i|\mathbf{n}} \equiv \left(\frac{\partial \log W(\mathbf{n})}{\partial n_i}\right)_{n_{j\neq i}}. \tag{2.24}$$

The subscript $_{i|\mathbf{n}}$ indicates that the derivative depends not only on the cluster mass i but also on the entire distribution to which this cluster belongs. This is an important point to which we will frequently return. The differential of $\log W(\mathbf{n})$ is

$$d \log W(\mathbf{n}) = \sum_i \log w_{i|\mathbf{n}} dn_i. \tag{2.25}$$

If we now take the differential of (2.23) with respect to both n_i and $\log w_{i|\mathbf{n}}$ and use Eq. (2.25), the result is

$$\sum_i n_i d \log w_{i|\mathbf{n}} = 0. \tag{2.26}$$

This is the companion Gibbs-Duhem equation to (2.25) and expresses the fact that homogeneity imposes the restriction that the derivatives of $\log W$ may not all vary independently.

The functional W and its derivatives are central elements of the theory and deserve their own names. We will refer to W as *selection functional* or *selection bias*, to reflect its main function, which is to bias the selection of a distribution of the ensemble. We will refer to $w_{i|\mathbf{n}}$ as *cluster bias*, *cluster weight*, or *cluster function*. Both W, $w_{i|\mathbf{n}}$, and their logarithms will appear throughout the theory. We will not introduce separate names for the logarithms but will let the context determine which form we are referring to.

Note 2.3 (Probability of Configuration)

We have biased the probability of distribution by a functional that depends on the distribution. If we wish to assign probabilities to configurations we need an additional rule to distribute $P(\mathbf{n})$ among the $\mathbf{n}!$ permutations that represent the distribution. We adopt the rule that all permutations of a configuration are equally probable. Then the probability of configuration $\mathbf{m}$ with distribution $\mathbf{n}$ is

$$P(\mathbf{m}|\mathbf{n}) = \frac{P(\mathbf{n})}{\mathbf{n}!}. \tag{2.27}$$

Returning to Eq. (2.23), which expresses the log of the selection functional as a summation over the distribution, we may equivalently express it as a summation over the cluster masses in the configuration,

$$\log W(\mathbf{m}|\mathbf{n}) = \sum_{j=1}^{N} \log w_{m_j|\mathbf{n}}, \tag{2.28}$$

where j goes over all masses m_j of configuration $\mathbf{m} = (m_1, m_2 \cdots m_N)$. We substitute this expression into (2.27) and use (2.20) for the probability of distribution to write the probability of configuration in the form

$$P(\mathbf{m}|\mathbf{n}) = \frac{1}{\Omega_{M,N}} \prod_{j=1}^{N} w_{m_j|\mathbf{n}}. \tag{2.29}$$

Here, as well as in Eq. (2.27), the bias $w_{m_j|\mathbf{n}}$ of cluster mass m_j is calculated within the distribution $\mathbf{n}$ of that configuration and will generally vary between configurations with different distributions.

It is possible to distribute the probability $P(\mathbf{n})$ among its configurations in many different ways, not necessarily uniform. As long as the probability of distribution that is calculated from such reapportionment is the same, the precise probability of configuration is immaterial. Effectively we are declaring distribution to be the observable quantity and configuration to be a non-observable. As a non-observable, the probability of configuration cannot be ascertained by the observer and may be set arbitrarily as long as it does not conflict with the probabilities of observable quantities. In this view the assignment of probabilities to configurations is a problem of inference. The uniform distribution represents the simplest choice that agrees with the observable distribution.

★

2.3.1 The Linear Selection Functional

In general the partial derivatives of $\log W$ depend on the entire distribution $\mathbf{n}$ but an important special case is when these derivatives are independent of n_i and functions of i only[4]:

$$\left(\frac{\partial \log W(\mathbf{n})}{\partial n_i}\right)_{n_{j\neq i}} = \log w_i. \tag{2.30}$$

[4]We use the notation w_i as opposed to $w_{i|\mathbf{n}}$ to indicate that w_i is the same for all $\mathbf{n}$.

The Euler relationship for $\log W$ is

$$\log W(\mathbf{n}) = \sum_i n_i \log w_i, \tag{2.31}$$

from which we obtain W in the equivalent form

$$W(\mathbf{n}) = \prod_i (w_i)^{n_i}. \tag{2.32}$$

We will call selection functionals that satisfy (2.31) *linear* with the understanding that it is the logarithm of the selection functional to which the linear condition applies. Using Eq. (2.29), the probability of configuration in this case becomes

$$P(\mathbf{m}|\mathbf{n}) = \frac{1}{\Omega_{M,N}} \prod_{j=1}^{N} w_{m_j}, \tag{2.33}$$

where w_{m_j} is a function of cluster mass, the same in all configurations.

Linearity is an important but special case. In general the selection functional will be nonlinear. As an example of a nonlinear functional consider the one below,

$$\log W(\mathbf{n}) = \sum_i c_i n_i F(n_i/N),$$

where c_i are constants and $F(x)$ is such that the concave and homogeneous properties are satisfied but otherwise arbitrary. The cluster bias of this functional is

$$\log w_{i|\mathbf{n}} = \left(\frac{\partial \log W(\mathbf{n})}{\partial n_i}\right)_{n_{j \neq i}} =$$

$$c_i F\left(n_i/N\right) + c_i \left(\frac{n_i}{N}\right) F'(n_i/N) - \sum_{i=1}^{\infty} c_i \left(\frac{n_i}{N}\right)^2 F'\left(n_i/N\right)$$

where $F'(x) = dF(x)/dx$. It is nonlinear because the cluster bias of cluster size i is not the same in all distributions. The distinguishing feature of linear functionals is that the cluster bias is a pure function of cluster mass whereas that of nonlinear functionals is a function of the cluster mass whose form depends on the distribution to which the cluster mass belongs.

Note 2.4 (Redundancy of the Selection Functional)

Consider the functional of the form

$$W'(\mathbf{n}) = e^{Nf(M/N)}, \tag{2.34}$$

where $f(z)$ is some function that satisfies the concavity requirement. This functional is a pure function[5] of M and N. It evaluates to the same value for all distributions in the ensemble and therefore adds no bias to the selection of distributions. All members of the infinite family of functionals defined by Eq. (2.34) assign identical probabilities and in this sense they are *redundant*. The simplest functional in this family is

$$W(\mathbf{n}) = 1. \tag{2.35}$$

This is the *uniform bias* and the resulting ensemble will be called *unbiased*.[6]

Redundancy extends to biased functionals. If $W(\mathbf{n})$ is an arbitrary selection functional, the new functional W' defined as

$$W'(\mathbf{n}) = W(\mathbf{n})e^{Nf(M/N)}$$

assigns identical probabilities as W. Since f can be chosen arbitrarily, the above equation defines an entirely family of functional redundant with W. In Chap. 7 we will introduce a rule that allows us to reduce redundant functionals in a manner consistent with the reduction of Eq. (2.34) into (2.35).

2.4 Connection to the Multinomial Distribution

Using the Euler equation, Eq. (2.23), we express the selection functional in the form

$$W(\mathbf{n}) = e^{\sum_i n_i \log w_{i|\mathbf{n}}} = \prod_i \left(w_{i|\mathbf{n}}\right)^{n_i}, \tag{2.36}$$

then, inserting this result into Eq. (2.23) and expanding the multinomial coefficient, the microcanonical probability becomes

[5] Since moments are functionals of the distribution, it is perfectly acceptable to have selection functionals that reduce to pure functions of certain moments. What makes N and M special is that these are the only two moments that evaluate to the same number in every distribution of the ensemble.

[6] Even though Eq. (2.34) also produces uniform bias, the term will be used exclusively to refer to the selection functional in Eq. (2.35).

$$P(\mathbf{n}) = \frac{N!}{\Omega_{M,N}} \prod_i \frac{(w_{i|\mathbf{n}})^{n_i}}{n_i!}. \tag{2.37}$$

This is an equivalent expression for the microcanonical probability and has the form of the multinomial probability distribution. It is not the multinomial distribution because the $w_{i|\mathbf{n}}$ are not normalized probabilities. It is, however, closely related to the multinomial distribution, as we will see below.

Suppose we have a fair die with a large number of sides, each labeled with an integer number $1, 2 \cdots$. We toss it N times, record the outcomes, and retain the result only if the sum of the N tosses is exactly M.[7] Each realization of this stochastic experiment produces an ordered partition of integer M into N parts, i.e., a configuration of the (M, N) ensemble. The total number of distinct outcomes is the volume of the ensemble, $\mathcal{V}_{M,N}$. The probability that the die produces distribution $\mathbf{n}$ is given by the multinomial distribution,

$$P(\mathbf{n}) = \frac{N!}{n_1! n_2 \cdots} p_1^{n_1} p_2^{n_2} \cdots ,$$

subject to the conditions

$$\sum_i n_i = N, \quad \sum_i i n_i = M.$$

Since the die is fair, all ordered partitions are equally probable. Then $p_i = p_*$ for all i and the probability of $\mathbf{n}$ is

$$P(\mathbf{n}) = \frac{N!}{n_1! n_2! \cdots} p_*^N .$$

The value of p_* is fixed by the normalization condition,

$$p_*^N \sum_{\mathbf{n} \in (M,N)} \mathbf{n}! = 1,$$

from which we find $p_*^N = 1/\mathcal{V}_{M,N}$. Then for the probability of distribution $\mathbf{n}$ we have

$$P(\mathbf{n}) = \frac{1}{\mathcal{V}_{M,N}} \frac{N!}{n_1! n_2 \cdots} = \frac{\mathbf{n}!}{\mathcal{V}_{M,N}}.$$

[7]It is sufficient to work with a die that has $M - N + 1$ sides, since this is the maximum possible number in N tosses that add up to M.

This is the probability of distribution in the unbiased ensemble. Indeed with $W(\mathbf{n}) = 1$ for all $\mathbf{n}$ and using $\Omega_{M,N} = \mathcal{V}_{M,N}$, Eq. (2.20) produces the same result.

Now we impose an additional rule: once we have a feasible sequence of N tosses, namely a sequence that adds up to M, we decide whether to accept it with probability proportional to $W(\mathbf{n})$, where $\mathbf{n}$ is the distribution of the received outcomes. The probability of distribution $\mathbf{n}$ in this case is proportional to the probability to obtain $\mathbf{n}$ after N rolls, times the weight factor $W(\mathbf{n})$,

$$P(\mathbf{n}) = \frac{1}{C}\frac{\mathbf{n}!W(\mathbf{n})}{\mathcal{V}_{M,N}},$$

and C is given by the normalization condition,

$$C = \frac{1}{\mathcal{V}_{M,N}}\sum_{\mathbf{m}} \mathbf{n}!W(\mathbf{n}) = \frac{\Omega_{M,N}}{\mathcal{V}_{M,N}}.$$

With this the probability of distribution $\mathbf{n}$ becomes

$$P(\mathbf{n}) = \mathbf{n}!\frac{W(\mathbf{n})}{\Omega_{M,N}}, \qquad [2.23]$$

which is the same as the microcanonical probability Eq. (2.20).

Note 2.5 (Entropy, Selection Functional, and the Reference Ensemble)

The multinomial derivation of the microcanonical probability highlights the relationship between multiplicity of distribution and selection functional. Multiplicity arises from the independence condition between successive tosses and represents the sampling frequency of a distribution in the ensemble. The selection functional is an additional rule that skews the acceptance of a sampled distribution. To assign probabilities to the distributions of the ensemble we only need the product $\mathbf{n}!W(\mathbf{n})$, not $\mathbf{n}!$ and $W(\mathbf{n})$ individually; we could have used a biased die that does not require a separate step to decide whether to accept or reject a feasible sequence. By splitting the microcanonical weight into entropy and bias we have effectively chosen the unbiased ensemble the "reference ensemble" for assigning probabilities, such that $W = 1$ defaults to selecting distributions in proportion to their multiplicity.[8] This is equivalent to assigning equal probability to all configurations.

★

[8] We will revisit this idea in Chap. 7.

2.5 Entropies in the Cluster Ensemble

In the cluster ensemble we are dealing with three related but distinct stochastic variables: the configuration of cluster masses $\mathbf{m}$ with probability $P(\mathbf{m})$; the distribution of cluster masses $\mathbf{n}$ with probability $P(\mathbf{n})$; and the cluster mass i within distribution $\mathbf{n}$, whose probability distribution is $P(i|\mathbf{n})$.[9] Each of these probability distributions is characterized by its entropy; we will obtain the relationship between these entropies. We use the following notation:

$S[P(\mathbf{m})]$	entropy of probability distribution $P(\mathbf{m})$ (entropy of configurations)
$S[P(\mathbf{n})]$	entropy of probability distribution $P(\mathbf{n})$ (entropy of distributions)
$S[P(i|\mathbf{n})]$	entropy of probability distribution $P(i|\mathbf{n})$ (entropy of cluster masses in distribution $\mathbf{n}$)

We begin by applying the entropy functional to the probability of configuration $P(\mathbf{m})$. Using Eq. (2.29) for $P(\mathbf{m})$ we have[10]

$$S[P(\mathbf{m})] = -\sum_{\mathbf{m}} P(\mathbf{m}) \log \frac{W(\mathbf{m})}{\Omega_{M,N}} = \log \Omega_{M,N} - \langle \log W \rangle ,$$

where $\langle \log W \rangle$ is the ensemble average of the selection bias among all distributions of the ensemble. Next we calculate the entropy of the probability of distribution $P(\mathbf{n})$. Using Eq. (2.20) we have

$$S[P(\mathbf{n})] = -\langle \log W \rangle - \langle \log \mathbf{n}! \rangle + \log \Omega_{M,N}.$$

where $\log \mathbf{n}! = NS(P[S[P(i|\mathbf{n})]])$ is the entropy of distribution $\mathbf{n}$. Its ensemble average is

$$\langle \log \mathbf{n}! \rangle = \sum_{\mathbf{n}} P(\mathbf{n}) \log \mathbf{n}! = N \langle S[P(i|\mathbf{n})] \rangle .$$

Combining these results we obtain

$$S[P(\mathbf{m})] = S[P(\mathbf{n})] + N \langle S[P(i|\mathbf{n})] \rangle . \tag{2.38}$$

[9] We previously used p_i for the cluster mass probability. Here we use $P(i|\mathbf{n})$ to emphasize that this probability refers to specific distribution $\mathbf{n}$.

[10] As previously, we are using the notation $W(\mathbf{m})$ to refer to the bias of configuration $\mathbf{m}$; this is equal to $W(\mathbf{n})$, where $\mathbf{n}$ is the cluster distribution in $\mathbf{m}$.

The entropy of configurations is the sum of the entropy of distributions plus the extensive entropy of cluster masses. We call $S(P(\mathbf{m}))$ *entropy of the ensemble*.

2.6 The Binary Exchange Reaction

Suppose we take a configuration $\mathbf{m}$ of the ensemble $\mu C(M, N)$ and by some algorithm convert it into a new configuration $\mathbf{m}'$ in the same ensemble. This transformation is represented by the transition

$$\mathbf{m} \to \mathbf{m}' \tag{2.39}$$

We define the *equilibrium constant* of this transition as

$$K_{\mathbf{m}\to\mathbf{m}'} \equiv \frac{P(\mathbf{m}')}{P(\mathbf{m})} = \frac{W(\mathbf{m}')}{W(\mathbf{m})}, \tag{2.40}$$

with the result on the far right obtained by application of Eq. (2.29). One particularly simple algorithm for transforming configurations is this: choose two clusters in configuration $\mathbf{m}$ at random, say m_i and m_j, and transfer a number of members from one cluster to the other, choosing the number of members to transfer and the direction of the transfer randomly among all possible transfers between the two clusters. We call this process binary exchange reaction, a term that alludes to a chemical reaction between physical clusters (Fig. 2.2). The binary reaction systematically transitions between all configurations of the microcanonical ensemble. To complete the process we accept the result of the reaction with probability $p_{\mathbf{m}\to\mathbf{m}'}$ given by

$$p_{\mathbf{m}\to\mathbf{m}'} = \begin{cases} K_{\mathbf{m}\to\mathbf{m}'} & \text{if } K_{\mathbf{m}\to\mathbf{m}'} \le 1, \\ 0 & \text{otherwise.} \end{cases} \tag{2.41}$$

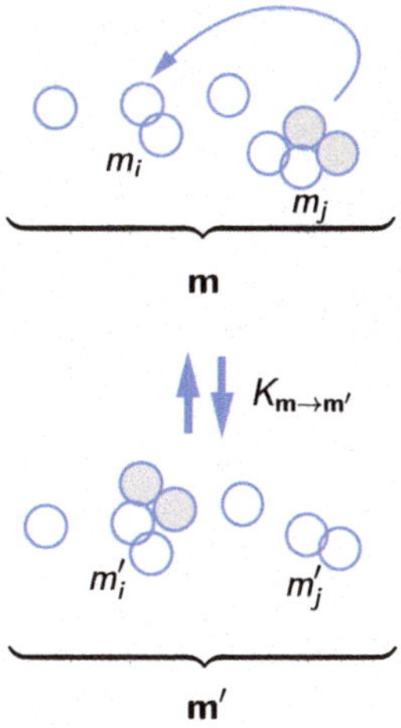

Fig. 2.2 In binary exchange reaction clusters i and j exchange a random number of members δ. The process transforms the parent configuration into a new configuration of the same ensemble. The process replaces cluster masses m_i and m_j with $m_i' = m_i + \delta$, $m_j' = m_j - \delta$

This is the acceptance probability of the Metropolis method and produces transition probabilities that satisfy the equilibrium constant in Eq. (2.40). The sequence of configurations that is produced by the method described here converges to the microcanonical probability $P(\mathbf{m})$. To see why, we take a closer look at the binary exchange reaction.

Given a configuration with N clusters and M total units, the number of units available for transfer is $M-N$, since a cluster with m units can transfer at most $m-1$ units (no cluster may be left empty after the transfer). Excluding the cluster from which transferred members originate, members can be moved to any of the other $N-1$ clusters. Therefore the total number of exchange reactions is $(M-N)(N-1)$ and is the same in all configurations of the microcanonical ensemble. As an example, with $M = 6$, $N = 3$, we have 6 possible exchange reactions in every configuration. These are enumerated below for the configuration $\mathbf{m} = (1, 3, 2)$:

$$(1,3,2) \to \begin{cases} (2,2,2) \\ (3,1,2) \\ (1,4,1) \\ (1,2,3) \\ (1,1,4) \\ (2,3,1) \end{cases}$$

Every exchange reaction that originates from the same configuration produces a unique configuration, as we confirm in this example. This generalizes to any M, N. It is a simple matter to show that we can convert through a series of binary exchange reactions any configuration of the microcanonical ensemble into any other configuration.[11] That is, this Markov process is reducible. If we form a graph whose nodes are configurations and its edges represent exchange reactions, this will be an undirected connected graph in which every node is connected to $z = (M-N)(N-1)$ neighbors. The transition probability $P(\mathbf{m} \to \mathbf{m}')$ for the exchange reaction $\mathbf{m} \to \mathbf{m}'$ with acceptance probability from Eq. (2.41) is

$$P(\mathbf{m}'|\mathbf{m}) = \begin{cases} \frac{1}{z} K_{\mathbf{m}\to\mathbf{m}'}, & \text{if } K_{\mathbf{m}\to\mathbf{m}'} \le 1 \\ \frac{1}{z}, & \text{otherwise.} \end{cases}$$

This transition probability satisfies detailed balance,

$$P(\mathbf{m})P(\mathbf{m} \to \mathbf{m}') = P(\mathbf{m}')P(\mathbf{m}' \to \mathbf{m}),$$

[11] For example, perform a series of transfers to a single cluster until its mass is $M - N + 1$ (and all other clusters are monomers), then transfer from that cluster to the rest until the desired configuration is reached.

and guarantees that the limiting distribution of configurations obtained by a sequence of binary exchange reactions is the microcanonical probability $P(\mathbf{m})$. It follows that the distributions of the sampled configurations is equal to the microcanonical probability $P(\mathbf{n})$.[12]

The binary exchange reaction provides a convenient method to sample configurations of the ensemble and is by far the simplest to implement numerically. It is also important in a different respect: it establishes a mapping between a static ensemble of configurations and one in which configurations undergo reactions that establish a stationary state of equilibrium. The two ensembles share the same distribution of configurations. The ensemble produced by the exchange reaction has direct analogies to a system of *physical* particles that undergo reversible clustering and fragmentation reactions. Let us return to the equilibrium constant in Eq. (2.40) and assume for the moment that the selection functional is linear and is given by Eq. (2.31). If the binary reaction involves cluster masses m_i and m_j as reactants, and m'_i and m'_j as products, the equilibrium constant becomes

$$K_{\mathbf{m}\to\mathbf{m}'} = \frac{w_{m'_i} w_{m'_j}}{w_{m_i} w_{m_j}}, \tag{2.42}$$

since the number of clusters and the cluster bias of all other clusters remain unaffected by the reaction.

$$m_i + m_j \overset{K}{\leftrightharpoons} m'_i + m'_j.$$

For nonlinear functionals we may write a similar equation but under more restrictive conditions. If all n_i on the reactant side are large, the change in $\mathbf{n}$ due to the exchange reaction (-1 for the n_i of the reactants, $+1$ for the products, and no change for all other n_i) represents a differential change and we may use the differential of $\log W$ in Eq. (2.25) to calculate it:

$$\log W(\mathbf{n}') - \log W(\mathbf{n}) = -\log w_{m_i|\mathbf{n}} - \log w_{m_j|\mathbf{n}} + \log w_{m'_i|\mathbf{n}} + \log w_{m'_j|\mathbf{n}}.$$

The equilibrium constant now becomes

$$K_{\mathbf{m}\to\mathbf{m}'} = \frac{w_{m'_i|\mathbf{n}} w_{m'_j|\mathbf{n}}}{w_{m_i|\mathbf{n}} w_{m_j|\mathbf{n}}}. \tag{2.43}$$

The result has the form of (2.42) but now the w's depend on the entire distribution, not only on the clusters that participate in the reaction. Moreover, the validity of Eq. (2.43) is restricted to distributions whose n_i are large, whereas Eq. (2.42) is generally true without restriction.

---

[12]See Kelly (2011) or Rozanov (1977).

Example 2.6 (Exchange Reactions)

We demonstrate the calculation of probabilities and sampling by binary exchange reaction with a numerical calculation in the ensemble $M = 6$, $N = 3$. We perform calculations for a linear selection bias with cluster bias $\log w_i = -ai^2$, with $a = 0.02$. The corresponding bias is

$$\log W(\mathbf{n}) = -a \sum_i i^2 n_i \quad \text{or} \quad W(\mathbf{n}) = e^{-a(n_1+4n_2+9n_3+\cdots)}.$$

As a linear bias, W may also be written in terms of configuration masses,

$$W(\mathbf{m}) = e^{-a(m_1^2+m_2^2\cdots+m_N^2)},$$

where m_i are the cluster masses in configuration $\mathbf{m}$.

First we compute the properties of the ensemble by direct calculation. The ensemble contains three distributions:

$$\mathbf{n}_A = (2, 0, 0, 1) \qquad \mathbf{n}_B = (1, 1, 1, 0) \qquad \mathbf{n}_C = (0, 3, 0, 0).$$

The corresponding values of the selection functional and multiplicity factors are

	$\mathbf{n}_A$	$\mathbf{n}_B$	$\mathbf{n}_C$
$W(\mathbf{n})$	0.697676	0.755784	0.786628
$\mathbf{n}!$	3	6	1

The partition function is

$$\Omega = \mathbf{n}_A!W(\mathbf{n}_A) + \mathbf{n}_B!W(\mathbf{n}_B) + \mathbf{n}_C!W(\mathbf{n}_C) = 7.41436.$$

Finally, the probabilities of all distributions are calculated from Eq. (2.20):

$$P(\mathbf{n}_A) = \mathbf{n}_A!\frac{W(\mathbf{n}_A!)}{\Omega} = 0.282294$$
$$P(\mathbf{n}_B) = \mathbf{n}_B!\frac{W(\mathbf{n}_B!)}{\Omega} = 0.611611$$
$$P(\mathbf{n}_C) = \mathbf{n}_C!\frac{W(\mathbf{n}_C!)}{\Omega} = 0.106095.$$

The mean cluster distribution is

$$\bar{\mathbf{n}} = \mathbf{n}_A P(\mathbf{n}_A) + \mathbf{n}_B P(\mathbf{n}_B) + \mathbf{n}_C P(\mathbf{n}_C) = (1.176, 0.930, 0.612, 0.282).$$

We now sample this ensemble by Monte Carlo simulation of exchange reactions. Starting with an initial configuration we randomly pick two clusters. Let i and j be

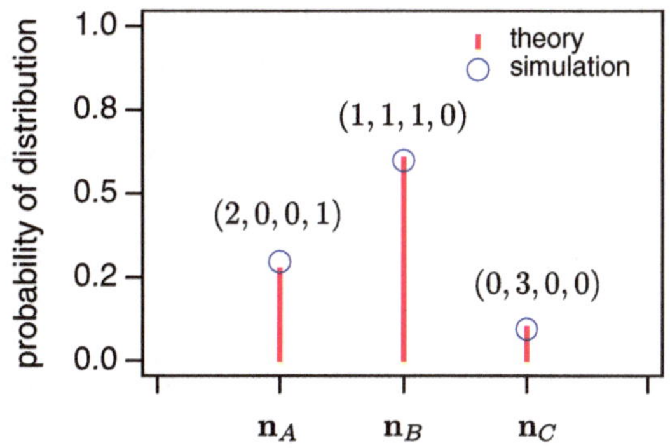

Fig. 2.3 Numerical simulation of cluster ensemble with $M = 6$, $N = 3$, $\log W = -0.1 \sum_i i^2 n_i$. The bars are exact calculations and the symbols are the results of Monte Carlo simulation by binary exchange reactions

their masses. We draw an integer random number k from a uniform distribution in the interval $[1, i + j - 1]$, and replace the original two clusters by two new ones with mass k and $i + j - k$, respectively. If $W(\mathbf{m})/W(\mathbf{m}) \geq 1$, the transition is accepted; if $W(\mathbf{m})/W(\mathbf{m}) < 1$ we draw a random number r in the interval (0, 1) and accept the move if $r < W(\mathbf{m})/W(\mathbf{m})$. These steps are repeated until enough data are collected for analysis. Figure 2.3 shows the results of this process after 5000 steps. The vertical sticks are the exact probabilities and the circles are the results of the simulation, and the two agree, as they should.

★

2.7 Linear Ensembles

The microcanonical ensemble will be called *linear* if $\log W(\mathbf{n})$ is a linear functional of $\mathbf{n}$; equivalently, if the cluster function is an intrinsic function of i for all distributions:

$$w_{i|\mathbf{n}} = w_i \quad (\text{for all } \mathbf{n}). \tag{2.44}$$

The fact that the cluster bias is the same function of cluster size in all distributions introduces significant mathematical simplifications that allow us to obtain several results in closed form. We begin by writing the microcanonical probability of distribution in the form

$$P(\mathbf{n}) = \frac{N!}{\Omega_{M,N}} \prod_i \frac{w_i^{n_i}}{n_i!}. \tag{2.45}$$

For the partition function then we have

$$\Omega_{M,N} = N! \sum_{\mathbf{n}} \prod_i \frac{w_i^{n_i}}{n_i!}. \tag{2.46}$$

We take the derivative of the partition function with respect to w_k. Noting that the term w_k appears once in every distribution of the ensemble, this derivative is

$$\frac{\partial \Omega_{M,N}}{\partial w_k} = N! \sum_{n_i} \cdots n_k \frac{w_k^{n_k-1}}{n_k!} \cdots . \tag{2.47}$$

where the $\cdots$ refer to all $w_i^{n_i}/n_i!$ with $i \neq k$. This result is also written as

$$\frac{\partial \Omega_{M,N}}{\partial w_k} = N \underbrace{\left\{ (N-1)! \sum_{n_i} \cdots \frac{w_k^{n_k-1}}{(n_k-1)!} \cdots \right\}}_{\Omega_{M-k,N-1}} . \tag{2.48}$$

The quantity in braces amounts to removing one cluster of mass k from the ensemble $\mu C(M, N)$. The summation then produces the partition function of the microcanonical ensemble $\mu C(M-k, N-1)$:

$$\frac{\partial \Omega_{M,N}}{\partial w_k} = N \Omega_{M-k,N-1}. \tag{2.49}$$

We return to Eq. (2.47), which we write in the form,

$$\frac{\partial \Omega_{M,N}}{\partial w_k} = \frac{\Omega_{M,N}}{w_k} \underbrace{\left\{ \frac{1}{\Omega_{M,N}} \sum_{n_i} \cdots n_k \frac{(w_k)^{n_k}}{n_k!} \cdots \right\}}_{\langle n_k \rangle} .$$

The quantity inside the braces is the mean value of n_k in the ensemble, therefore,

$$\frac{\partial \Omega_{M,N}}{\partial w_k} = \frac{\langle n_k \rangle}{w_k} \Omega_{M,N}. \tag{2.50}$$

The result is easily generalized to derivatives of higher order in w_k. By recursive application of Eqs. (2.49) and (2.50), the derivative of order j is

$$\begin{aligned} \frac{\partial^j \Omega_{M,N}}{\partial w_k^j} &= N(N-1)\cdots(N-j+1)\, \Omega_{M-jk,N-i} \\ &= \frac{\langle n_k(n_k-1)\cdots(n_k-j+1)\rangle}{w_k^j} \Omega_{M,N}, \end{aligned} \tag{2.51}$$

Eliminating the derivative of the partition function between the two results we also have,[13]

[13]In writing $\Omega_{M-jk,N-j}$ we understand the subscripts to satisfy the conditions $M - jk \geq N - j \geq 1$.

$$\langle n_k(n_k-1)\cdots(n_k-j+1)\rangle = w_k^j \frac{N!}{(N-j)!}\frac{\Omega_{M-jk,N-j}}{\Omega_{M,N}}. \tag{2.52}$$

Comparing this result with Eq. (2.51) we obtain the following identity involving the partition function and the cluster bias:

$$\frac{\partial^j \log \Omega_{M,N}}{\partial w_k^j} = \frac{N!}{(N-j)!}\frac{\Omega_{M-jk,N-j}}{\Omega_{M,N}}, \tag{2.53}$$

which is a generalization of Eq. (2.49).

All integer moments are obtained from Eq. (2.52). With $i = 1$ we obtain the mean cluster distribution in the ensemble:

$$\langle n_k\rangle = N w_k \frac{\Omega_{M-k,N-1}}{\Omega_{M,N}}. \tag{2.54}$$

The second moment is obtained from (2.52) with $i = 2$,

$$\left\langle n_k^2\right\rangle = (w_k)^2 N(N-1)\frac{\Omega_{M-2k,N-2}}{\Omega_{M,N}} + N w_k \frac{\Omega_{M-k,N-1}}{\Omega_{M,N}}. \tag{2.55}$$

The variance $\mathrm{Var}(n_k) = \left\langle n_k^2\right\rangle - \langle n_k\rangle^2$ is,

$$\mathrm{Var}(n_k) = w_k^2 N(N-1)\frac{\Omega_{M-2k,N-2}}{\Omega_{M,N}} - w_k N \frac{\Omega_{M-k,N-1}}{\Omega_{M,N}}\left(N w_k \frac{\Omega_{M-k,N-1}}{\Omega_{M,N}} - 1\right). \tag{2.56}$$

It is more convenient to express the variance as a ratio relative to $(\overline{n_k})^2$:

$$\frac{\mathrm{Var}(n_k)}{\langle n_k\rangle^2} = \frac{N-1}{N}\left(\frac{\Omega_{M-2k,N-2}}{\Omega_{M,N}}\right) + 1 - \frac{1}{\langle n_k\rangle}. \tag{2.57}$$

Higher order moments can be calculated by the same method.

Note 2.7 (Linear ensembles in the literature)
Given an integer partition $\mathbf{m} = \{m_1, m_2 \cdots m_N\}$ of M into N parts, the linear ensemble assigns a probability to that partition that is proportional to the product

$$w_{m_1} w_{m_2} \cdots w_{m_N},$$

where the w_k are factors that depend on the size k of the elements of the partition. Such probability spaces are called *Gibbs partitions* (Pitman 2006). One of the earliest references to this type of probability space was made by Darwin and Fowler (1922) in their proof of the equality between the mean and most probable distribution in statistical mechanics. In this case the factors w_i are introduced as a mathematical convenience but are eventually set to 1 to obtain the canonical distribution of statistical thermodynamics. A tutorial discussion of the method is given in Schrödinger (1989). Gibbs partitions have been studied in relationship to stationary stochastic Markov processes (Kelly 2011) and also with reference to aggregation/fragmentation process (Berestycki and Pitman 2007; Durrett et al. 1999). We discuss population balances in Chaps. 8, 9, and 10 where we will see that linear ensembles represent an important but rather special case.

★

2.8 The Unbiased Ensemble

The simplest linear ensemble is when W is the unbiased functional:

$$W(\mathbf{n}) = 1 \qquad [2.35]$$

with cluster bias

$$w_i = 1. \tag{2.58}$$

for all distributions. The uniform bias assigns probabilities to distributions in proportion to their multiplicity; equivalently, it assigns uniform distribution to all configurations. Since all configurations are equally probable, the partition function is equal to the volume of the ensemble in Eq. (2.7):

$$\Omega_{M,N} = \mathcal{V}_{M,N} = \binom{M-1}{N-1}. \tag{2.59}$$

It follows that the probability of distribution $\mathbf{n}$ in the unbiased ensemble is

Fig. 2.4 Surface plot of the mean distribution of the unbiased ensemble for $M = 12, N = 1, \cdots 12$

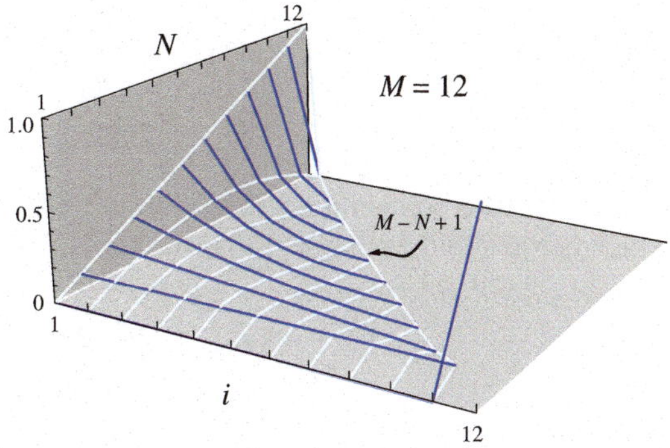

$$P(\mathbf{n}) = \mathbf{n}! \Big/ \binom{M-1}{N-1}. \tag{2.60}$$

The mean cluster distribution is obtained from Eq. (2.54):

$$\frac{\langle n_k \rangle}{N} = \binom{M-k-1}{N-2} \Big/ \binom{M-1}{N-1}. \tag{2.61}$$

This distribution is exact for all finite M and N (Fig. 2.4).

Example 2.8 (Unbiased Ensemble)

We demonstrate the unbiased ensemble with a calculation for $M = 12$, $N = 7$. First we collect the distributions of the ensemble by constructing the integer partitions of $M = 12$ into $N = 7$ parts. The ensemble consists of 7 distributions, represented in the table below by a sorted configuration of cluster masses:

m
2 2 2 2 2 1 1
3 2 2 2 1 1 1
3 3 2 1 1 1 1
4 2 2 1 1 1 1
4 3 1 1 1 1 1
5 2 1 1 1 1 1
6 1 1 1 1 1 1

The first configuration represents distribution

$$\mathbf{n} = (2, 5, 0, 0, 0, 0)$$

whose multiplicity is

$$\mathbf{n}! = \frac{(5+2)!}{5!\,2!} = 21.$$

To calculate the probability of distribution we first calculate the partition function, which is equal to the volume of the ensemble,

$$\Omega_{12,7} = \binom{11}{6} = 462.$$

The probability of distribution $\mathbf{n} = (2, 5, 0, 0, 0, 0)$ is

$$P_{2,5,0,0,0,0} = \frac{21}{462} = 0.04545.$$

The table below summarizes the results for all distributions in this ensemble.

$\mathbf{n}$	$\mathbf{n}!$	$P(\mathbf{n})$
2 5 0 0 0 0	21	21/462 = 0.045454
3 3 1 0 0 0	140	140/462 = 0.303030
4 1 2 0 0 0	105	105/462 = 0.227273
4 2 0 1 0 0	105	105/462 = 0.227273
5 0 1 1 0 0	42	42/462 = 0.090909
5 1 0 0 1 0	42	42/462 = 0.090909
6 0 0 0 0 1	7	7/462 = 0.015151
	462	1.000000

Each distribution is listed in rows (first element is number monomers, second element number of dimers, etc.). The table defines the statistics of the ensemble completely. As an example, we calculate the mean number of dimers:

$$\begin{aligned}\langle n_2 \rangle &= \sum_{\mathbf{n}} n_{2|\mathbf{n}} P(\mathbf{n}) \\ &= (5)\frac{21}{462} + (3)\frac{140}{462} + (1)\frac{105}{462} + (1)\frac{105}{462} + (0)\frac{42}{462} + (1)\frac{42}{462} + (0)\frac{7}{462} = \frac{21}{11}\end{aligned}$$

The same result is obtained from Eq. (2.61):

$$\bar{n}_2 = N\binom{M-3}{N-2} \Big/ \binom{M-1}{N-1} = 7\binom{9}{5} \Big/ \binom{11}{6} = \frac{21}{11}.$$

The second moment of the dimers is

$$\left\langle n_2^2 \right\rangle = (5)^2 \frac{21}{462} + (3)^2 \frac{140}{462} + (1)^2 \frac{105}{462} + \\ (1)^2 \frac{105}{462} + (0)^2 \frac{42}{462} + (1)^2 \frac{42}{462} + (0)^2 \frac{7}{462} = \frac{56}{11}.$$

Alternatively, using Eq. (2.55) we find

$$\overline{n_2^2} = 7 \cdot 6 \cdot \binom{7}{4} \Big/ \binom{11}{6} + 7 \binom{9}{5} \Big/ \binom{11}{6} = \frac{56}{11},$$

in agreement with the exact calculation.

★

2.9 Canonical Ensemble

We start with a microcanonical configuration $\mathbf{m}$ of the (M, N) ensemble and select N' clusters $(0 < N' < N)$ in the order they appear in $\mathbf{m}$. This produces a *canonical* configuration $\mathbf{m}'$ with N' ordered elements (Fig. 2.5). The remaining $N - N'$ elements of $\mathbf{m}$ form the complement $\mathbf{m}''$ of canonical configuration $\mathbf{m}'$. This complement is also a canonical configuration. The canonical phase space is therefore a sub-partitioning of the microcanonical space into two complementary partitions.

The relationship between the canonical and microcanonical phase spaces can be illustrated using the microcanonical table (Table 2.2). The table has N columns and $\mathcal{V}_{M,N}$ rows that contain every integer partition of M into N parts in all possible permutations. Reordering the columns amounts to a permutation in the order of clusters in all configurations but since the ensemble contains all possible permutation, reordering columns produces an identical set of microcanonical configurations. Reordering the rows has also no effect. That is, the ordering of the rows and columns in the microcanonical table is immaterial and can be set arbitrarily. Each row of the microcanonical table is a microcanonical configuration, a set of N integers whose sum is M. Each column constitutes a *canonical slice*. Canonical slices have several important properties:

1. *All canonical slices contain* $\mathcal{V}_{M,N}$ *elements.*

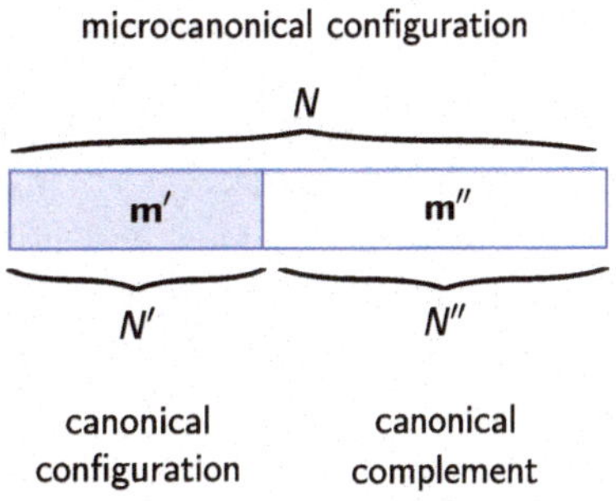

Fig. 2.5 A canonical configuration is constructed by selecting the first N' elements selected from a microcanonical configuration

Table 2.2 Canonical configurations produced from two columns of the microcanonical ensemble with 5 clusters and total mass 8, i.e., $N' = 2$, $M = 8$, $N = 5$ (configurations are reordered and grouped by the mass in the canonical configuration.). The shaded columns of cluster masses are the canonical ensemble and the unshaded are its microcanonical complement. Horizontal lines mark the microcanonical slices of the canonical ensemble

#	m_1	m_2	m_3	m_4	m_5	M'
1	1	1	2	2	2	2
2	1	1	1	2	3	2
3	1	1	1	3	2	2
4	1	1	2	1	3	2
5	1	1	2	3	1	2
6	1	1	3	1	2	2
7	1	1	3	2	1	2
8	1	1	1	1	4	2
9	1	1	1	4	1	2
10	1	1	4	1	1	2
11	1	2	1	2	2	3
12	1	2	2	1	2	3
13	1	2	2	2	1	3
14	2	1	1	2	2	3
15	2	1	2	1	2	3
16	2	1	2	2	1	3
17	1	2	1	1	3	3
18	1	2	1	3	1	3
19	1	2	3	1	1	3
20	2	1	1	1	3	3
21	2	1	1	3	1	3
22	2	1	3	1	1	3
23	2	2	1	1	2	4
24	2	2	1	2	1	4
25	2	2	2	1	1	4
26	1	3	1	1	2	4
27	1	3	1	2	1	4
28	1	3	2	1	1	4
29	3	1	1	1	2	4
30	3	1	1	2	1	4
31	3	1	2	1	1	4
32	2	3	1	1	1	5
33	3	2	1	1	1	5
34	1	4	1	1	1	5
35	4	1	1	1	1	5

This is a straightforward consequence of the fact that the number of rows in the microcanonical table is equal to the total number of configurations, $\mathcal{V}_{M,N}$.

2. *All canonical slices contain every cluster mass m from* 1 *to* $M - N + 1$; *cluster mass m appears* $\mathcal{V}_{M-m,N-1}$ *times in every canonical slice.*

The microcanonical ensemble contains every cluster mass from 1 up to the maximum cluster mass $M - N + 1$ and since the microcanonical table contains every permutation of cluster masses, each mass appears in every column of the table.

3. *The multiplicity of cluster mass m in every canonical slice is $\mathcal{V}_{M-m,N-1}$. This is the number of times mass* m *appears in the slice.*

   The multiplicity of mass m is equal to the number of complementary configurations, which is easier to calculate. The complement of cluster mass m is a configuration with $N-1$ clusters and total mass $M-m$; there are $\mathcal{V}_{M-m,N-1}$ such configurations.[14] Accordingly, the multiplicity of mass m is equal to the microcanonical volume of its complement,

$$\mathcal{V}_{M-m,N-1} = \binom{M-m-1}{N-2}, \qquad m = 1, 2, \cdots M-N+1. \tag{2.62}$$

4. *All canonical slices contain the same distribution of clusters masses.* This follows from the fact that all slices contain the same cluster masses and each mass has the same multiplicity. The canonical slice then is a homogeneous subset of the cluster masses contained in the microcanonical ensemble: it contains a fraction $1/N$ of the total clusters in the same proportions as in the entire ensemble.
5. *The following identities are true:*

$$\sum_{m=1}^{M-N+1} \mathcal{V}_{M-m,N-1} = \mathcal{V}_{M,N}, \tag{2.63}$$

$$\sum_{m=1}^{M-N+1} m \frac{\mathcal{V}_{M-m,N-1}}{\mathcal{V}_{M,N}} = \frac{M}{N}. \tag{2.64}$$

   The left-hand side in Eq. (2.63) is the sum of the multiplicities of all cluster masses in the canonical slice; this is equal to the total number of clusters in the canonical slice. The left-hand side of Eq. (2.64) is the mean cluster mass in the canonical slice; this is equal to the mean cluster mass in the entire ensemble, which is M/N.

Example 2.9 (Canonical Slices)

Canonical slices of the microcanonical ensemble ($M = 8, N = 5$) contain cluster masses between $m_{\text{min}} = 1$ and $m_{\text{max}} = M - N + 1 = 4$. The multiplicity of the cluster masses in the canonical slice is calculated below:

m	$\mathcal{V}_{M-m,N-1}$
1	20
2	10
3	4
4	1

[14] A mass m in a given column of the microcanonical table is paired with every permutation of every configuration that contains $N-1$ clusters with total mass $M-m$; their number is the volume of the $(M-m, N-1)$ ensemble.

All canonical slices contains 20 monomers, 10 dimers, 4 trimers, and 1 tetramer. The total number of clusters is

$$20 + 10 + 4 + 1 = 35,$$

equal to $\mathcal{V}_{8,5}$, the volume of the microcanonical ensemble $\mu C(8, 5)$. The average cluster mass in the canonical slice is

$$\frac{20(1) + 10(2) + 4(3) + 1(4)}{35} = 1.6.$$

This is the same as $M/N = 8/5 = 1.6$.

★

2.9.1 Canonical Configuration

A *canonical configuration* is an ordered list of N' cluster masses ($N > N' \geq 1$) picked randomly without replacement from a microcanonical configuration of the ensemble $\mu C(M, N)$. Since every cluster appears the same number of times in every column of the microcanonical table, we may construct the canonical configuration by picking the first N' elements of a configuration. Then, the first N' columns of the microcanonical table constitute the canonical ensemble associated with the parent microcanonical ensemble. The complement of the canonical configuration is also a canonical configuration with $N - N'$ clusters and mass $M - M'$, where M' is the total mass in the canonical configuration.

The total mass M' in a canonical configuration is a fluctuating quantity. The smallest canonical mass is when all members are monomers. A canonical configuration with all monomers is always present in the ensemble. To see why, recall that the microcanonical ensemble contains one configuration with $N - 1$ monomers plus one cluster of size $M - N + 1$, which is the maximum cluster that exist in the (M, N) ensemble. The permutations of this configuration ensure that any canonical ensemble with $1 \leq N' < N$ will contain one configuration (and possibly more) with N' monomers. Therefore the smallest mass of configuration in the canonical ensemble is

$$M'_{\text{min}} = N'. \tag{2.65}$$

The largest mass of canonical configuration is when the complementary configuration consists of monomers only. In this case the mass of the complement is $N - N'$, therefore the mass of the canonical configuration

$$M'_{\text{max}} = M - N + N'. \tag{2.66}$$

As a corollary, the bounds of the canonical mass satisfy the condition,

$$M'_{\max} - M'_{\min} = M - N, \tag{2.67}$$

which states that the difference between the largest and smallest canonical mass is the same in all configurations and equal to the difference between the maximum and minimum cluster masses in the microcanonical ensemble.[15]

In contrast to the microcanonical ensemble where configurations appear only once, configurations in the canonical ensemble generally appear multiple times (Table 2.1). The multiplicity of canonical configuration is equal to the total number of complementary configurations with which it combines to form a microcanonical configuration in $\mu C(M, N)$. All complementary configurations to canonical configuration $\mathbf{m}'$ contain $N - N'$ clusters and have total mass $M - M'$. The set of complements to canonical configuration $\mathbf{m}'$ constitutes the complete microcanonical ensemble $\mu C(M - M', N - N')$. This implies that the complement contains every configuration of $\mu C(M-M', N-N')$, exactly once. This is because the combination of configuration $\mathbf{m}'$ with *any* configuration that produces N clusters with total mass M is a configuration in $\mu C(M, N)$ and appears in the ensemble exactly once. Therefore, the multiplicity of configuration $\mathbf{m}'$ in the canonical ensemble $C(M', N')$ is

$$\left(\text{multiplicity of } \mathbf{m}'\right) = \mathcal{V}_{M-M',N-N'}.$$

The multiplicity of the corresponding cluster distribution $\mathbf{n}'$ is

$$\left(\text{multiplicity of } \mathbf{n}'\right) = \mathbf{n}'!\,\mathcal{V}_{M-M',N-N'}, \tag{2.68}$$

since the cluster masses of configuration $\mathbf{m}'$ appear in $\mathbf{n}'!$ permutations. Within distributions with the same mass M', the canonical multiplicity is proportional to the microcanonical multiplicity, which is given by the multinomial factor. The subset of canonical configurations with the same mass constitutes a *microcanonical slice*. This subset contains all configurations of the microcanonical ensemble $\mu C\left(M', N'\right)$, each present $\mathcal{V}_{M-M,N-N'}$ times, equal to the volume of the complementary microcanonical ensemble. Just as the canonical ensemble is a union of canonical slices, the canonical ensemble is a union of microcanonical slices.

Example 2.10 (Canonical Configuration)

Table 2.2 shows the canonical ensemble with two clusters ($N' = 2$) extracted from the microcanonical ensemble $M = 8$, $N = 10$.

[15] In the microcanonical ensemble the largest and smallest masses are $M-N+1$ and 1, respectively; their difference is $M - N$.

Consider the configuration $\mathbf{m}' = (3, 1)$ with mass $M' = 4$. All of its complementary configurations have mass $M - M' = 4$ and contain $N - N' = 3$ clusters. The multiplicity of the configuration is

$$\mathcal{V}_{M-M',N-N'} = \binom{3}{2} = 3.$$

As we confirm in Table 2.2, this configuration appears in positions 29, 30, and 31. All canonical configurations with $M' = 4$ and $N' = 2$ appear 3 times because they all have the same set of complements. These configurations are

$$(1, 3),\ (3, 1),\ (2, 2)$$

and constitute the complete microcanonical ensemble $\mu C\,(4, 3)$. The corresponding distributions are:

$$\mathbf{m}'_1 = (1, 3) \to \mathbf{n}'_A = (1, 0, 1)$$
$$\mathbf{m}'_2 = (3, 1) \to \mathbf{n}'_A = (1, 0, 1)$$
$$\mathbf{m}'_3 = (2, 2) \to \mathbf{n}'_B = (0, 1, 0)$$

Let us consider the canonical configuration $\mathbf{m}' = (1, 3)$ with total mass $M' = 4$. The multiplicity of this configuration is equal to the volume of the microcanonical ensemble $\mu C(M - M', N - N') = \mu C(4, 3)$:

$$\mathcal{V}_{4,3} = \binom{3}{2} = 3.$$

Indeed, configuration (3,1) appears three times, in rows 29, 30 and 31.

The complements of $\mathbf{m}' = (3, 1)$ are

$$(1, 1, 2),\ (1, 2, 1),\ (2, 1, 1)$$

and represent the complete microcanonical ensemble of all ordered partitions of 4 into 3 parts.

The cluster distribution of configuration $\mathbf{m}' = (3, 1)$ is $\mathbf{n}'_A = (1, 0, 1)$ and its multiplicity is

$$\mathbf{n}'_A!\,\mathcal{V}_{4,3} = \frac{2!}{1!\,1!}\binom{3}{2} = 6.$$

We can confirm that distribution $\mathbf{n}'_A!$ is represented 6 times in the canonical ensemble, 3 times corresponding to configuration $(3, 1)$, and 3 times corresponding to its permutation $(1, 3)$.

Canonical Reconstruction

The canonical slice represents a condensed form of the microcanonical table that contains all information about the complete set of microcanonical configurations. Conversely, the microcanonical table can be constructed by systematically assembling canonical slices together. We call this procedure canonical reconstruction and build it recursively, assuming that M and N are known.

We begin by observing that all canonical slices contain the same integers $m = 1, 2, \cdots M - N + 1$, and that each integer m appears $\mathcal{V}_{M-m,N-1}$ times, according to Eq. (2.62). This produces a list of $\mathcal{V}_{M,N}$ integers. The first canonical slice then is some permutation of this list whose order does not matter and can be set arbitrarily. Suppose we have placed N' columns in the table and are in the process of adding slice $j = N' + 1$. The current table consists of $\mathcal{V}_{M,N}$ configurations each with N' clusters. The problem now is how to pair each mass of the j canonical slice with an existing configuration in the table. A configuration $\mathbf{m}'$ with mass M' can be paired with any cluster mass m of the canonical slice provided that the mass of the complement, $M - M' - m$, is at least equal to the number of clusters in the complement, $N - N' - 1$ (we cannot have clusters with zero or negative mass). Therefore, the cluster mass m that can be paired with a configuration whose mass is M' are all those that satisfy the condition[16]

$$M - M' + N' - N + 1 \geq m \geq 1. \tag{2.69}$$

Now we know which cluster masses can be paired with a given configuration and we are only left to determine the number of times a mass m can be paired with the same configuration. This number is equal to the microcanonical volume of the complement,

$$\mathcal{V}_{M-M'-m,N-N'-1}.$$

This process is continued until the table contains $N - 1$ columns. The last column is obtained by the condition that the total mass of configuration is M. This construction is illustrated with a calculation below.

Example 2.11 (Canonical Reconstruction)

In this example we begin with one microcanonical slice and systematically reconstruct the entire microcanonical table for $M = 6$, $N = 4$.

First slice: The canonical slice contains cluster masses from 1 up to $M - N + 1 = 3$. The multiplicity of each mass is calculated from Eq. (2.62),

$$\mathcal{V}_{M-m,N-1} = \binom{M - m - 1}{N - 2},$$

[16]We must also have $N' \leq N - 1$ because the complement must contain at least one cluster. Therefore, the condition in (2.69) applies to all slices except the last one. The last column will be obtained by mass balance.

which gives

m	1 2 3
$\mathcal{V}_{6-m,3}$	6 3 1

There are 6 ones, 3 twos, and 1 three, a total of 10 cluster masses. Placing these clusters in a list (the order is immaterial, here we chose ascending), we obtain the first slice

m_1	1 1 1 1 1 1 2 2 2 3
m_2	
m_3	
m_4	

For economy of space we have transposed the table so that slices are shown as rows.

Second slice: We calculate an auxiliary table that shows the masses that can be paired with the existing masses in each configuration, along with their multiplicity, which we calculate from Eq. (2.62) using $N = N' = 1$ (recall that N' refers to the number of slices that have been placed in the table at the start of the current iteration):

M'	m	multiplicity
1	1	$\mathcal{V}_{4,2} = 3$
1	2	$\mathcal{V}_{3,2} = 2$
1	3	$\mathcal{V}_{2,2} = 1$
2	1	$\mathcal{V}_{3,2} = 2$
2	2	$\mathcal{V}_{2,2} = 1$
3	1	$\mathcal{V}_{2,2} = 1$

Here is how we read this table: configurations with total mass $M' = 1$ will be paired with cluster mass 3 times, with cluster mass 2 twice, and with cluster mass 3 once; configurations with mass $M' = 2$ will be paired with cluster mass 1 twice, and with cluster mass 2 once; configurations with mass $M' = 3$ will be paired with cluster mass 1 once. This produces the second slice:

m_1	1 1 1 1 1 1 2 2 2 3
m_2	1 1 1 2 2 3 1 1 2 1
m_3	
m_4	

No other pairings are possible. If we try to pair configuration mass $M' = 1$, say, with cluster mass $m = 4$, we will be left with a complementary mass $M - M' = 2$ to be distributed into 3 clusters, which is not possible.

Third slice: The auxiliary table with $N' = 2$ is

M'	m	multiplicity
2	1	$\mathcal{V}_{3,1} = 1$
2	2	$\mathcal{V}_{2,1} = 1$
2	3	$\mathcal{V}_{1,1} = 1$
3	1	$\mathcal{V}_{2,1} = 1$
3	2	$\mathcal{V}_{1,1} = 1$
4	1	$\mathcal{V}_{1,1} = 1$

Configuration $(1, 1)$, whose mass is $M' = 2$, will be paired once with each of the three cluster masses, 1, 2, and 3. Configurations $(1, 2)$ and $(2, 1)$ with mass $M' = 3$ will each be paired once with cluster mass 1 and 2. Configurations $(2, 2)$, $(1,3)$, and $(3, 1)$, all of which have $M' = 4$, will be paired with cluster mass 1, once each. This yields the third slice of the table:

m_1	1	1	1	1	1	1	2	2	2	3
m_2	1	1	1	2	2	3	1	1	2	1
m_3	1	2	3	1	2	1	1	2	1	1
m_4										

Fourth slice: The final slice cannot be constructed from the properties of canonical slices because these only apply to canonical configurations with a nonzero complement. It will be constructed instead by adding the cluster mass that makes the total mass of the configuration equal to $M = 6$:

m_1	1	1	1	1	1	1	2	2	2	3
m_2	1	1	1	2	2	3	1	1	2	1
m_3	1	2	3	1	2	1	1	2	1	1
m_4	3	2	1	2	1	1	2	1	1	1

This table contains the complete set of ordered integer partitions of 6 into 4 parts. We can confirm that all slices contain the same distribution of clusters: 6 monomers, 3 dimers, and 1 trimer.

Exercise Complete the microcanonical table starting with the canonical slice that contains the following masses in the multiplicities shown:

$$1, 2, 3, 4, 5$$

(Confirm first that this is indeed a canonical slice.)

★

2.9.2 Canonical Probability

We now wish to calculate the probability of canonical distribution $\mathbf{n}'$ in the canonical ensemble $C(N'; M; N)$, given that the microcanonical probability is known. Distribution $\mathbf{n}'$ and its complement $\mathbf{n}''$ form a microcanonical distribution $\mathbf{n}$ that is represented by $\mathbf{n}!$ microcanonical configurations.[17] There are $\mathbf{n}!$ configurations with probability $P(\mathbf{n})$ and a fraction $\mathbf{n}'!\mathbf{n}''!/\mathbf{n}!$ of them contain canonical distribution $\mathbf{n}'$. The probability of $\mathbf{n}'$ within this sub-ensemble is

$$P(\mathbf{n}' + \mathbf{n}'')\frac{\mathbf{n}'!\mathbf{n}''!}{(\mathbf{n}' + \mathbf{n}'')!} \tag{2.70}$$

The overall probability of $\mathbf{n}'$ is obtained by summing over all complements of $\mathbf{n}'$:

$$P(\mathbf{n}'|N'; M, N) = \sum_{\mathbf{n}''} P(\mathbf{n}' + \mathbf{n}'')\frac{\mathbf{n}'!\mathbf{n}''!}{(\mathbf{n}' + \mathbf{n}'')!}. \tag{2.71}$$

Using $P(\mathbf{n}) = \mathbf{n}!W(\mathbf{n})/\Omega_{M,N}$ we obtain,

$$P(\mathbf{n}'|N'; M, N) = \frac{\mathbf{n}'!}{\Omega_{M,N}} \sum_{\mathbf{n}''} W(\mathbf{n}' + \mathbf{n}'')\mathbf{n}''!. \tag{2.72}$$

As a corollary we also have

$$\Omega_{M,N} = \sum_{\mathbf{n}'}\sum_{\mathbf{n}''} W(\mathbf{n}' + \mathbf{n}'')\mathbf{n}'!\mathbf{n}''!, \tag{2.73}$$

which gives the microcanonical partition function in terms of canonical distributions and their complement. In all of the above, $\mathbf{n}' + \mathbf{n}''$ is understood to be a member of the microcanonical ensemble $\mu C(M, N)$ with M and N fixed. Since a distribution $\mathbf{n}'$ can be associated with any number of microcanonical ensembles, the probability $P(\mathbf{n}')$ depends not only on $\mathbf{n}'$ but also on the associated microcanonical ensemble.

[17] A note on nomenclature: Here we use $\mathbf{n}'$ to indicate the canonical distribution, $\mathbf{n}''$ to indicate its microcanonical complement, and $\mathbf{n} = \mathbf{n}' + \mathbf{n}''$ to indicate the microcanonical distribution that is formed by combining the canonical distribution with its complement.

Table 2.3 Setup for the calculation of the probability of distribution $\mathbf{n}' = (1, 1)$ in the canonical ensemble $C(2; 8, 5)$. The distribution has two complements, $\mathbf{n}''_A = (2, 0, 1)$ and $\mathbf{n}''_B = (1, 2, 0)$. The canonical multiplicity of $\mathbf{n}'$ is equal to the sum $\mathbf{n}''_A! + \mathbf{n}''_B! = 6 + 6 = 12$

#	m_1	m_2	m_3	m_4	m_5
1	1	1	2	2	2
2	1	1	1	2	3
3	1	1	1	3	2
4	1	1	2	1	3
5	1	1	2	3	1
6	1	1	3	1	2
7	1	1	3	2	1
8	1	1	1	1	4
9	1	1	1	4	1
10	1	1	4	1	1
11	1	2	1	2	2
12	1	2	2	1	2
13	1	2	2	2	1
14	2	1	1	2	2
15	2	1	2	1	2
16	2	1	2	2	1
17	1	2	1	1	3
18	1	2	1	3	1
19	1	2	3	1	1
20	2	1	1	1	3
21	2	1	1	3	1
22	2	1	3	1	1
23	2	2	1	1	2
24	2	2	1	2	1
25	2	2	2	1	1
26	1	3	1	1	2
27	1	3	1	2	1
28	1	3	2	1	1
29	3	1	1	1	2
30	3	1	1	2	1
31	3	1	2	1	1
32	2	3	1	1	1
33	3	2	1	1	1
34	1	4	1	1	1
35	4	1	1	1	1

Example 2.12 (Canonical Probabilities)

We consider the canonical ensemble $C(2; 8, 5)$ that is formed by selecting the first two clusters from all microcanonical configurations in $\mu C(8, 5)$ (Table 2.3). The canonical ensemble contains the following configurations with their corresponding distributions:

	m_1	m_2	n'_1	n'_2	n'_3	n'_4	multiplicity
$\mathbf{n}'_A$	1	1	2	0	0	0	10
$\mathbf{n}'_B$	1	2	1	1	0	0	12
$\mathbf{n}'_C$	1	3	1	0	1	0	6
$\mathbf{n}'_D$	1	4	1	0	0	1	2
$\mathbf{n}'_E$	2	2	0	2	0	0	3
$\mathbf{n}'_F$	2	3	0	1	1	0	2

The rows labeled m_1 and m_2 show the canonical configurations (in a single permutation); the rows n'_1 through n'_4 show the corresponding canonical distributions; the last column shows the multiplicity of the canonical distributions, namely, the number of times the distribution appears in the microcanonical table. For example, configurations with one monomer and one dimer appear 12 times, this is the canonical multiplicity of distribution $\mathbf{n}'_B = (1, 1, 0, 0)$. In all, this canonical ensemble contains six distributions. Notice that we obtain identical ensembles regardless of which two rows we select.

We will calculate the canonical probability of distribution $\mathbf{n}'_B$. This distribution has two complements in the ensemble. One complement is

$$\mathbf{n}''_{B1} = (2, 0, 1, 0)$$

and corresponds to permutations of configuration $\mathbf{m}'' = (1, 1, 3)$. When combined with this complement, $\mathbf{n}'_B$ produces the microcanonical distribution

$$\mathbf{n}'_B + \mathbf{n}''_{B1} = (1, 1, 0, 0) + (2, 0, 1, 0) = (3, 1, 1, 0)$$

The second complement is

$$\mathbf{n}''_{B1} = (1, 2, 0, 0)$$

and together with $\mathbf{n}'_B$ forms the microcanonical distribution

$$\mathbf{n}'_B + \mathbf{n}''_{B2} = (1, 1, 0, 0) + (1, 2, 0, 0) = (2, 3, 0, 0).$$

The canonical probability of $\mathbf{n}'$ from Eq. (2.71) is

$$P[(1, 1, 0, 0)] =$$

$$P[(3, 1, 1, 0)]\frac{(1, 1, 0, 0)!\,(2, 0, 1, 0)!}{(3, 1, 1, 0)!} + P[(2, 3, 0, 0)]\frac{(1, 1, 0, 0)!\,(1, 2, 0, 0)!}{(2, 3, 0, 0)!}$$

$$= P[(3, 1, 1, 0)]\frac{6}{20} + P[(2, 3, 0, 0)]\frac{6}{10}.$$

The ratio 6/20 is the fraction of configurations with distribution (3, 1, 1, 0) that contain canonical distribution (1, 1, 0, 0), also equal to the fraction of permutations of configuration $\mathbf{m} = (1, 1, 1, 2, 3)$ whose first two elements are one 1 and one 2, in any order. This fraction is multiplied by the microcanonical probability of the containing configuration and summed over all complements to give the probability of the canonical distribution.

For the microcanonical probabilities we have

$$P[(3, 1, 1, 0)] = W(3, 1, 1, 0)\frac{(3, 1, 1, 0)!}{\Omega_{8,5}}$$

and

$$P[(2, 3, 0, 0)] = W(2, 3, 0, 0)\frac{(2, 3, 0, 0)!}{\Omega_{8,5}},$$

and with these results the canonical probability is

$$P[(1, 1, 0, 0)] = W(3, 1, 1, 0)\frac{6}{\Omega_{8,5}} + W(2, 3, 0, 0)\frac{6}{\Omega_{8,5}}.$$

Unbiased ensemble: In the special case of unbiased ensemble, $W = 1$ for all distributions and $\Omega_{8,5} = V8, 5 = 35$. The canonical probability then is

$$P[(1, 1, 0, 0)] = \frac{12}{35}.$$

Indeed, distribution $\mathbf{n}'_B = (1, 1, 0, 0)$ appears 12 out of 35 times in Table 2.3.

★

2.10 M'-Canonical Ensemble

The M'-canonical ensemble is a sub-ensemble of the microcanonical in which all configurations have the same total mass M'. We construct it as follows: given a microcanonical configuration of the $\mu C(M, N)$ ensemble, we pick consecutive clusters starting at position 1 until the total mass is M'; if it is not possible to obtain a contiguous sequence of clusters with total mass M' from a given microcanonical configuration, we reject that configuration and repeat the process with the next configuration in the microcanonical table. We call the ensemble produced by this method *M'-canonical* to indicate that the mass of all configurations is fixed. This implies that the number of clusters N' fluctuates. This is analogous to the canonical ensemble in which the number of clusters is fixed and the total mass fluctuates.[18] Table 2.4 shows an example for $M = 8$, $N = 5$. We start with the microcanonical table and extract M'-canonical configurations with mass $M' = 5$. One feature that jumps out is that the M'-canonical ensemble lacks the neat organization of the N-canonical ensemble. In particular, the M'-canonical subset contains fewer configurations than the microcanonical parent. This is because it is not always possible to find a microcanonical configuration with successive elements that add up to M'. Therefore, not all microcanonical configurations are represented in the M'-canonical ensemble. Roughly, the fraction of microcanonical configurations that contain an M'-canonical configuration with the specified mass is N/M.[19] We will see in Chap. 3 that this is the correct limit when M', M, and N are large.

To calculate the probability of distribution in the M'-canonical ensemble we use Table 2.4 as a guide. An M'-canonical distribution $\mathbf{n}'$ and its complement $\mathbf{n}''$ form a microcanonical distribution $\mathbf{n} = \mathbf{n}'+\mathbf{n}''$. Distribution $\mathbf{n}$ appears $\mathbf{n}!$ times in the table, and $\mathbf{n}'$ appears in a fraction $\mathbf{n}'!\mathbf{n}''!/\mathbf{n}!$ of them. The M'-canonical probability then is proportional to the microcanonical probability of $\mathbf{n}$ times the fraction represented by $\mathbf{n}'$. That is,

$$\frac{1}{C'}\frac{\mathbf{n}'!\mathbf{n}''!W(\mathbf{n}'+\mathbf{n}'')}{(\mathbf{n}'+\mathbf{n}'')!} = \frac{1}{C'}\frac{\mathbf{n}'!\mathbf{n}''!W(\mathbf{n}'+\mathbf{n}'')}{\Omega_{M,N}} = \frac{\mathbf{n}'!\mathbf{n}''!W(\mathbf{n}'+\mathbf{n}'')}{C}, \tag{2.74}$$

with $C = C'\Omega_{M,N}$ to be determined by normalization. The M'-canonical probability of distribution $\mathbf{n}'$ is obtained by summing this result over all complements of $\mathbf{n}'$:

$$P(\mathbf{n}'|M';\, M, N) = \frac{\mathbf{n}'!}{C}\sum_{\mathbf{n}''} W(\mathbf{n}'+\mathbf{n}'')\mathbf{n}''!. \tag{2.75}$$

The normalization constant is

$$C = \sum_{\mathbf{n}'} \mathbf{n}'! \sum_{\mathbf{n}''} \mathbf{n}''! W(\mathbf{n}'+\mathbf{n}''). \tag{2.76}$$

with the outer summation going over all M'-canonical configurations. Equation (2.75) is the same as Eq. (2.72) for canonical probability except for the normalization constant. The different constants arise from the fact that the M'-canonical

[18] We may rename the canonical ensemble to N'-canonical, to be consistent with the nomenclature used in this section. We choose not do this and will continue to refer to the N'-canonical ensemble simply as canonical. We will use the term N'-canonical only where needed to avoid possible confusion when discussing both canonical ensembles together.

[19] The fraction of M'-canonical configurations with $M' = 5$ in this example is $20/35 = 0.57$, fairly close to $N/M = 5/8 = 0.625$.

Table 2.4 Configurations of the M'-canonical ensemble with mass $M' = 5$, extracted from the microcanonical ensemble $M = 8$, $N = 5$ (shown by the shaded cells). The number of clusters N' is shown in the last column. This table is grouped by permutations of the microcanonical configuration

i	m_1	m_2	m_3	m_4	m_5	N'
1	4	1	1	1	1	2
2	1	4	1	1	1	2
3	1	1	4	1	1	
4	1	1	1	4	1	
5	1	1	1	1	4	
6	3	2	1	1	1	2
7	3	1	2	1	1	
8	3	1	1	2	1	3
9	3	1	1	1	2	3
10	2	3	1	1	1	2
11	2	1	3	1	1	
12	2	1	1	3	1	
13	2	1	1	1	3	4
14	1	3	2	1	1	
15	1	3	1	2	1	3
16	1	3	1	1	2	3
17	1	2	3	1	1	
18	1	2	1	3	1	
19	1	2	1	1	3	4
20	1	1	3	2	1	3
21	1	1	3	1	2	3
22	1	1	2	3	1	
23	1	1	2	1	3	4
24	1	1	1	3	2	
25	1	1	1	2	3	4
26	2	2	2	1	1	
27	2	2	1	2	1	3
28	2	2	1	1	2	3
29	2	1	2	2	1	3
30	2	1	2	1	2	
31	2	1	1	2	2	
32	1	2	2	2	1	3
33	1	2	2	1	2	3
34	1	2	1	2	2	
35	1	1	2	2	2	

ensemble has the same number of configurations as the microcanonical ensemble from which it is extracted, while the M'-canonical ensemble has fewer.

Note 2.13 (Microcanonical, Canonical, and M'-Canonical)

The terms "canonical" and "microcanonical" come to us from Gibbs. Gibbs identified the exponential distribution of microstates in energy as the "most simple case conceivable" of distributions that satisfy the governing equation of an ensemble of interacting particles (Gibbs 1902, p 34) and named it canonical. Microstates in the canonical ensemble of statistical mechanics are distributed with respect to energy, in complete analogy to canonical configurations in our development, which are distributed in cluster "mass." Gibbs envisioned the ensemble of fixed energies as a differential slice of the canonical ensemble (in our nomenclature this is a *microcanonical slice*, see p. 53). Hence "microcanonical" implies that this ensemble is a differential constituent of the canonical. In the same spirit grand canonical encompasses the canonical, which in turn encompasses the microcanonical ensemble.

Our terminology is based on Gibbs's but some conflicts arise because Gibbs is dealing with a rather specific problem. We use *microcanonical* to refer to an ensemble that is completely defined by extensive variables, in our case M and N. We use *canonical* to refer to subsets of the microcanonical ensemble in which one or more extensive variables are allowed to fluctuate. In the finite phase space, this makes the canonical probability a function of not only the canonical distribution but also the microcanonical distribution from which it has been extracted. By contrast, the microcanonical probability does not depend on variables external to the ensemble. This elevates the microcanonical ensemble as the fundamental ensemble of the theory. In Chap. 3 we will add more extensive variables, which may represent different components or other attributes. We may construct canonical ensembles for any of these variables and their combinations. Variable N is intrinsically different. While we can have multiple variables M_i to represent different attributes of the population, we can only have one variable N that represents the number of clusters. It might seem that the M'-canonical ensemble should be more appropriately called *grand canonical*. The correspondence between the two ensembles is only partial. Gibbs's grand ensemble fluctuates with respect to both energy and number of particles, and as such it encompasses the canonical ensemble which allows fluctuations in energy only. This is not the case with the M'-canonical ensemble in which N is the sole fluctuating variable. We prefer the generic term *canonical family* for all ensembles that allow fluctuations in any combination of the extensive variables that define the microcanonical ensemble.

★

Chapter 3
Thermodynamic Limit (ThL)

If we increase the size of the cluster ensemble at fixed mean, the ensemble becomes a container of every possible cluster distributions $\mathbf{n}$ with fixed mean. We call this *thermodynamic limit (ThL)* and define it by the condition

$$M > N \gg 1, \quad M/N = \text{constant}. \tag{3.1}$$

We write $M > N \gg 1$ rather than $M > N \to \infty$ because we are interested in the asymptotic behavior as the ensemble increases in size. In the thermodynamic limit the microcanonical probability becomes sharply peaked around one particular distribution of the ensemble that we call *most probable distribution* (MPD). In this chapter we examine the convergence of the ensemble into a narrow region, and ultimately a single point of phase space and obtain the most probable distribution in terms of M, N, and the selection functional W.

3.1 Convergence in the ThL

The microcanonical probability is proportional to the microcanonical weight $\mathbf{n}!W(\mathbf{n})$, which, as a concave function of $\mathbf{n}$, must have a maximum. The distribution that maximizes the microcanonical weight is the *most probable distribution* (MPD) and in the ThL is also the most important distribution in the ensemble. It satisfies the condition

$$\left.\frac{\partial \log \mathbf{n}!W(\mathbf{n})}{\partial n_i}\right|_{\tilde{\mathbf{n}}} = 0, \tag{3.2}$$

T. Matsoukas, *Generalized Statistical Thermodynamics*,
Understanding Complex Systems, https://doi.org/10.1007/978-3-030-04149-6_3

under the two constraints that define the ensemble,

$$\sum_i n_i = N, \tag{3.3}$$

$$\sum_i i n_i = M. \tag{3.4}$$

We expand the logarithm of the microcanonical weight in power series of **n** in the vicinity of the most probable distribution:

$$\log \mathbf{n}!W(\mathbf{n}) = \log \tilde{\mathbf{n}}!W(\tilde{\mathbf{n}}) + \frac{1}{2}\sum_{i,j} \left.\frac{\partial^2 \log \mathbf{n}!W(\mathbf{n})}{\partial n_i \partial n_j}\right|_{\tilde{\mathbf{n}}} \delta n_i \delta n_j, +O(1/N^2), \tag{3.5}$$

with $\delta\mathbf{n} = \mathbf{n} - \tilde{\mathbf{n}}$. The linear term is missing by virtue of Eq. (3.2); the logarithm of the microcanonical weight is homogeneous in **n** with degree 1; its k-order derivative with respect to n_i is homogeneous with degree $1 - k$. Accordingly, the second derivative vanishes as $1/N$ and higher derivatives vanish even faster. Since $\log \tilde{\mathbf{n}}!W(\tilde{\mathbf{n}})$ is the maximum of $\log \mathbf{n}W(\mathbf{n})$, we must have

$$\sum_{i,j} \left.\frac{\partial^2 \log \mathbf{n}!W(\mathbf{n})}{\partial n_i \partial n_j}\right|_{\tilde{\mathbf{n}}} \delta n_i \delta n_j \le 0,$$

which we write more explicitly as

$$\sum_i \left.\frac{\partial^2 \log \mathbf{n}!W(\mathbf{n})}{\partial n_i^2}\right|_{\tilde{\mathbf{n}}} \delta n_i^2 + \sum_{i,j\neq i} \left.\frac{\partial^2 \log \mathbf{n}!W(\mathbf{n})}{\partial n_i \partial n_j}\right|_{\tilde{\mathbf{n}}} \delta n_i \delta n_j \le 0. \tag{3.6}$$

By the concave property of $\log \mathbf{n}!W(\mathbf{n})$,

$$\frac{\partial^2 \log \mathbf{n}!W(\mathbf{n})}{\partial n_i^2} \le 0, \tag{3.7}$$

which implies that the first summation on the right-hand side of Eq. (3.6) is always negative. The second summation may be positive or negative because δn_i and δn_j can be varied essentially independently when M and N are large, even though they are constrained by Eqs. (3.3) and (3.4). To satisfy the inequality in Eq. (3.6) for arbitrary δn_i, δn_j, the summation with the mixed derivatives must vanish faster than the summation of the second derivatives in n_i. The expansion in Eq. (3.5) then becomes

$$\log \mathbf{n}!W(\mathbf{n}) \to \log \tilde{\mathbf{n}}!W(\tilde{\mathbf{n}}) + \frac{1}{2}\sum_i \left.\frac{\partial^2 \log \mathbf{n}!W(\mathbf{n})}{\partial n_i^2}\right|_{\tilde{\mathbf{n}}} \delta n_i^2, \tag{3.8}$$

and the microcanonical weight is given by the Gaussian function,

$$\mathbf{n}!W(\mathbf{n}) \to \tilde{\mathbf{n}}!W(\tilde{\mathbf{n}}) \exp\left(-\sum_{i,j} \frac{\delta n_i^2}{2\sigma_i^2}\right), \tag{3.9}$$

with

$$\sigma_i^2 = -1 \bigg/ \left(\frac{\partial^2 \log \mathbf{n}!W(\mathbf{n})}{\partial n_i^2}\right)_{\tilde{\mathbf{n}}} .$$

As the inverse of the second derivative, the variance σ_{ij}^2 scales as $\tilde{n}_i$, which in turn scales as N. For large N the microcanonical weight converges sharply about the most probable distribution. We sum both sides of Eq. (3.9) over all $\mathbf{n}$, then take the logarithm:

$$\log \sum_{\mathbf{n}} \mathbf{n}!W(\mathbf{n}) \quad \to \quad \log \tilde{\mathbf{n}}!W(\tilde{\mathbf{n}}) + \underbrace{\log \int \exp\left(-\sum_{i,j} \frac{\delta n_i^2}{2\sigma_i^2}\right) d\delta n_1\, d\delta n_2 \cdots}_{O(\log N)} \tag{3.10}$$

On the left-hand side we have the log of the microcanonical partition function; on the right-hand side we have the log of the maximum microcanonical weight plus the log of an integral that scales as $\sim N$. The result reduces to

$$\log \Omega_{M,N} \to S(\tilde{\mathbf{n}}) + \log W(\tilde{\mathbf{n}}) + O(\log N),$$

The term of order $\log N$ may be dropped relative to $S(\tilde{\mathbf{n}}) + \log W(\tilde{\mathbf{n}})$ (both S and $\log W$ scale as $\sim N$), therefore we have the asymptotic relationship

$$\log \Omega_{M,N} = S(\tilde{\mathbf{n}}) + \log W(\tilde{\mathbf{n}}). \tag{3.11}$$

The fact that the microcanonical weight (and therefore the microcanonical probability) peaks very sharply around the most probable distribution implies that the entire ensemble converges into a narrow region of phase space populated by distributions that are differential variations of the MPD. All ensemble averages then reduce to averages over a single distribution, the MPD. If $F = F(\mathbf{n})$ is a property of distribution, its ensemble average in the ThL is equal to its value at the most probable distribution:

$$\langle F \rangle \to F(\tilde{\mathbf{n}}). \tag{3.12}$$

Table 3.1 Common homogeneous properties and their degree of homogeneity

Property	Homogeneity	Common name
$\bar{x} = \sum_i i n_i \Big/ \sum_i n_i$	$\nu = 0$	Mean size
$N = \sum_i n_i$	$\nu = 1$	Number of clusters
$M = \sum_i i n_i$	$\nu = 1$	Mass
$J_k[\mathbf{n}] = \sum_i i^k n_i$	$\nu = 1$	Moment of order k
$S(\mathbf{n}) = -\sum_k n_i \log \frac{n_i}{N}$	$\nu = 1$	Entropy
$A(\mathbf{n}) = \frac{1}{2} \sum_i \sum_j i(i-1) n_i$	$\nu = 1$	Number of intra-cluster pairs
$B(\mathbf{n}) = \frac{1}{2} \sum_i \sum_j n_i (n_j - \delta_{ij})$	$\nu = 2$	Number of cluster pairs

If F is a homogeneous function of $\mathbf{n}$ with degree ν its ensemble average is also homogeneous in $\tilde{\mathbf{n}}$ with the same degree (see Table 3.1 for some examples of homogeneous properties). The preservation of homogeneity from a function of $\mathbf{n}$ to its ensemble average is only possible because the ensemble converges into a single distribution. In general, if F is homogeneous in $\mathbf{n}$ the ensemble average

$$\langle F \rangle = \sum_{\mathbf{n}} P(\mathbf{n}) F(\mathbf{n}) \tag{3.13}$$

is *not*, because $P(\mathbf{n})$ is not homogeneous (its logarithm is). In the special case that $P(\mathbf{n})$ is a delta function at $\mathbf{n} = \tilde{\mathbf{n}}$, we obtain $\langle F \rangle = F(\tilde{\mathbf{n}})$. Then, and then only, does the ensemble average inherit the homogeneity of F.

3.2 Most Probable Distribution (MPD)

We obtain the MPD by maximizing the logarithm of the microcanonical weight under the constraints that define the ensemble,

$$\sum_i n_i = N, \qquad [3.3]$$

$$\sum_i i n_i = M. \qquad [3.4]$$

We perform the maximization by the method of Lagrange multipliers. We construct the function to be maximized by incorporating the constraints,

$$\mathcal{F} = \log \mathbf{n}! + \log W(\mathbf{n}) - \alpha \left(\sum_i n_i - N \right) - \beta \left(\sum_i i n_i - M \right)$$

where α and β are Lagrange multipliers and all n_i are treated as independent variables. Using the Stirling approximation $\log x! \approx x \log x - 1$, we differentiate with respect to n_i and set the derivative equal to zero:

$$\left.\frac{\partial \mathcal{F}}{\partial n_i}\right|_{\tilde{n}_i} = 0 = -\log\frac{\tilde{n}_i}{N} - \left.\frac{\partial \log W(\mathbf{n})}{\partial n_i}\right|_{\tilde{n}_i} - \alpha - \beta i.$$

The derivative that appears on the right-hand side is the cluster bias $\log \tilde{w}_i$ evaluated at the MPD[1]:

$$\log \tilde{w}_i = \left.\frac{\partial \log W(\mathbf{n})}{\partial n_i}\right|_{\tilde{n}_i}. \tag{3.14}$$

Setting $\alpha = \log q$ and solving for $\tilde{n}_i$ we obtain the MPD in the form

$$\frac{\tilde{n}_i}{N} = \tilde{w}_i \frac{e^{-\beta i}}{q}. \tag{3.15}$$

We obtain the Lagrange multipliers β and q by back-substitution of the MPD into the constraints. The first constraint gives

$$q = \sum_{i=1}^{M-N+1} \tilde{w}_i e^{-\beta i}, \tag{3.16}$$

and from the second constraint we obtain,

$$\frac{\sum_{i=1}^{M-N+1} i\tilde{w}_i e^{-\beta i}}{\sum_{i=1}^{M-N+1} \tilde{w}_i e^{-\beta i}} = \frac{M}{N}, \tag{3.17}$$

with both summations running from $i = 1$ to $i = i_{\max} = M - N + 1$.

Equation (3.15) along with (3.14), (3.16), and (3.17) define the most probable distribution and its parameters in terms of M, N, and W. If the selection functional is linear the cluster weights are known functions of cluster size and this set of equations is solved sequentially to obtain the MPD: calculate β by solving Eq. (3.17); with β known calculate q from (3.16); finally obtain the distribution from (3.15). In the general case these equations must be solved simultaneously for $\tilde{n}_i$, β, and q.

---

[1] We use $\tilde{w}_i$ as shorthand for $w_{i|\tilde{\mathbf{n}}}$. As a general notational convention, variables decorated with the tilde, ˜, refer to the MPD.

3.2.1 The $\beta q \bar{x}$ Relationship

Equations (3.16) and (3.17) define an implicit relationship between β, q, and $\bar{x}$. We will now obtain this relationship in explicit form. We start with Eq. (3.16) and take the derivative of $\log q$ with respect to β:

$$\frac{d \log q}{d\beta} = \frac{1}{q} \sum \left(-i \tilde{w}_i e^{-\beta i} + \frac{d\tilde{w}_i}{d\beta} e^{-\beta i} \right).$$

The first summation on the right-hand side is

$$\frac{1}{q} \sum_i \left(-i \tilde{w}_i e^{-\beta i} \right) = -\sum_i \frac{i \tilde{n}_i}{N} = -\frac{M}{N}.$$

For the second summation we have

$$\sum_i \frac{d\tilde{w}_i}{d\beta} e^{-\beta i} = q \sum_i \left(\frac{d \log \tilde{w}_i}{d\beta} \right) \left(\frac{w_i e^{-\beta i}}{q} \right) = \frac{q}{N} \sum_i \tilde{n}_i \frac{d \log \tilde{w}_i}{d\beta} = 0.$$

The last result on the right is identically equal to zero by virtue of the Gibbs-Duhem equation (see Eq. (2.26)). Therefore,

$$\frac{d \log q}{d\beta} = -\bar{x}. \tag{3.18}$$

The result establishes a relationship between the three intensive properties of the ensemble, β, q, and $\bar{x} = M/N$.

3.2.2 Linearized Ensemble

Except for the special case of linear bias, the cluster weight w_i depends on the distribution to which the cluster belongs. The notation $\tilde{w}_i$ signifies that the cluster weight refers specifically to the most probable distribution, i.e.,

$$\tilde{w}_i \equiv w_{i|\tilde{\mathbf{n}}}. \tag{3.19}$$

In the ThL the only distributions with nonvanishing probability are those in the immediate vicinity of the MPD and may be treated as differential variations of the MPD of the form

$$\mathbf{n} \equiv \tilde{\mathbf{n}} + \delta\mathbf{n}.$$

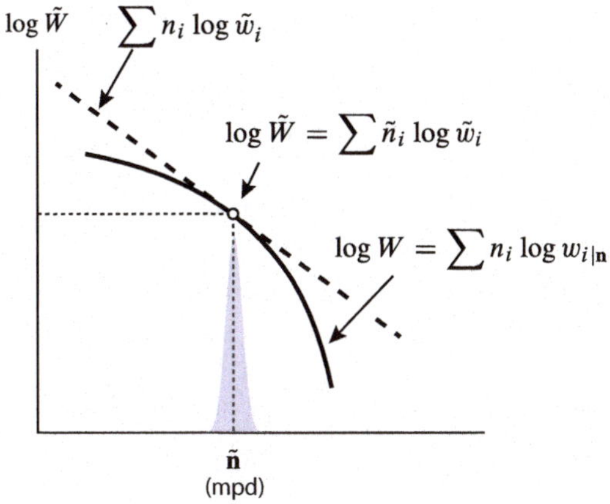

Fig. 3.1 In the vicinity of the MPD the ensemble is linear with cluster weight equal to the cluster weight of the MPD, $\tilde{w}_i$

The logarithm of the selection functional can be expressed as a linear function in $\delta\mathbf{n}$:

$$\log W(\tilde{\mathbf{n}} + \delta\mathbf{n}) = \log W(\tilde{\mathbf{n}}) + \sum_i \left(\frac{\partial \log W}{\partial n_i}\right)_{\tilde{\mathbf{n}}} \delta\mathbf{n} + O(\delta\mathbf{n})^2.$$

Writing $\log W(\tilde{\mathbf{n}})$ in Euler form,

$$\log W(\tilde{\mathbf{n}}) = \sum_i \tilde{n}_i \log \tilde{w}_i,$$

the selection bias in the vicinity of the most probable distribution is[2]

$$\log W(\mathbf{n}) = \sum_i n_i \log \tilde{w}_i. \tag{3.20}$$

In the ThL all distributions with nonvanishing probability have the same cluster bias as the MPD. In geometric language, $\log W(\mathbf{n})$ is a hypersurface above the plane of distributions, $\log \tilde{W}$ is the point on this surface at $\mathbf{n} = \tilde{\mathbf{n}}$, and $\sum_i n_i \log \tilde{w}_i$ is the hyperplane that is tangent to $\log W$ at $\tilde{\mathbf{n}}$ and has slope $\log \tilde{w}_i$ with respect to the i axis of the phase space (Fig. 3.1). In the vicinity of the MPD the selection functional is fully described by its linearized form and all distributions in this region have the same cluster weights. This should not be confused with the linear ensemble, in which case $\log W$ is linear over the entire phase space. Rather, it means that the ensemble converges into a linearized region of $\log W$ around the most probable distribution.

[2]Recall that the derivatives of $\log W(\mathbf{n})$ of order k scale as $1/N^{k-1}$ and vanish asymptotically as the size of the system increases. Therefore, all terms of order $\delta\mathbf{n}^2$ or higher can be dropped.

3.3 The Microcanonical Equations

We apply the entropy functional to the MPD in Eq. (3.15):

$$\tilde{S} \equiv S(\tilde{\mathbf{n}}) = -\sum_i \tilde{n}_i \log \frac{n_i}{N} = -\sum_i \tilde{n}_i \log \tilde{w}_i + \beta \sum_i i\tilde{n}_i + \log q \sum_i \tilde{n}_i.$$

The first summation on the right-hand side is $\log W(\tilde{\mathbf{n}})$, the second summation is the total mass M, and the last summation is the number of clusters; therefore,

$$\tilde{S} = -\log \tilde{W} + \beta M + (\log q)N,$$

Using $\tilde{S} = \log \Omega_{M,N} - \log \tilde{W}$ from Eq. (3.11) we obtain:

$$\log \Omega_{M,N} = \beta M + (\log q)N. \tag{3.21}$$

This result of remarkable simplicity connects the microcanonical partition function to its formal extensive variables M and N and to the intensive parameters β and q that appear in the MPD. Suppose we multiply both M and N by the same factor λ. From Eq. (3.11) we have

$$\log \Omega_{\lambda M,\lambda N} = \lambda S(\tilde{\mathbf{n}}) + \lambda \log W(\tilde{\mathbf{n}}) = \lambda \log \Omega_{M,N}.$$

The result states that $\log \Omega_{M,N}$ in the ThL is homogeneous in M and N with degree 1. By Euler's theorem $\log \Omega$ is expressed as

$$\log \Omega_{M,N} = M \left(\frac{\partial \log \Omega_{M,N}}{\partial M}\right)_N + N \left(\frac{\partial \log \Omega_{M,N}}{\partial N}\right)_M. \tag{3.22}$$

Direct comparison with Eq. (3.21) leads to the following identifications:

$$\beta = \left(\frac{\partial \log \Omega}{\partial M}\right)_N, \tag{3.23}$$

$$\log q = \left(\frac{\partial \log \Omega}{\partial N}\right)_M. \tag{3.24}$$

The parameters β and q, which were introduced as Lagrange multipliers, are now seen to represent the partial derivatives of the microcanonical partition function with respect to the primary extensive variables of the ensemble, M and N. It follows that both β and q are intensive, i.e., homogeneous in M and N with degree 0. The differential of $\log \Omega$ can now be written as

$$d \log \Omega_{M,N} = \beta dM + (\log q) dN. \tag{3.25}$$

Differentiating Eq. (3.21) with respect to all four variables on the right-hand side, and combining with (3.25) we obtain the associated Gibbs-Duhem equation,

$$M d\beta + N d \log q = 0. \tag{3.26}$$

We may express this result in the equivalent form,

$$\frac{d \log q}{d\beta} = -\frac{M}{N} = \bar{x}, \qquad [3.18]$$

which we recognize as Eq. (3.18). Thus we have recovered the relationship between q, β, and $\bar{x}$ through an independent derivation.

Equation (3.21) is the central relationship of the cluster ensemble. Along with Eqs. (3.15), (3.23), (3.24), and (3.18), this set of equations establishes the interdependencies between the main variables of the ensemble:

$$M, \ N, \ \log \Omega, \ \beta, \ \log q, \ \tilde{w}_i.$$

The selection functional does not appear in the fundamental equation itself. It is present only in Eq. (3.15) for the MPD, which is given in terms of the cluster weights $\tilde{w}_i$. Entropy does not appear either, except in Eq. (3.11); it is the log of the microcanonical partition function whose variational properties determine the state of the system.

3.3.1 Special Case: Linear Ensemble

In Chap. 2 we derived the mean cluster distribution of the linear ensemble in the form,

$$\langle n_k \rangle = N w_k \frac{\Omega_{M-k,N-1}}{\Omega_{M,N}}. \qquad [2.54]$$

We will obtain its relationship to the MPD. First we write

$$\log \frac{\Omega_{M-k,N-1}}{\Omega_{M,N}} = \log \Omega_{M-k,N-1} - \log \Omega_{M,N}.$$

The difference on the right-hand side represents the change in $\log \Omega_{M,N}$ when M is changed by $\delta M = -k$ and N by $\delta N = -1$. Treating δM and δN as differential quantities relative to M and N we use Eq. (3.25) to obtain the corresponding change in $\log \Omega$:

$$\log \Omega_{M-k,N-1} = (M-k)\beta + (N-1)\log q. \tag{3.27}$$

We then have

$$\frac{\Omega_{M-k,N-1}}{\Omega_{M,N}} = e^{-\beta k - \log q}. \tag{3.28}$$

The mean cluster distribution then takes the form

$$\frac{\langle n_k \rangle}{N} = w_k \frac{e^{-\beta k}}{N} = \frac{\tilde{n}_i}{N}. \tag{3.29}$$

Therefore, the mean cluster distribution is equal to the most probable distribution. This is the expected result: when the ensemble converges into a single distribution, the mean and the most probable distributions converge to each other. This is true in all ensembles, linear or not, but the linear ensemble allows us to show this explicitly.

3.4 Canonical Probability in the Thermodynamic Limit

In Chap. 2 we defined the canonical configuration as N' clusters sampled from a microcanonical configuration of a larger microcanonical ensemble $\mu C(M, N)$ and obtained its probability in the form

$$P(\mathbf{n}'|N'; M, N) = \frac{\mathbf{n}'!}{\Omega_{M,N}} \sum_{\mathbf{n}''} W(\mathbf{n}' + \mathbf{n}'')\mathbf{n}''!. \qquad [2.72]$$

Here $\mathbf{n}'$ is the canonical distribution and $\mathbf{n}''$ is its microcanonical complement such that $\mathbf{n}' + \mathbf{n}''$ is a microcanonical distribution; the summation goes over all complements of distribution $\mathbf{n}'$. We make the further assumption that the enclosing microcanonical ensemble is much larger than the canonical ensemble from which it is extracted. That is, $N' \ll N$ and $M' \ll M$, which further imply $\mathbf{n}' \ll \mathbf{n}''$. We treat $\mathbf{n}'$ as a differential variation on $\mathbf{n}''$ to write

$$\log W(\mathbf{n}' + \mathbf{n}'') = \log W(\mathbf{n}'') + \delta \log W$$

with

$$\delta \log W = \log W(\mathbf{n}') = \sum_i n_i' \log \tilde{w}_i$$

It follows that

$$\log W(\mathbf{n}' + \mathbf{n}'') = \log W(\mathbf{n}') + \log W(\mathbf{n}''),$$

and finally,

$$W(\mathbf{n}' + \mathbf{n}'') = W(\mathbf{n}')W(\mathbf{n}'').$$

We substitute into Eq. (2.72) to obtain

$$P(\mathbf{n}'|N'; M, N) = \frac{\mathbf{n}'!W(\mathbf{n}'!)}{\Omega_{M,N}} \sum_{\mathbf{n}''} \mathbf{n}''!W(\mathbf{n}''). \tag{3.30}$$

The summation is over all complements of distribution $\mathbf{n}'$, all of which contain $N - N'$ clusters with total mass $M - M'$, where M' is the mass in the canonical configuration $\mathbf{n}'$. We recognize the summation on the right-hand side as the microcanonical partition function $\Omega_{M-M',N-N'}$:

$$P(\mathbf{n}'|N'; M, N) = \mathbf{n}'!W(\mathbf{n}'!)\frac{\Omega_{M-M',N-N'}}{\Omega_{M,N}}. \tag{3.31}$$

We treat M' and N' as differential changes on M and N, respectively and use Eq. (3.25) to write

$$\log \Omega_{M,N} - \log \Omega_{M-M',N-N'} = \beta M' + \log q N', \tag{3.32}$$

where β and $\log q$ are the parameters of the associated microcanonical ensemble. This leads to the following result for the canonical probability,

$$P(\mathbf{n}'|N'; M, N) = \mathbf{n}'!W(\mathbf{n}'!)\frac{e^{-\beta M'}}{q^{N'}}. \tag{3.33}$$

Notice that the mass M and number N of clusters of the parent microcanonical ensemble are not present in this result. Instead, the microcanonical ensemble is represented by its intensive properties β and q. We drop the primes (they are needed only when we must distinguish between same quantities of the canonical and microcanonical ensemble) and write the final result as

$$P(\mathbf{n}|N; \beta, q) = \mathbf{n}!W(\mathbf{n}!)\frac{e^{-\beta M}}{q^N}, \tag{3.34}$$

with the understanding that M and N now refer to the canonical ensemble. It is instructive to derive this result by a simpler, more intuitive argument. The canonical configuration consists of N clusters picked from a microcanonical configuration. In the thermodynamic limit and under the condition that the canonical configuration is much smaller than the microcanonical configuration from which it is extracted, this amounts to drawing N clusters from an infinite pool of clusters with microcanonical MPD. The probability to sample distribution $\mathbf{n}$ is given by the multinomial distribution:

$$P(\mathbf{n}|N, \mathrm{MPD}) = \mathbf{n}! \prod_i \tilde{p}_i^{n_i},$$

with

$$\tilde{p}_i = \tilde{w}_i \frac{e^{-\beta i}}{q}.$$

Substituting $\tilde{p}_i$ into the canonical probability we have

$$P(\mathbf{n}|N, \mathrm{MPD}) = \mathbf{n}! \left(\prod \tilde{w}_i^{n_i}\right) \frac{e^{-\beta M}}{q^N}$$

and finally,

$$P(\mathbf{n}|N, \mathrm{MPD}) = \mathbf{n}! W(\mathbf{n}) \frac{e^{-\beta M}}{q^N},$$

which is the same as Eq. (3.34). In this interpretation the canonical ensemble is a *homogeneous sample* of the microcanonical ensemble, namely a sample with the same intensive properties as the microcanonical ensemble from which it is extracted.

3.4.1 Canonical Partition Function

The part of the canonical probability that depends explicitly on the distribution is the canonical weight of that distribution. Since q and N are constants this weight is

$$(\text{canonical weight}) = \mathbf{n}! W(\mathbf{n}) e^{-\beta M}, \tag{3.35}$$

where $M = \sum_i i n_i$ is the mass of the canonical configuration. The canonical weight therefore is equal to the microcanonical weight times the factor $e^{-\beta M}$. The summation of the canonical weights over all canonical distributions defines the canonical partition function Q. Since the canonical distribution is properly normalized,

$$\sum_{\mathbf{n}} \frac{\mathbf{n}! W(\mathbf{n}) e^{-\beta M}}{q^N} = 1$$

from which we obtain

$$Q \equiv \sum_{\mathbf{n}} \mathbf{n}!\, W(\mathbf{n})\, e^{-\beta M} = q^N. \tag{3.36}$$

In terms of their logarithms the relationship between Q and q is

$$\log Q = N \log q, \tag{3.37}$$

and identifies $\log Q$ as the extensive form of $\log q$. In combination with Eq. (3.21) and (3.23) we also have

$$\log Q = \log \Omega - M \left(\frac{\partial \log \Omega}{\partial M} \right)_N, \tag{3.38}$$

which identifies the canonical partition function as the Legendre transform of the microcanonical partition function with respect to M. As an immediate consequence of Legendre properties, the differential of $\log Q$ is

$$d \log Q = -M d\beta + (\log q)\, dN, \tag{3.39}$$

with partial derivatives

$$M = -\left(\frac{\partial \log Q}{\partial \beta} \right)_N, \quad \log q = \left(\frac{\partial \log Q}{\partial N} \right)_\beta. \tag{3.40}$$

These equations establish that the canonical partition function $Q = Q(\beta, N)$ is a formal function of β and N. Recall that in the canonical ensemble the mass M of configuration is not fixed but is a fluctuating quantity; the mass that appears in Eqs. (3.39) and (3.40) is the *mean* mass. We establish this in Sect. 3.4.3 where we discuss fluctuations.

3.4.2 The Canonical MPD

The canonical MPD is the distribution that maximizes the canonical weight

$$\log \left(\mathbf{n}!\, W(\mathbf{n})\, e^{-\beta M} \right)$$

under the conditions that define the ensemble:

$$\sum_i n_i = N, \quad \beta = \text{const.}, \quad \tilde{w}_i = \text{const.}$$

Both β and $\tilde{w}_i$ are held constant because they are fixed by the microcanonical MPD. We express M and $\log W$ in terms of the elements n_i of the canonical distribution to write

$$M = \sum_i i n_i \quad \text{and} \quad \log W(\mathbf{n}) = \sum_i n_i \log \tilde{w}_i,$$

and seek the unconstrained maximum of

$$\log\left(\mathbf{n}!\, W(\mathbf{n})\, e^{-\beta M}\right) - \log q' \left(\sum_i n_i - N\right) =$$

$$-N \sum_i n_i \log \frac{n_i}{N} + \sum_i n_i \log \tilde{w}_i - \beta \sum_i i n_i - \log q' \left(\sum_i n_i - N\right), \tag{3.41}$$

with respect to n_i, where $\log q'$ is the Lagrange multiplier for the constraint of fixed N. The MPD that results from this maximization is

$$\frac{\tilde{n}_i}{N} = \tilde{w}_i \frac{e^{-\beta i}}{q'},$$

subject to the normalization

$$\sum_i \tilde{w}_i e^{-\beta i} = q'.$$

This summation is identical with the normalization of the microcanonical MPD in Eq. (3.16). This implies $q' = q$, and the canonical MPD becomes

$$\left(\frac{\tilde{n}_i}{N}\right)_{\text{canonical}} = \tilde{w}_i \frac{e^{-\beta i}}{q} = \left(\frac{\tilde{n}_i}{N}\right)_{\text{microcanonical}}. \tag{3.42}$$

This establishes that the MPD of the canonical ensemble is the same as that of the microcanonical ensemble.

Note 3.1 (On the Canonical MPD)

The fact that the canonical and microcanonical MPDs are the same should not come as a surprise considering that the canonical configuration is a sample of

N clusters drawn randomly from the microcanonical MPD. For large enough N the two distributions converge to each other. We can also view this result from the stand point of maximizing the canonical weight. Notice that even though the constraint of constant mass is no longer present, the canonical objective function in Eq. (3.41) does contain the term $-\beta \sum_i in_i$ as part of the canonical weight. This term plays the same role as the Lagrange multiplier for the constraint of fixed mass in the microcanonical maximization. Mathematically then, we are maximizing the *same* quantity in both ensembles, although we construct this quantity differently in each case. We return to this point with a more detailed discussion in Sect. 3.6.

★

3.4.3 Canonical Fluctuations

Fluctuations of the canonical distribution about its MPD can be calculated by straightforward application of the properties of the multinomial distribution.[3] The mean number of clusters of size i is

$$\langle n_i \rangle = N\tilde{p}_i = N\tilde{w}_i \frac{e^{-\beta i}}{q} = \tilde{n}_i, \tag{3.43}$$

and is equal to the most probable value. The variance is

$$\mathrm{Var}(n_i) = N\tilde{p}_i(1 - \tilde{p}_i), \tag{3.44}$$

and scales as $\sim N$. The standard deviation scales as $\sim N^{1/2}$ and vanishes in relationship to $\bar{n}_i$ when N becomes large.

[3]Two elementary properties of the multinomial distribution

$$P(n_1, n_2 \cdots) = N! \frac{p_1^{p_1}}{n_1!} \frac{p_2^{n_2}}{n_2!} \cdots$$

are the mean value of n_k,

$$\bar{n}_k = Np_k$$

and its variance

$$\mathrm{Var}(n_k) = Np_k(1 - p_k).$$

The mass of the microcanonical configuration is a fluctuating quantity. To obtain the mean mass we start with the canonical partition function in Eq. (3.36) and calculate the derivative of its log with respect to β[4]:

$$\left(\frac{\partial \log Q}{\partial \beta}\right)_N = -\frac{1}{Q}\left(\sum_{\mathbf{n}} \mathbf{n}!W(\mathbf{n}!)Me^{-\beta M}\right) = -\bar{M}. \tag{3.45}$$

Comparing this with Eq. (3.40) we now see that the mass that appears in that equation, as well as in (3.39), is the mean mass of the canonical ensemble. The mean mass per cluster is

$$\frac{\bar{M}}{N} = -\frac{1}{N}\left(\frac{\partial \log Q}{\partial \beta}\right)_N = -\frac{d \log q}{d\beta}. \tag{3.46}$$

The left-hand side is the mean cluster in the canonical configuration, while the term on the far right is according to Eq. (3.18) equal to the mean size in the associated microcanonical ensemble. The result reaffirms that the two are equal. The variance of the canonical mass is

$$\mathrm{Var}(M) = \frac{\partial^2 \log Q}{\partial \beta^2}, \tag{3.47}$$

which can be shown by means of straightforward calculus by differentiation of Eq. (3.45). The log of the canonical partition function is extensive, which means that the variance scales as N. The standard deviation of the canonical mass scales as $\sim N^{1/2}$ and vanishes relative to the mean mass, which is extensive. The conclusion from the study of fluctuations is that all intensive properties are the same in both the microcanonical and the canonical ensemble.

3.5 M-Canonical Probability in the ThL

The M'-canonical configuration,[5] defined in Chap. 2 (Sect. 2.10), consists of clusters drawn from a microcanonical ensemble such that the total mass of the configuration is fixed, while the number of clusters is variable. The M'-canonical probability is

$$P(\mathbf{n}'|M'; M, N) = \frac{\mathbf{n}'!}{C} \sum_{\mathbf{n}''} W(\mathbf{n}' + \mathbf{n}'')\mathbf{n}''!, \tag{2.75}$$

[4] As functionals of $\mathbf{n}$, both $\mathbf{n}!$ and $W(\mathbf{n})$ are independent of β and are treated as constants in this differentiation.

[5] We use the primes in M' to distinguish the M'-canonical subs ensemble from the microcanonical ensemble (M, N) from which it is extracted. We will drop the prime in the thermodynamic limit.

where $\mathbf{n}'$ is the M'-canonical configuration, $\mathbf{n}''$ is its microcanonical complement, and C is fixed by normalization. We obtain the ThL limit of the M'-canonical probability by the same method we used for the canonical probability. Under the condition $\mathbf{n}' \ll \mathbf{n}''$ we have

$$W(\mathbf{n}' + \mathbf{n}'') = W(\mathbf{n}')W(\mathbf{n}''),$$

and the M'-canonical probability in Eq. (2.75) becomes

$$P(\mathbf{n}'|M';\, M, N) = \frac{\mathbf{n}'!W(\mathbf{n}')}{C} \sum_{\mathbf{n}''} W(\mathbf{n}'')\mathbf{n}''! = \frac{\mathbf{n}'!W(\mathbf{n}')}{C} \Omega_{M-M',N-N'},$$

with the result on the far right obtained by recognizing the summation of the microcanonical weights $\mathbf{n}''!W(\mathbf{n}'')$ of the complementary configurations as the microcanonical partition function $\Omega_{M-M',N-N'}$. We rewrite this as

$$P(\mathbf{n}'|M';\, M, N) = \frac{\mathbf{n}'!W(\mathbf{n}')}{C'} \frac{\Omega_{M-M',N-N'}}{\Omega_{M,N}},$$

where $C' = C/\Omega_{M,N}$ is a new normalization constant. Using Eq. (3.32) to write

$$\frac{\Omega_{M-M',N-N'}}{\Omega_{M,N}} = e^{-\beta M' - \log q N'},$$

the M'-canonical probability becomes

$$P(\mathbf{n}'|M';\, M, N) = \frac{\mathbf{n}'!W(\mathbf{n}')}{C'} \frac{e^{-\beta M'}}{q^{N'}}. \tag{3.48}$$

The normalization constant is given by the condition

$$C' = \sum_{\mathbf{n}'} \mathbf{n}'!W(\mathbf{n}') \frac{e^{-\beta M'}}{q^{N'}} \tag{3.49}$$

with the summation going over distributions $\mathbf{n}'$ whose total mass is exactly M'. The summand has a multinomial interpretation, it is the probability to sample distribution $\mathbf{n}'$ by drawing clusters from the microcanonical MPD[6]

[6] Use $M' = \sum_i i n_i'$, $N' \sum_i n_i'$, $\log W(\mathbf{n}') = \sum_i n_i' \log \tilde{w}_i$, $\tilde{p}_i = \tilde{w}_i e^{-\beta i}/q$, to write

$$\mathbf{n}'!W(\mathbf{n}') \frac{e^{-\beta M'}}{q^{N'}} = \mathbf{n}'! \prod \tilde{p}_i^{n_i'} \equiv \mathrm{Prob}\left(\sum_i m_i = M' | N'\right),$$

$$\mathbf{n}'!W(\mathbf{n}')\frac{e^{-\beta M'}}{q^{N'}}.$$

The factor C' in Eq. (3.49) is then the probability to sample a configuration with total mass M' by drawing clusters from the MPD, regardless of the number of clusters that must be drawn. To calculate this probability we consider the following equivalent problem: We imagine clusters of size i as chains with i beads. We draw a large sample of clusters and line them up in the order they were sampled. We then look at the M'th bead and if it is the *last* bead of its chain ("terminal" bead), then the configuration of clusters up to that bead has the desired mass M'. For large enough M' the probability to find a terminal bead in a long string of chains does not depend on the total number of beads before it, i.e., it is independent of M'. In this limit the probability to find a terminal bead is found easily: in a string of N^* clusters with total mass M^*, there are N^* terminal beads (one per cluster). The probability to choose a terminal bead is $N^*/M^* = 1/\bar{x}$, where $\bar{x}$ is the mean cluster size of the MPD:

$$C' = 1/\bar{x}.$$

With this the M'-canonical probability becomes

$$P(\mathbf{n}'|M'; M, N) = \mathbf{n}'!W(\mathbf{n}')\left(\bar{x}\frac{e^{-\beta M'}}{q^{N'}}\right). \tag{3.50}$$

Dropping the primes (the only extensive properties on the right-hand side are those of the M-canonical ensemble) we rewrite this in the form

$$P(\mathbf{n}|M; \beta, q) = \mathbf{n}!W(\mathbf{n})\left(\bar{x}\frac{e^{-\beta M}}{q^{N}}\right). \tag{3.51}$$

Except for the factor $\bar{x}$, the result is the same as the canonical probability in Eq. (3.34). This additional factor arises from the fact that the M-canonical phase space contains fewer configurations than the canonical phase space.[7] To satisfy

---

which is the multinomial probability to sample distribution $\mathbf{n}'$ in N' draws from a pool of clusters with distribution $\tilde{p}_i$. The summation in Eq. (3.49), which can be expressed as

$$\sum_{N'} \text{Prob}\left(\sum_i m_i = M'|N'\right),$$

is the probability that the sampled distribution has mass M' regardless of the number of clusters.

[7]See discussion in Sect. 2.10 on p. 61.

normalization, the M-canonical probability must be multiplied by a factor larger than 1; this factor turns out to be $\bar{x}$. An alternative explanation for this factor and its relationship to the size of the phase space is given in the example below.

Example 3.2 (Fluctuations in Mass Versus Fluctuations in Number)

We investigate the relationship between fluctuations in mass and number with a calculation. First, consider the following process: Draw N^* random samples of random variable m with distribution $p(m)$ and calculate the sum

$$M = \sum_{i=1}^{N^*} m_i.$$

Suppose that the probability distribution of M is $f_M(M)$. According to the central limit theorem, this probability converges to a Gaussian with mean and variance equal to $\bar{M} = N^*\bar{x}$, where $\bar{x}$ is the mean of m_i. This is shown in Fig. 3.2a for $N^* = 100$, drawn from a distribution with mean $\bar{m} = 3$.

The process we just described samples canonical configurations of length N^*. Consider now a modification that allows us to sample N-canonical configurations: Draw consecutive samples but save them only if their sum is exactly M^*; if a draw causes the total mass to exceed M^*, reject the sample and start again. In this process the number of clusters N, defined implicitly by the condition

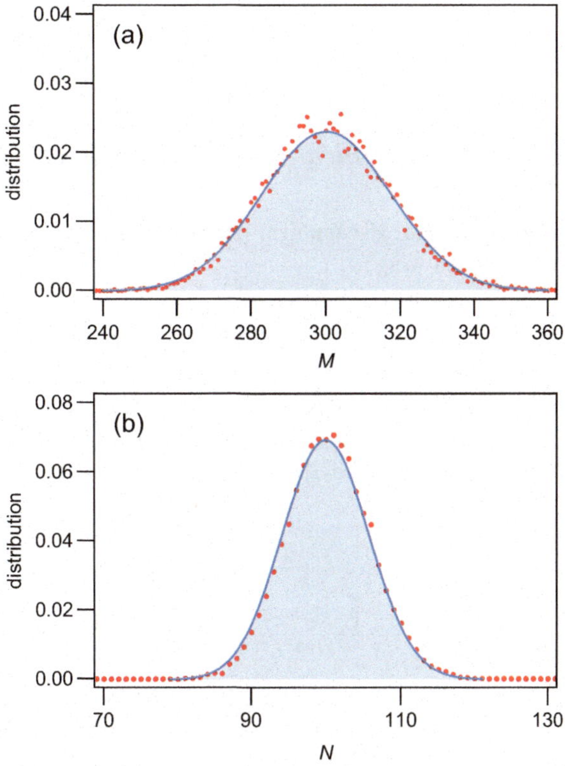

Fig. 3.2 Distribution of the mass and number of clusters in samples drawn from a distribution of clusters with mean size $\bar{x} = 3$. (**a**) Distribution of the mass of $N^* = 100$ random samples. (**b**) Distribution of the number of clusters in samples whose total mass is exactly $M^* = \bar{x}N^* = 300$. Solid lines are Gaussians with parameters given in Example 3.2. The variance of the N-distribution is smaller than the variance of the M-distribution by a factor $\bar{x} = M^*/N^*$

$$\sum_{i=1}^{N} m_i = M^*,$$

is a stochastic variable. We will relate its probability distribution to that of M. Let $f_M(M)$ be the probability distribution of M at fixed number of draws N^*, and $f_N(N)$ the distribution of the number of clusters at fixed total mass M^*. We assume that both distributions can be treated as continuous functions of their argument. We have

$$f_N(N)dN = f_M(N)\frac{dM}{dN}.$$

Setting $M/N = M^*/N^* = \bar{x}$ we have $dM/dN = \bar{x}$. The result for F_N is

$$f_N(N) = \bar{x} f_M(\bar{x}N).$$

Therefore F_N is also Gaussian with mean and variance equal to $\bar{N} = \bar{M}/\bar{x} = N^*$. This variance is less than the variance of $f_M(M)$ by a factor $\bar{x}$ (Fig. 3.2).

A final comment on this calculation: By fixing the ratio M/N to $M^*/N^* = \bar{x}$ we have tacitly assumed that both f_N and f_M are narrow (strictly, this equality applies only between the means of the two distributions). Since both distributions are Gaussians with variance equal to their mean, they become increasingly narrower relative to their means as M and N increase. The assumption therefore holds.

★

3.5.1 *M*-Canonical Partition Function

The part of the M-canonical probability that depends explicitly on **n** defines the M-canonical weight:

$$(M\text{-canonical weight}) = \frac{\mathbf{n}!W(\mathbf{n})}{q^N},$$

($\bar{x}$, M, and β are all constant). The summation of the weights defines the M-canonical partition function and since the M-canonical probability in Eq. (3.51) is properly normalized, the partition function is

$$\Phi = \frac{e^{\beta M}}{\bar{x}}. \tag{3.52}$$

It is a function of q, which appears explicitly in the statistical weight, and of M, which constrains the phase space. Combining with Eq. (3.21) we obtain its relationship to the microcanonical partition function:

$$\log(\bar{x}\Phi) = \log \Omega_{M,N} - (\log q)N = \log \Omega_{M,N} - N\left(\frac{\partial \log \Omega_{M,N}}{\partial N}\right)_M. \tag{3.53}$$

The result identifies $\log(\bar{x}\Phi)$ as the Legendre transformation of $\log \Omega_{M,N}$ with respect to N. By Legendre properties the derivatives of $\log(\bar{x}\Phi)$ are

$$\left(\frac{\partial \log(\bar{x}\Phi)}{\partial M}\right)_q = \beta, \quad \left(\frac{\partial \log(\bar{x}\Phi)}{\partial q}\right)_M = -N. \tag{3.54}$$

The number of clusters is a fluctuating quantity in the M-canonical ensemble; it is the mean number of clusters that appears in the second of these equations. We also have

$$d\log(\bar{x}\Phi) = \beta dM - Nd\log q, \tag{3.55}$$

which follows immediately from Eq. (3.54). Using

$$\log(\bar{x}\Phi) = \log \bar{x} + \log \Phi = \beta M,$$

and noticing that $\bar{x}$ is an intensive property, as is β, while M is extensive, for large M at fixed β and $\bar{x}$ we have

$$\log(\bar{x}\Phi) \approx \log \Phi.$$

Therefore all the relationships written for $\log \bar{x}\Phi$ may be written asymptotically for $\log \Phi$ as well.

3.5.2 M-Canonical MPD

We obtain the M-canonical MPD by maximizing the logarithm of the M-canonical weight,

$$\log \frac{\mathbf{n}!W(\mathbf{n}!)}{q^N}$$

under the conditions

$$q = \text{const}; \quad \sum_i i n_i = M.$$

The equivalent quantity to maximize is

$$\log \frac{\mathbf{n}!W(\mathbf{n}!)}{q^N} - b\left(\sum_i i n_i - M\right)$$

where b is the Lagrange multiplier for the constraint that fixes the mass of the configuration. The distribution that maximizes this functional is

$$\frac{\tilde{n}_i}{N} = \tilde{w}_i \frac{e^{-bi}}{q^N}. \tag{3.56}$$

Comparing with the microcanonical MPD we conclude $b = \beta$ since both distributions satisfy the same normalization condition.[8] Therefore,

$$\left(\frac{\tilde{n}_i}{N}\right)_{M\text{-canonical}} = \tilde{w}_i \frac{e^{-\beta i}}{q} = \left(\frac{\tilde{n}_i}{N}\right)_{\text{microcanonical}}. \tag{3.57}$$

Therefore the N-canonical MPD is the same as the microcanonical MPD.

3.6 Convergence of Ensembles

In the thermodynamic limit all ensembles, microcanonical, canonical, and M-canonical, converge to each other, as demonstrated by the fact that the most probable distribution is the same in all of them. This convergence is the result of a corresponding converges between the probabilities of distribution in each ensemble. The canonical probability of configuration can be expressed as

$$P(\mathbf{n}|\text{canonical}) = \mathbf{n}!W(\mathbf{n}!)e^{-\beta M + N\log q} = \frac{\mathbf{n}!W(\mathbf{n})}{\Omega_{M,N}},$$

which is the same as the microcanonical probability in the (M, N) ensemble, where M is the mass of the canonical configuration. In this expression $\Omega_{M,N}$ is not strictly constant because M varies, but is asymptotically constant because fluctuations

[8] Both b and β satisfy the normalization

$$\sum_i \tilde{w}_i e^{-bi} = q,$$

from which we obtain

$$\frac{d\log q}{db} = \frac{d\log q}{d\beta} = -\bar{x}.$$

This implies $b = \beta + c$ but since both distributions (the one in Eq. (3.56) and the microcanonical MPD) are normalized to unit area, we must have $c = 0$.

decay as the size of the canonical system increases. Similarly the probability of M-canonical configuration is

$$P(\mathbf{n}|M\text{-canonical}) = \bar{x}\mathbf{n}!W(\mathbf{n}!)e^{-\beta M+N\log q} = \bar{x}\frac{\mathbf{n}!W(\mathbf{n})}{\Omega_{M,N}},$$

which again is the microcanonical probability in the $\mu C(M, N)$ ensemble with M fixed and N variables (as we have discussed already, the factor $\bar{x}$ arises from the decreased size of the M-canonical space). In all three ensembles, the probability of distribution is the same. Accordingly, in all ensembles the MPD maximizes the microcanonical probability functional under the particular constraints that define each ensemble. These constraints can be summarized as follows:

ensemble	**constraints**	
microcanonical	$\sum_i n_i = N$	$\sum_i in_i = M$
canonical	$\sum_i N_i = N$	$\beta = \text{const}$
M-canonical	$q = \text{const.}$	$\sum_i in_i = M$

In the ThL all ensembles converge to a single distribution, the MPD. As the sole distribution of the ensemble with nonzero probability at this limit, the MPD has unit probability, $P(\tilde{\mathbf{n}}) = 1$,[9] and this leads to the fundamental relationship

$$\log \tilde{\mathbf{n}}!W(\tilde{\mathbf{n}}) \to \beta M + N\log q = \log \Omega_{M,N}, \tag{3.58}$$

which is a combination of Eqs. (3.11) and (3.21). This alternative view offers a concise summary of the cluster ensemble in the ThL. The quantity that is maximized in the cluster ensemble is the microcanonical functional—*not* entropy. Entropy is maximized only in the special case $W(\mathbf{n}) = 1$. This relegates entropy to a property of secondary significance relative to the microcanonical partition function.

3.7 Microcanonical Surface and the Quasistatic Process

We have obtained the differential of the microcanonical partition function in the form

$$d\log \Omega_{M,N} = \beta dM + (\log q)dN. \qquad [3.25]$$

The geometric interpretation is that β and $\log q$ are the slopes of the surface $\log \Omega$ over the (M, N) plane (Fig. 3.3). The surface is constructed by allowing M and

[9]More precisely, $P(\mathbf{n})$ converges to a delta function at $\mathbf{n} = \tilde{\mathbf{n}}$.

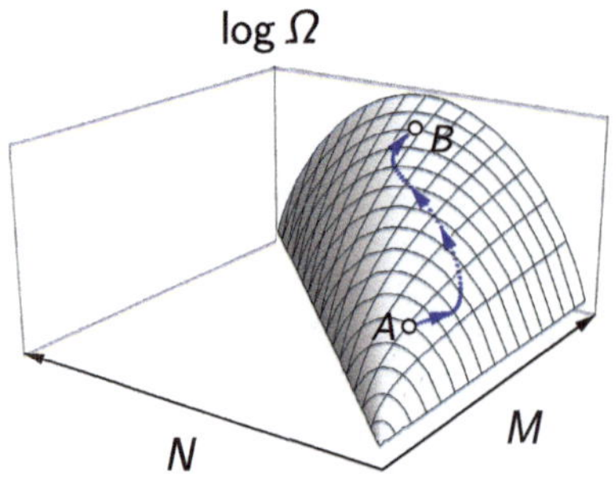

Fig. 3.3 A point on the $\log \Omega$ surface represents a state of the cluster ensemble with β and $\log q$ given by the slopes of the surface along M and N, respectively. A quasistatic MN process is any path on this surface. Along this path β and $\log q$ vary smoothly and are given by the local slopes of the surface. (The partition function in this illustration is that of the unbiased ensemble)

N to vary while W remains fixed. A continuous line on this surface, such as line AB in Fig. 3.3, is a sequence of states whose β and $\log q$ values are given by the local slopes. This line represents a sequence of ensembles from initial state $A = (M_A, N_A)$ to final state $B = (M_B, N_B)$,

$$(M_A, N_A) \to \cdots (M, N) \ \to \ (M + dM, N + dN) \ \to \ \cdots (M_B, N_B),$$

or, equivalently, a sequence of most probable distributions,

$$\mathbf{n}_A \to \cdots \mathbf{n} \to \mathbf{n} + d\mathbf{n} \to \cdots \mathbf{n}_B.$$

We call this path *quasistatic* to indicate that it represents a sequence of ensembles that are completely determined by the local coordinates in the (M, N) plane. Evolution along this path is governed by the fundamental differential on Eq. (3.25). In Chaps. 8, 9, and 10 we discuss two specific examples of quasistatic process, clustering, and fragmentation.

3.8 Generalized Cluster Ensemble

The (M, N) ensemble is the simplest microcanonical ensemble we can construct and describes a population characterized by a single extensive attribute that we have called *mass*. We may generalize this to any number of extensive attributes. The extension is straightforward but the added dimensionality requires that we restate our definitions carefully. Suppose that the members of the population are distinguished by mutually exclusive attributes such that a member is of a single type $k = 1, 2, \cdots K$. By attribute we refer to any extensive variables, such as energy, volume, or other, that characterize the state of the population. The total number of members in the population with the particular attribute k is X_k, $k = 1, \cdots K$. We construct the generalized microcanonical ensemble $\mu C(X_1, \cdots X_K; N)$ as the collection of all possible ways to distribute the set of attributes $\{X_1, X_2, \cdots X_K\}$

among N clusters. A cluster of the generalized ensemble is an ordered sequence of nonnegative integers,

$$\mathbf{x} = (x_1, x_2, \cdots, x_K),$$

such that x_k is the number of members in the cluster with attribute k. A configuration is an ordered sequence of N clusters. The distribution $\mathbf{n}$ of configuration is the multidimensional vector

$$\mathbf{n} = \{n_{x_1,x_2,\cdots,x_K}\},$$

whose element $n_{x_1,x_2,\cdots,x_K} \equiv n_{\mathbf{x}}$ is the number of clusters in state $\mathbf{x} = (x_1, x_2, \cdots, x_K)$.

The total number of members in the distribution with attribute k is[10]

$$X_i = \sum_{\mathbf{x}} x_i \, n_{\mathbf{x}}, \tag{3.59}$$

We also have

$$N = \sum_{\mathbf{x}} n_{\mathbf{x}}, \tag{3.60}$$

which fixes the total number of clusters. The microcanonical ensemble consists of all distributions that satisfy Eqs. (3.59) and (3.60). There are K constraints corresponding to Eq. (3.60) plus one constraint on N. The generalized cluster ensemble is therefore defined by the set of $K+1$ variables $(X_1, \cdots X_K; N)$ with the X_k replacing M in the basic (M, N) ensemble.

---

[10] In expanded notation this is

$$X_i = \sum_{x_1,x_2,\cdots} x_i \, n_{x_1,x_2,\cdots,x_K}$$

with the summation going over all states $(x_1, x_2, \cdots)$ in $\mathbf{n}$. Since $n_{x_1,x_2\cdots}$ is zero for states that are not present in the distribution, we may take the limits of the summation to go from 0 to ∞ for all x_i:

$$\sum_{x_1,x_2,\cdots} = \sum_{x_1=0}^{\infty} \sum_{x_2=0}^{\infty} \cdots .$$

We will abbreviate this as

$$\sum_{\mathbf{x}}$$

The same convention applies to the product in Eq. (3.61).

3.8.1 Multiplicity, Entropy, and Selection Functional

Permutations in the order of clusters in a configuration produce new configurations with the same cluster distribution. The number of permutations is the multiplicity of distribution and is given by the multinomial coefficient

$$\mathbf{n}! = \frac{N!}{\prod_{\mathbf{x}} n_{\mathbf{x}}!}. \tag{3.61}$$

The entropy of distribution is

$$S(\mathbf{n}) = \log \mathbf{n}! = -N \sum_{\mathbf{x}} \frac{n_{\mathbf{x}}}{N} \log \frac{n_{\mathbf{x}}}{N} = -N S(\mathbf{p}), \tag{3.62}$$

where $\mathbf{p} = \mathbf{n}/N$ is the intensive distribution. The microcanonical probability of distribution $\mathbf{n}$ is

$$P(\mathbf{n}) = \mathbf{n}! \frac{W(\mathbf{n})}{\Omega},$$

where $W(\mathbf{n})$ is the selection functional and $\Omega = \Omega(X_1, \cdots X_K; N)$ is the partition function. As in the (M, N) ensemble, the log of the selection functional is required to be homogeneous in $\mathbf{n}$ with degree 1, and concave or linear. The homogeneity condition implies the Euler relationship

$$\log W(\mathbf{n}) = \sum_{\mathbf{x}} n_{\mathbf{x}} \log w_{\mathbf{x}|\mathbf{n}} \tag{3.63}$$

where $\log w_{\mathbf{x}|\mathbf{n}}$ is the partial derivative of $\log W$ with respect to element $(x_1, \cdots x_K)$ of distribution $\mathbf{n}$:

$$\log w_{\mathbf{x}|\mathbf{n}} = \frac{\partial \log W(\mathbf{n})}{\partial n_{\mathbf{x}}}, \tag{3.64}$$

In the special case that $\log W$ is linear, the cluster bias is independent of distribution $\mathbf{n}$ and a function of the $x_1, \cdots x_K$ only:

$$\left.\begin{aligned} \log W(\mathbf{n}) &= \textstyle\sum_x a(\mathbf{x}) n_{\mathbf{x}} \\ \log w_{\mathbf{x}|\tilde{n}} &= a(\mathbf{x}) \end{aligned}\right\} \quad \text{(linear functional).}$$

In general, the cluster bias $\log w_{\mathbf{x}|\mathbf{n}}$ depends not only on the cluster vector $\mathbf{x}$ but also on the distribution to which the cluster belongs.

3.8.2 MPD of the Generalized Ensemble

The most probable distribution maximizes the logarithm of the microcanonical weight $\mathbf{n}!W(\mathbf{N})$ with respect to $\mathbf{n}$ subject to the constraints in Eqs. (3.59) and (3.60). By standard Lagrange maximization generalized MPD is

$$\frac{\tilde{n}_{\mathbf{x}}}{N} = \tilde{w}_{\mathbf{x}} \frac{e^{-\beta_1 x_1 - \beta_2 x_2 \cdots}}{q} \equiv \tilde{p}_{\mathbf{x}}, \tag{3.65}$$

where $\tilde{w}_{\mathbf{x}}$ is the cluster weight evaluated at the MPD, β_i are the Lagrange multipliers corresponding to the K attributes, and q is the normalizing factor and arises from the constraint that fixes the value of N. This is of the same form as the MPD of the (M, N) ensemble, with each attribute introducing the multiplicative term $\exp(-\beta_i x_i)$. This close relationship allows us to extend the results of the (M, N) ensemble by inspection without going through the entire derivation again.

In the thermodynamic limit the distributions of the ensemble converge to the MPD and the logarithm of the partition function converges to the logarithm of the microcanonical weight:

$$\log \Omega = S(\tilde{\mathbf{n}}) + \log W(\tilde{\mathbf{n}}). \tag{3.66}$$

Applying the entropy functional to the generalized MPD and carrying out the summations we obtain the following expression for the entropy

$$S(\mathbf{n}) = \beta_1 X_1 + \beta_2 X_2 \cdots + N \log q - \log W(\tilde{\mathbf{n}}). \tag{3.67}$$

We combine with Eq. (3.66) to write the fundamental equation of the generalized ensemble:

$$\log \Omega = \beta_1 X_1 + \beta_2 X_2 \cdots + (\log q) N. \tag{3.68}$$

This is Euler's theorem applied to $\log \Omega$ and leads to the identification of all partial derivatives,

$$\beta_i = \left(\frac{\partial \log \Omega}{\partial X_i} \right), \quad \log q = \left(\frac{\partial \log \Omega}{\partial N} \right). \tag{3.69}$$

Here, all partial derivatives of $\log \Omega$ are understood to be within the microcanonical set of independent variables $(X_1, X_2, \cdots, N)$. As a corollary we obtain the differential of $\log \Omega$ in the form

$$d \log \Omega = \beta_1 dX_1 + \beta_2 dX_2 \cdots + (\log q) dN. \tag{3.70}$$

This differential may be interpreted as the governing equation of the quasistatic process.

Upon increasing the scale of the ensemble by a factor λ all extensive properties in the ThL are increased by the same factor while all intensive properties remain constant. It is useful then to express all variables in intensive form. The intensive ratios

$$\bar{x}_k = \frac{X_k}{N}, \tag{3.71}$$

are the ensemble average values of X_k per cluster. We define the intensive microcanonical partition function as

$$\log \omega = \frac{\log \Omega}{N}. \tag{3.72}$$

As an intensive quantity, $\omega = \omega(\bar{x}_1, \bar{x}_2 \cdots)$ is a function of the intensive set of $\bar{x}_k$. Accordingly, the intensive state is defined by the K-dimensional vector $(\bar{x}_1, \bar{x}_2 \cdots)$. This is to be contrasted with the $K + 1$ dimensions that are needed to describe the extensive state. Dividing both sides of Eq. (3.68) by N we obtain the intensive form fundamental equation:

$$\log \omega - \log q = \beta_1 \bar{x}_1 + \beta_2 \bar{x}_2 \cdots, \tag{3.73}$$

The β_i, which were defined as derivatives of the extensive $\log \Omega$, can also be expressed as derivatives of the intensive $\log \omega$:

$$\beta_i = \left(\frac{\partial \log \Omega}{\partial X_i} \right)_{X_j, N} = \left(\frac{\partial \log \omega}{\partial \bar{x}_i} \right)_{\bar{x}_j}. \tag{3.74}$$

Since $\log \omega$ is a function of the $\bar{x}_i$ with partial derivatives β_i, its differential is

$$d \log \omega = \beta_1 d\bar{x}_1 + \beta_2 d\bar{x}_2 + \cdots. \tag{3.75}$$

While $\log \Omega$ is homogeneous in X_i with degree 1, no similar homogeneity condition exists between $\log \omega$ and $\bar{x}_i$, unless $\log q = 0$, in which case $\log \omega$ is homogeneous with degree 1 in all x_k.[11] Homogeneity between $\log \omega$ and $\bar{x}_i$ is possible but not necessary; when it exists, it is because of the particulars of the problem, not because the cluster ensemble requires it.

Solving Eq. (3.73) for $\log q$ we obtain

$$\log q = \log \omega - \bar{x}_1 \beta_1 - \bar{x}_2 \beta_2 \cdots = \log \omega - \bar{x}_1 \left(\frac{\partial \log \omega}{\partial \bar{x}_1} \right) - \bar{x}_2 \left(\frac{\partial \log \omega}{\partial \bar{x}_2} \right) \cdots$$

The result identifies $\log q$ as the Legendre transform of $\log \omega$ with respect to all $\bar{x}_i$. By application of the Legendre properties its differential is

$$d \log q = -\bar{x}_1 d\beta_1 - \bar{x}_2 d\beta_2 - \cdots . \tag{3.76}$$

By inspection of this result we recognize the partial derivatives,

$$\bar{x}_i = -\left(\frac{\partial \log q}{\partial \beta_i} \right)_{\beta_{j \neq i}} . \tag{3.77}$$

There are K such relationships and they provide a connection between β_i, the corresponding average $\bar{x}_i$, and $\log q$.

3.8.3 Generalized Canonical Ensemble

We define the generalized canonical ensemble in relationship to its associated microcanonical. We form a canonical configuration by sampling N clusters from a microcanonical ensemble characterized by the intensive set

$$(\beta_1, \cdots \beta_K; \log q)$$

[11] With $q = 0$, Eq. (3.73) becomes

$$\log \omega = \beta_1 \bar{x}_1 + \beta_2 \bar{x}_2 \cdots ,$$

which in combination with Eq. (3.74) states that $\log \omega$ is homogeneous in all x_k with degree 1.

we form a canonical configuration by sampling N clusters. The size of the microcanonical ensemble is immaterial as long as the length of the microcanonical configuration is much larger than N. The probability of canonical distribution is

$$P(\mathbf{n}|\text{canonical}) = \mathbf{n}! \prod_i \tilde{p}_i^{n_i},$$

where $\tilde{p}_\mathbf{x}$ is the microcanonical MPD from Eq. (3.65). By analogy to Eq. (3.34) the final result is

$$P(\mathbf{n}|\text{canonical}) = \mathbf{n}!\, \tilde{W}(\mathbf{n}) \frac{e^{-\beta_1 X_1 - \beta_2 X_2 \cdots}}{q^N}, \tag{3.78}$$

where X_k is the total number of clusters with trait k and $\tilde{W}(\mathbf{n})$ is the linearized selection bias

$$\log \tilde{W}(\mathbf{n}) = \exp \sum_\mathbf{x} n_\mathbf{x} \log \tilde{w}_\mathbf{x},$$

constructed using the cluster weights $\tilde{w}_\mathbf{x}$ of the microcanonical MPD. The normalization condition in Eq. (3.78) defines the canonical partition function Q,

$$q^N = \sum_\mathbf{n} \mathbf{n}!\, \tilde{W}(\mathbf{n}) e^{-\beta_1 X_1 - \beta_2 X_2 \cdots} \equiv Q. \tag{3.79}$$

The summation of canonical weights defines the canonical partition function Q. Let $\bar{X}_k$ be the mean number of attribute k in the canonical configuration,

$$\bar{X}_k = \frac{1}{N} \sum_k X_k.$$

Then we have

$$\bar{X}_k = -\left(\frac{\partial \log Q}{\partial \beta_k}\right)_{\beta_{j\neq k}, N} = -N \left(\frac{\partial \log q}{\partial \beta_k}\right)_{\beta_{j\neq k}}. \tag{3.80}$$

In combination with Eq. (3.77) this reads

$$\bar{X}_k = N \bar{x}_k, \tag{3.81}$$

and states that the mean number of members with attribute k in the canonical configuration is the same as the mean number in the microcanonical configuration. The canonical MPD is obtained by maximization of the canonical weight

$$\mathbf{n}!\,\tilde{W}(\mathbf{n})e^{-\beta_1 X_1 - \beta_2 X_2 \cdots}$$

under the constraints that N, q, and all β_i are constant. It is a simple exercise to show that this is the same as the MPD of the microcanonical ensemble.

3.8.4 The Canonical Family

We have constructed the canonical ensemble in such a way that all X_i are left free to fluctuate. It is possible to fix the value of some of them and let the others float to produce ensembles whose configurations are subsets of the full canonical ensemble. Suppose we let attributes $i = 1, \cdots L$ float and fix the values of all other attributes to their expected values $X_i = N\bar{x}_i$ for all $i > L$. With $L = K$ we recover the basic case discussed in the previous section. By choosing which variables to fix and which to let float we may form a family of canonical ensembles.[12]

The final results can be written down in formulaic fashion without the need to repeat the derivations. The probability of distribution is

$$P(\mathbf{n}) = \mathbf{n}!\tilde{W}(\mathbf{n})\frac{e^{-\beta_1 X_1 \cdots - \beta_L X_L}}{Q_L} \tag{3.82}$$

and the MPD is

$$\frac{\tilde{n}_{\mathbf{x}}}{N} = \tilde{w}_{\mathbf{x}}\frac{e^{-\beta_1 x_1 - \cdots \beta_L x_L}}{q_L} \tag{3.83}$$

with[13]

$$Q_L = q_L^N. \tag{3.84}$$

The log of the L-canonical partition function is the Legendre transformation of the log of the microcanonical partition function with respect to the L variables that fluctuate:

$$\log q_L = \log \omega - \sum_{i=1}^{L} \beta_i \bar{x}_i. \tag{3.85}$$

---

[12]We can always sort attributes so that the first L of them are those that fluctuate and the remaining $K - L$ are those that are fixed.

[13]Using the notation of this section, q and Q in Sect. 3.4 would be notated as q_K and Q_K, respectively.

Its differential is

$$d\log q_L = -\sum_{i=1}^{L} \bar{x}_i d\beta_i + \sum_{i=L+1}^{K} \beta_i d\bar{x}_i \tag{3.86}$$

with

$$\bar{x}_i = -\left(\frac{\partial \log q_L}{\partial \beta_i}\right), \quad (i = 1, L) \tag{3.87}$$

$$\beta_i = \left(\frac{\partial \log q_L}{\partial \bar{x}_i}\right), \quad (i = L+1, K). \tag{3.88}$$

These derivatives are understood to be taken within the set of variables

$$(\beta_1, \cdots \beta_L, \bar{x}_{L+1}, \cdots \bar{x}_K).$$

The generalized cluster ensembles are summarized in Tables 3.2 and 3.3.

Table 3.2 Summary of generalized microcanonical cluster ensemble

Generalized microcanonical ensemble	Equation
Probability of distribution	
$P(\mathbf{n}) = \mathbf{n}!\dfrac{W(\mathbf{n})}{\Omega}$	(2.20)
Most probable distribution	
$\dfrac{\tilde{n}_i}{N} = \tilde{w}_i \dfrac{e^{-\beta_1 x_{i1} - \beta_2 x_{i2} \cdots}}{q}$	(3.65)
Entropy	
$S(\tilde{\mathbf{n}}) = \log\Omega - \log W(\tilde{\mathbf{n}})$	(3.66)
Microcanonical partition function: $\Omega(X_1, X_2, \cdots; N)$	
$\log\Omega = \beta_1 X_1 + \beta_2 X_2 + \cdots + (\log q)N$	(3.68)
$d\log\Omega = \beta_1 dX_1 + \beta_2 dX_2 + \cdots + (\log q)dN$	(3.70)
$\beta_i = \left(\dfrac{\partial \log\Omega}{\partial X_i}\right), \quad \log q = \left(\dfrac{\partial \log\Omega}{\partial N}\right)$	(3.69)
Intensive microcanonical partition function: $\omega(\bar{x}_1, \bar{x}_2, \cdots)$	
$\log\omega \equiv \log\Omega/N$	
$= \beta_1\bar{x}_1 + \beta_2\bar{x}_2 \cdots + \log q$	(3.73)
$d\log\omega = \beta_1 d\bar{x}_1 + \beta_2 d\bar{x}_2 + \cdots$	(3.75)
$\beta_i = \left(\dfrac{\partial \log\omega}{\partial \bar{x}_i}\right)$	(3.74)
Cluster partition function: $q(\beta_1, \beta_2 \cdots)$	
$d\log q = -\bar{x}_1\, d\beta_1 - \bar{x}_2\, d\beta_2 \cdots$	(3.76)
$\bar{x}_i = -\left(\dfrac{\partial \log q}{\partial \beta_i}\right)$	(3.77)

Table 3.3 Summary of generalized canonical ensemble

Generalized canonical family	Equation
Probability of distribution	
$P(\mathbf{n}) = \mathbf{n}!\tilde{W}(\mathbf{n})\dfrac{e^{-\beta_1 X_1 \cdots - \beta_L X_L}}{Q_L}$	(3.82)
Most probable distribution	
$\dfrac{\tilde{n}_i}{N} = \tilde{w}_i \dfrac{e^{-\beta_1 x_{1i} \cdots - \beta_L x_{Li}}}{q_L}$	(3.83)
$\log q_L = \dfrac{\log Q_L}{N}$	(3.84)
Partition function $Q_L(\beta_1, \beta_2, \cdots, \beta_L, X_{L+1}, \cdots X_K; N)$	
$\log Q_L = \log \Omega - \beta_1 X_1 \cdots - \beta_L X_L$	(†)
$d \log Q_L = -\sum_{i=1}^{L} \bar{X}_i d\beta_i + \sum_{i=L+1}^{K} \beta_i d\bar{X}_i$	(†)
$\bar{X}_i = -\left(\dfrac{\partial \log Q_L}{\partial \beta_i}\right), \quad (i = 1, \cdots L)$	(†)
$\beta_i = \left(\dfrac{\partial \log Q_L}{\partial \bar{X}_i}\right), \quad (i = L+1, \cdots K)$	(†)

† Derivations left as exercise.

Intensive partition function $q_L(\beta_1, \beta_2, \cdots, \beta_L, \bar{x}_{L+1}, \cdots \bar{x}_K)$	
$\log q_L = \log \omega - \beta_1 \bar{x}_1 \cdots - \beta_L \bar{x}_L$	(3.85)
$d \log q_L = -\sum_{i=1}^{L} \bar{\bar{x}}_i d\beta_i + \sum_{i=L+1}^{K} \beta_i d\bar{\bar{x}}_i$	(3.86)
$\bar{\bar{x}}_i = -\left(\dfrac{\partial \log q_L}{\partial \beta_i}\right), \quad (i = 1, \cdots L)$	(3.87)
$\beta_i = \left(\dfrac{\partial \log q_L}{\partial \bar{x}_i}\right), \quad (i = L+1, \cdots K)$	(3.88)

Chapter 4
The Most Probable Distribution in the Continuous Limit

The cluster ensemble is inherently discrete but when the characteristic cluster size is much larger than the unit of the ensemble, the MPD may be treated as a continuous variable. We define the continuous limit by the condition

$$\text{continuous limit:} \quad M \gg N \gg 1. \tag{4.1}$$

It is distinct from the thermodynamic limit in that it requires the M/N to be large relative to the unit of the population. In practice though we will be operating mostly where the two limits overlap. In the continuous limit we will replace size by the continuous variable x and the normalized (intensive) MPD by the continuous function $f(x)$:

$$i \to x, \quad M/N \to \bar{x}, \quad \tilde{n}_i/N \to f(x)dx.$$

In the continuous limit all extensive properties increase proportionally to N and all intensive properties are functions of the mean size $\bar{x} = M/N$. As an intensive property, the MPD $f(x)$ also depends on the average size and a more accurate notation would be $f(x;\bar{x})$. Nonetheless we will continue to notate the MPD more simply as $f(x)$ but will occasionally use the notation $f(x;\bar{x})$ when necessary to emphasize the dependence on $\bar{x}$. The continuous MPD satisfies the normalizations

$$\int f(x)dx = 1, \tag{4.2}$$

$$\int x f(x)dx = \bar{x}. \tag{4.3}$$

T. Matsoukas, *Generalized Statistical Thermodynamics*,
Understanding Complex Systems, https://doi.org/10.1007/978-3-030-04149-6_4

These correspond to the two fundamental constraints of the cluster ensemble and are obtained by dividing both constraints by M and replacing the summations by integrals. In this chapter we examine the behavior and properties of the MPD in the continuous limit and reach the conclusion that *any* distribution that satisfies the integral constraints in Eqs. (4.2) and (4.2) can be obtained as the MPD of the cluster ensemble under an appropriate selection bias.

Note 4.1 (On the Limits of Integration)
The domain for all integrals is the same as that for the corresponding summations, i.e., from 1 to ∞. The lower limit can be shifted to zero as long as the analytic continuation of $f(x)$ at $x \to 0$ is such that the integrals converge (the summations always converge by virtue of the homogeneous construction of the ensemble). It is possible to encounter distributions for which some of these integrals do not converge in $(0, \infty)$ and we will point them out explicitly. As a rule we assume the limits to go from 0 to ∞ and for simplicity we will not write them out.

★

4.1 Properties in the Continuous Limit

The entropy of the continuous MPD is[1]

$$S[f] = -\int f(x) \log f(x) dx, \tag{4.4}$$

and is the generalization Eq. (2.17). The cluster bias $\tilde{w}_i$ of the MPD an intensive property, becomes in the continuous limit a function of x and $\bar{x}$:

$$\tilde{w}_i \to w(x; \bar{x}). \tag{4.5}$$

For the logarithm of the selection functional we have

$$\begin{aligned} \log \tilde{W} = \sum_i \tilde{n}_i \log \tilde{w}_i \to \int N f(x) \log w(x; \bar{x}) dx \\ = \log W[Nf] = N \log W[f], \end{aligned} \tag{4.6}$$

[1]We use the notation $J[f]$ to indicate J is a functional of function f. The calculus of functionals is reviewed briefly in the appendix of Chap. 7.

or

$$\log W[f] = \int f(x) \log w(x; \bar{x}) dx. \tag{4.7}$$

The associated Gibbs-Duhem equation is

$$\int f(x) \left(\frac{\partial \log w(x; \bar{x})}{\partial \bar{x}} \right) dx = 0, \tag{4.8}$$

and follows from the corresponding discrete result in Eq. (2.26)[2]. Using $\log \omega = \log \Omega / N$ for the intensive microcanonical partition function we also have

$$\log \omega = \log W[f] + S[f], \tag{4.9}$$

which is the intensive form of Eq. (3.11). The intensive microcanonical partition function satisfies the fundamental relationship,[3]

$$\log \omega = \beta \bar{x} + \log q. \tag{4.10}$$

This is the application of Eq. (3.73) to the cluster ensemble with one distributed variable M. From Eq. (3.74) we have

$$\beta = \frac{d \log \omega}{d \bar{x}}, \tag{4.11}$$

and from (3.77) we obtain

$$\bar{x} = -\frac{d \log q}{d \beta}. \tag{4.12}$$

We have replaced in these equations the partial derivatives with simple derivatives because in the case of the simple ensemble (M, N) all intensive properties are functions of a single intensive variable, the average size $\bar{x} = M/N$. For the MPD itself we have

$$f(x) = w(x; \bar{x}) \frac{e^{-\beta x}}{q}. \tag{4.13}$$

[2]This derivation is discussed in the Appendix of Chap. 7 (see Eq. 7.143).

[3]The properties of the intensive partition function for the generalized ensemble $(M_1, M_2, \cdots; N)$ were discussed in Sect. 3.8.2. Here we adapt the results of that section to the simple microcanonical ensemble $\mu C(M, N)$ with $M = \bar{x} N$.

Table 4.1 Discrete properties of the cluster ensemble and their continuous counter parts

Discrete		Continuous
$\tilde{n}_i/N$	$\rightarrow$	$f(x) = w(x;\bar{x})\dfrac{e^{-\beta x}}{q}$
$S(\tilde{\mathbf{n}})/N$	$\rightarrow$	$S[f] = -\int f(x)\log f(x)dx$
$\log W(\tilde{\mathbf{n}})/N$	$\rightarrow$	$\log W[f] = \int f(x)\log w(x;\bar{x})dx$
$\log\Omega/N$	$\rightarrow$	$\log\omega = \beta\bar{x} + \log q = \log W[f] + S[f]$
		$\beta = \dfrac{d\log\omega}{d\bar{x}}$
		$q = \int w(x;\bar{x})e^{-\beta x}dx$

We insert this result into the normalization condition in Eq. (4.2) to obtain the following expression for q:

$$q = \int w(x;\bar{x})e^{-\beta x}dx. \tag{4.14}$$

This form expresses q as the Laplace transform of $w(x;\bar{x})$ with respect to x, with β as the conjugate variable in Laplace space:

$$q = \mathcal{L}\{w; x; \beta\}, \tag{4.15}$$

where $F = \mathcal{L}\{f; t; s\}$ indicates that $F[s]$ is the Laplace transform of $f(t)$. These mappings between the variables of the discrete and continuous domain are summarized in Table 4.1.

4.2 Unbiased Ensemble

In linear ensembles the cluster bias is a pure function of size. We notate this function as

$$w(x;\bar{x}) = w(x).$$

In Chap. 2 we derived the main results of linear ensembles in the discrete domain, as well as of the unbiased ensemble, which is the most basic case of linear ensemble. Here and in the next section we will reformulate the solution to linear ensembles in the continuous domain. We discuss two examples: The first one is the unbiased ensemble and we will obtain its results by taking the continuous limit on the results obtained in Chap. 2. In the second example (Sect. 4.3) we discuss a linear selection functional with cluster bias that is power-law function of size and use to demonstrate the treatment of linearity in the continuous domain. The unbiased ensemble ($W[f] = w(x) = 1$) is the simplest cluster ensemble that is fully solvable analytically. In Sect. 2.7 we obtained the partition function and mean distribution for discrete finite M and N. Here we obtain the properties of the ensemble in the ThL.

The partition function of the unbiased ensemble is equal to the microcanonical volume,

$$\Omega_{M,N} = \binom{M-1}{N-1}. \tag*{[2.6]}$$

Using Stirling's asymptotic formula for the factorial, this becomes

$$\Omega_{M,N} = \frac{\left(\bar{x}^{\bar{x}}(\bar{x}-1)^{-(\bar{x}-1)}\right)^N}{\sqrt{2\pi N\bar{x}(\bar{x}-1)}}, \tag{4.16}$$

with $\bar{x} = M/N$. This is the continuous representation of the partition function. Taking the log of this expression and dividing by N we obtain the intensive partition function $\log\omega = \log\Omega_{M,N}/N$ in the form

$$\log\omega = \bar{x}\log\bar{x} - (\bar{x}-1)\log(\bar{x}-1), \tag{4.17}$$

where we have dropped terms that grow logarithmically.[4] The parameter β is according to Eq. (4.11) the derivative of $\log\omega$ with respect to $\bar{x}$,

$$\beta = \log\frac{\bar{x}}{\bar{x}-1}, \tag{4.18}$$

and q is calculated as $\log q = \log\omega - \beta\bar{x}$,

$$q = \bar{x} - 1, \tag{4.19}$$

Alternatively, these results can be obtained by first calculating β and q in the discrete finite domain and then passing to the continuous limit. Using the finite difference representation for the derivative of $\log\Omega_{M,N}$ with respect to M, the parameter β is

$$\beta = \log\frac{\Omega_{M+1,N}}{\Omega_{M,N}} = \log\frac{M}{M-N+1} = \log\frac{\bar{x}}{\bar{x}-1+1/N} \to \log\frac{\bar{x}}{\bar{x}-1}. \tag{4.20}$$

Similarly, for $\log q$ we have

$$\log q = \log\frac{\Omega_{M,N+1}}{\Omega_{M,N}} = \log\frac{M-N}{N} = \log(\bar{x}-1). \tag{4.21}$$

[4]The term that is dropped is $\log\sqrt{2\pi N\bar{x}(\bar{x}-1)}$. This amounts to using the simpler Stirling formula, $\log x! = x\log x - x$.

Table 4.2 Summary of unbiased ensemble

Property	Discrete finite		ThL
$\log \omega$	$\frac{1}{N} \log \binom{M-1}{N-1}$	$\rightarrow$	$\log \bar{x} + 1$
β	$\log \frac{M}{M-N+1}$	$\rightarrow$	$\frac{1}{\bar{x}}$
q	$\log \frac{M-N}{M}$	$\rightarrow$	$\bar{x}$
Mean distribution	$\frac{1}{N} \log \binom{M-k-1}{N-2} \Big/ \log \binom{M-1}{N-1}$	$\rightarrow$	$\frac{e^{-x/\bar{x}}}{\bar{x}}$
MPD	$\frac{1}{\bar{x}-1}\left(1-\frac{1}{\bar{x}}\right)^x$	$\rightarrow$	$\frac{e^{-x/\bar{x}}}{\bar{x}}$

Both results agree with Eqs. (4.18) and (4.19). With β and q known, and $w(x) = 1$, the MPD (Eq. 4.13) is

$$f(x) = \frac{1}{\bar{x}-1}\left(1-\frac{1}{\bar{x}}\right)^x. \tag{4.22}$$

If $\bar{x} \gg 1$, we can put these results into a simpler form. We find[5]

$$\log \omega \rightarrow 1 + \log \bar{x} \tag{4.23}$$

$$\beta \rightarrow \frac{1}{\bar{x}}, \tag{4.24}$$

$$q \rightarrow \bar{x}, \tag{4.25}$$

$$f(x) \rightarrow \frac{e^{-x/\bar{x}}}{\bar{x}}. \tag{4.26}$$

These results are summarized in Table 4.2.

As we see from Eq. (4.26), the MPD of the unbiased ensemble in the continuous limit is the exponential distribution. Recall that the MPD maximizes the log of

[5]When $\bar{x} \gg 1$ the right-hand side of Eq. (4.17) is asymptotically equal to the derivative of $\bar{x} \log \bar{x}$:

$$\log \omega \rightarrow \frac{d\bar{x} \log \bar{x}}{d\bar{x}} = \log \bar{x} + 1.$$

Then obtain β as $d \log \omega / d\bar{x}$ and $\log q$ as $\log \omega - \beta \bar{x}$.

the microcanonical weight, $\log W[F] + S[f]$, which in the case of the unbiased ensemble ($W = 1$) is equal to the entropy of the distribution. In this we have recovered the well-known result in the maximum entropy method, namely, that the exponential distribution maximizes entropy among all distributions with the same mean.

Note 4.2 (Entropy in the ThL)

In the unbiased ensemble entropy is equal to $\log \omega$ and this is always positive,

$$S[f] = \log \omega = 1 + \log \bar{x} \geq 0,$$

since $\bar{x} > 1$. The passage to the continuous limit does not destroy this fundamental property. This property arises from the fact that entropy is the log of multiplicity, and multiplicity is defined in a discrete distribution whose unit mass is 1. We may rescale the cluster mass to any other unit z_0 by defining the scaled variable $z/z_0 \equiv x$. The probability distribution of z is

$$h(z) = f(x)\frac{dx}{dz} = \frac{f(z/z_0)}{z_0}.$$

The entropy of $h(z)$ is

$$\begin{aligned} S[h] &= -\int h(z) \log h(z) dz \\ &= -\int \Big(-\log z_0 + \log f(z/z_0) \Big) f(z/z_0) \frac{dz}{z_0} \\ &= S[f] + \log z_0. \end{aligned}$$

The entropy of the rescaled function is shifted by $\log z_0$. This rescaling does not affect maximization (z_0 is constant) but entropy may now be positive or negative.

Note 4.3 (ThL and Continuity)

Equations (4.17), (4.18), (4.19), and (4.22) are ThL expressions for $\log \omega$, β, q, and the MPD in the transitional regime between the discrete and continuous domain. In this region we have ThL conditions but the mean cluster size is not sufficiently large to treat the distribution as a continuous function of size. We demonstrate this distinction by a numerical example. Figure 4.1a shows the sampled distribution for $M = 100$, $N = 50$, which corresponds to mean cluster size $\bar{x} = 2$. The ThL prediction given by Eq. (4.22) is in excellent agreement with the simulation despite the fact that the mean cluster size is of the order of 1. Clearly, the values of M and N in this example are large enough to ensure ThL. By contrast, the continuum prediction in Eq. (4.26) is clearly not appropriate. Figure 4.1b shows simulation results for $M = 1000$, $N = 50$, corresponding to average cluster size $M/N = 20$. In this case the system is well into the continuous regime and both Eqs. (4.22)

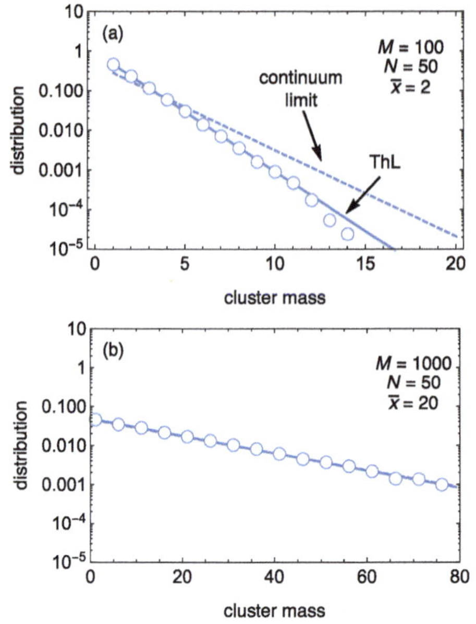

Fig. 4.1 MPDs of the unbiased ensemble. The solid line is the ThL (Eq. 4.22), the dashed line is the continuum limit (Eq. 4.26) and the symbols are results from MC simulation of the ensemble (in the bottom graph the dashed and solid lines coincide). The ThL equation represents the MPD correctly at all M/N. The continuum limit is the asymptotic form of the ThL result when $M/N \gg 1$

and (4.26) converge to the sampled distribution. The point of this calculation is to demonstrate that the thermodynamic limit is distinct from the continuous limit and is established regardless of whether the MPD is discrete or continuous, as long as the size of the population is large. The broader implication is that the theory of the cluster ensemble applies to both discrete and continuous distributions. Our interest, however, will be in distributions in the continuous domain.

★

4.3 Linear Ensemble with Power-Law Bias

The unbiased ensemble is an example where we were able to obtain the discrete finite partition function, and the ThL by taking the limit at $M, N \gg 1$ at fixed $M/N = \bar{x}$. In the general case of linear ensembles this is not possible and instead we obtain the ThL by applying the relationships of the continuum domain. We demonstrate the approach using as an example the power-law cluster bias,

$$w(x) = x^a, \tag{4.27}$$

where a is a constant. In the discrete domain this represents the selection functional

$$W(\mathbf{n}) = \prod_i i^{a n_i}, \tag{4.28}$$

with corresponding microcanonical probability

$$P(\mathbf{n}) = \frac{N!}{\Omega_{M,N}} \prod_i \frac{i^{an_i}}{n_i!}. \tag{4.29}$$

First we calculate q from Eq. (4.14), which expresses $q(\beta)$ as the Laplace transform of $w(x)$ with β as the transformed variable:

$$q = \int x^a e^{-\beta x} dx = \beta^{-a-1}\Gamma(a+1). \tag{4.30}$$

Next we obtain the average cluster size $\bar{x}$ from Eq. (4.12) by differentiation of $\log q(\beta)$,

$$\bar{x} = -\frac{d \log q}{d\beta} = \frac{a+1}{\beta}. \tag{4.31}$$

We express β and q in terms of $\bar{x}$. Solving Eq. (4.31) for β we have

$$\beta = \frac{a+1}{\bar{x}} \tag{4.32}$$

and substituting into Eq. (4.30) we findc

$$q = \left(\frac{\bar{x}}{a+1}\right)^{a+1} \Gamma(a+1). \tag{4.33}$$

Finally we construct the MPD by combining Eqs. (4.32), (4.33), and (4.27), the MPD is

$$f(x) = \frac{x^a}{\Gamma(1+a)} \left(\frac{1+a}{\bar{x}}\right)^{1+a} e^{-(1+a)x/\bar{x}}. \tag{4.34}$$

This is the gamma distribution with shape parameter $a+1$. It is easy to confirm that it is normalized to unit area and that its mean is $\bar{x}$. Figure 4.2 shows this distribution for $a = 2$, $\bar{x} = 20$, along with results from a Monte Carlo sampling of the ensemble using the method of the exchange reaction and the cluster bias function in Eq. (4.27). The last step is to construct the microcanonical partition function as $\log\omega = \beta\bar{x} + \log q$:

$$\log\omega = -(a+1)\log\left(\frac{a+1}{\bar{x}}\right) + \log\Gamma(a+1) + a + 1. \tag{4.35}$$

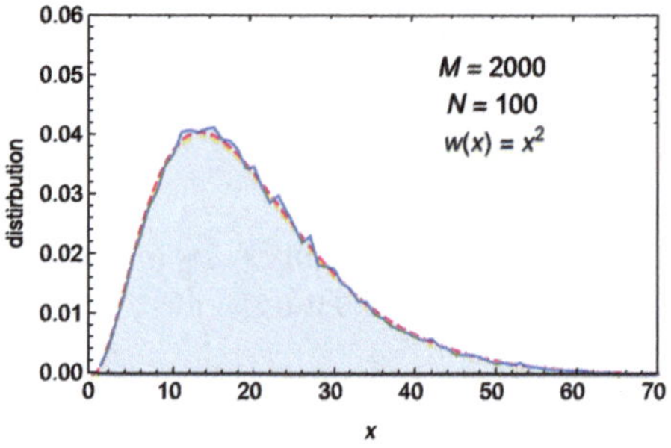

Fig. 4.2 MC simulation of the power-law cluster bias with $a = 2$, $\bar{x} = 20$. The dashed line is Eq. (4.34), the shaded curve is a Monte Carlo simulation of the ensemble with $M = 2000$, $N = 100$

The example demonstrates how to obtain the canonical form of the MPD and all the associated parameters of a linear ensemble starting with the cluster function $w(x)$. The entire ensemble is reconstructed from knowledge of the function $w(x)$. We will now show that it is also possible to reconstruct the ensemble starting from $\omega(\bar{x})$.

Suppose that the microcanonical partition function is given as a function of $\bar{x}$ by Eq. (4.35). β is obtained from Eq. (4.11) as the derivative of $\log \omega$ with respect to $\bar{x}$:

$$\beta = \frac{d \log \omega}{d\bar{x}} = \frac{1+a}{\bar{x}}. \tag{4.36}$$

Next we obtain $\log q$ as the Legendre transformation of $\log \omega(\bar{x})$:

$$\log q = \log \omega - \bar{x}\frac{d \log \omega}{d\bar{x}} = \log \Gamma(a+1) - (a+1)\log\left(\frac{a+1}{\bar{x}}\right), \tag{4.37}$$

and obtain q in the form:

$$q = \left(\frac{\bar{x}}{a+1}\right)^{a+1} \Gamma(a+1). \tag{4.38}$$

Finally, we obtain $w(x)$ by inverse Laplace transform of $q(\beta)$ with respect to β with transformed variable x.[6] To do this we first express $\bar{x}$ in terms of β by solving Eq. (4.36) and substituting into (4.37) to obtain q in terms of β. We find $\bar{x} = (1 + a)/\beta$ and

$$q = \beta^{-a-1}\Gamma(a+1). \tag{4.39}$$

[6]Start with Eq. (4.14)

$$q = \mathcal{L}\{w; x; \beta\},$$

and solve for w by inverting the transform:

$$w = \mathcal{L}^{-1}\{q; \beta; x\}.$$

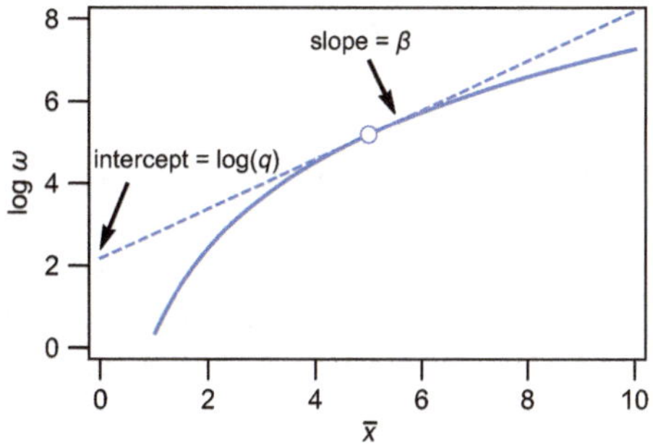

Fig. 4.3 Graphical representation of Legendre relationships of the partition function. β is the slope of $\log \omega(\bar{x})$ at given $\bar{x}$, log is the intercept, and w is the inverse Laplace transform of the $q(\beta)$ relationship. The partition function is Eq. (4.35) with $a = 2$

We then obtain $w(x)$ by inverse Laplace transform of this result:

$$w(x) = \mathcal{L}^{-1}\left\{\beta^{-a-1}\Gamma(a+1); \beta; x\right\} = x^a. \tag{4.40}$$

Thus we have recovered all the properties of the linear ensemble from the microcanonical partition function.

Note 4.4 (Encoding of Information in the Linear Ensemble)

The microcanonical partition function of the linear ensemble is a *complete representation* of the ensemble. If the partition function is a known function of the average cluster size, all properties of the ensemble can be obtained by mathematical manipulations of that function. This reconstruction is demonstrated graphically in Fig. 4.3. Given $\bar{x}$, we draw the tangent line at $\log \omega(\bar{x})$. The slope of this tangent is β and its intercept is $\log q$. By repeating this construction at all $\bar{x}$ we obtain a relationship between q and β. The only function that is not represented graphically in this figure is the cluster function $w(x)$. This is obtained as the inverse Laplace transform of $q(\beta)$ with respect to β. The ensemble may also be reconstructed from $w(x)$. This is always possible, since the ensemble is fully specified by the selection functional, and this, in the case of linear ensemble, is in turn fully specified by the cluster function $w(x)$. In the linear case then

$$W[f] \Longleftrightarrow \omega(\bar{x}),$$

which we take to mean that the linear ensemble in the ThL is fully encoded in the selection functional as well as in the partition function. From left to right we have the *forward* problem, i.e., the calculation of the ensemble starting from the selection functional. From right to left we have the *inverse* problem, i.e., inferring W given ω. In the linear case the flow of information is reversible. We will see that in the nonlinear case this is not the case.

★

4.4 Nonlinear Ensembles

The fundamental difference between linear and nonlinear ensembles is that the cluster functions of the linear ensemble are intrinsic functions of the cluster size while for nonlinear ensembles they also depend on the distribution to which a cluster belongs to. In the ThL the dependence on the distribution appears as a dependence on the average cluster, size since the most probable distribution is fixed once the mean cluster size is specified. Therefore we have

$$w = w(x) \qquad \text{linear ensemble}$$

$$w = w(x; \bar{x}) \qquad \text{nonlinear ensemble}$$

We will demonstrate the different behavior between linear and nonlinear ensembles through an example that allows us to calculate the MPD and all ThL properties exactly. We will do this using the "entropic" selection functional, which we define as

$$\log W(\mathbf{n}) = S(\mathbf{n}) = -\sum_i n_i \log \frac{n_i}{N}, \tag{4.41}$$

with cluster functions

$$\log w_{i|\mathbf{n}} = \frac{\partial S(\mathbf{n})}{\partial n_i} = -\log \frac{n_i}{N}. \tag{4.42}$$

The log of the entropic selection functional is equal to entropy, which is a nonlinear functional of distribution since clearly the cluster bias $w_{i|\mathbf{n}}$ depends not only on i but also on n_i. The corresponding microcanonical probability of distribution $\mathbf{n}$ is

$$P(\mathbf{n}) = \mathbf{n}! \frac{e^{S[\mathbf{n}]}}{\Omega_{M,N}}. \tag{4.43}$$

This functional samples distributions in proportion to the product of their natural multiplicity times $\exp S(\mathbf{n})$. This is *not* the same as the unbiased ensemble, which picks distributions in proportion to their multiplicity alone. The microcanonical probability of the entropic ensemble is functionally equivalent to[7]

$$P'(\mathbf{n}) = \frac{(\mathbf{n}!)^2}{\Omega_{M,N}}. \tag{4.44}$$

[7]The MPD maximizes the logarithm of the probability and since $\log \mathbf{n}! = S[\mathbf{n}]$ in the ThL, Eqs. (4.43) and (4.44) both produce the same MPD, even though their probabilities are not the same (their logarithms are).

The entropic ensemble essentially picks distributions in proportion to the *square* of their multiplicity, it applies therefore a nonuniform bias.

We could obtain the MPD by standard Lagrange maximization of the microcanonical weight but in this case we will get to the final answer more easily by noting that the cluster function of the entropic bias in Eq. (4.42) can be expressed as

$$w(x) = \frac{N}{n_i} = \frac{1}{f(x)}. \tag{4.45}$$

Using this in the canonical form of the MPD we have

$$f(x) = \frac{1}{f(x)} \frac{e^{-\beta x}}{q}, \tag{4.46}$$

which solved for f gives

$$f(x) = \frac{e^{-\beta x/2}}{q^2}. \tag{4.47}$$

The parameters β and q are obtained by back substitution of the MPD into the two constraints:

$$1 = \int_0^\infty f(x)dx = \frac{2}{\beta\sqrt{q}}, \tag{4.48}$$

$$\bar{x} = \int_0^\infty x f(x)dx = \frac{4}{\beta^2\sqrt{q}}. \tag{4.49}$$

We find

$$\beta = 2/\bar{x}, \tag{4.50}$$

$$q = \bar{x}^2. \tag{4.51}$$

Using these values in Eq. (4.47) and (4.42), the MPD is

$$f(x) = \frac{e^{-x/\bar{x}}}{\bar{x}}. \tag{4.52}$$

The cluster function is $w(x; \bar{x})$, or

$$w(x; \bar{x}) = \bar{x} e^{x/\bar{x}}. \tag{4.53}$$

The explicit dependence on $\bar{x}$ demonstrates the distinguishing feature of the nonlinear cluster function, a feature that is not present in the linear case. Finally, the microcanonical partition function is calculated as $\log\omega = \beta\bar{x} + \log q$ and we find

$$\log\omega = 2 + 2\log\bar{x}. \tag{4.54}$$

With this we have obtained all the results of the entropic ensemble in the ThL. The entropic ensemble has the same MPD as the unbiased ensemble but differs in all other properties.

Note 4.5 (Entropic Versus Unbiased Ensemble)

The entropic and unbiased ensembles share the same MPD. While the two ensembles are distinctly different, they are closely related. The microcanonical weight of the unbiased ensemble is

$$\mathbf{n}!W(\mathbf{n}) = \mathbf{n}!$$

and of the entropic ensemble,

$$\mathbf{n}!W(\mathbf{n}) = \mathbf{n}!\,e^{S(\mathbf{n})} \sim (\mathbf{n}!)^2$$

The unbiased ensemble picks distributions in proportion to their multiplicity, the entropic ensemble picks them in proportion to the square of multiplicity. The logarithms of these weights are maximized by the same distribution. Despite sharing the same MPD, the two ensembles are fundamentally different because they assign different probabilities to the distributions and configurations of the ensemble. While the unbiased ensemble samples all configurations with the same probability, the entropic ensemble is biased towards configurations with higher entropy. It is for this reason that the functions ω, q, β, and w are different, even though they produce the same MPD. We can see the difference between the two ensembles in fluctuations about the MPD. The entropic ensemble applies a nonuniform bias, it is more "choosy" than the unbiased ensemble and its fluctuations are narrower. We see this in simulation. Figure 4.4 shows the distribution of clusters by Monte Carlo using $M = 200$, $N = 50$, in the unbiased (top) and the entropic (bottom) ensembles. The MPD in both cases agrees with the exponential distribution. Figure 4.4 shows the fluctuation in the number of monomers, n_1. The mean in both cases is the same, equal to the number of monomers in the exponential MPD. It is clear, however, that fluctuations in the entropic ensemble are of narrower magnitude than those of the unbiased ensemble. In the thermodynamic limit the difference between the unbiased and the entropic ensemble vanishes and the two ensembles converge to the same point in the phase space, the exponential distribution.

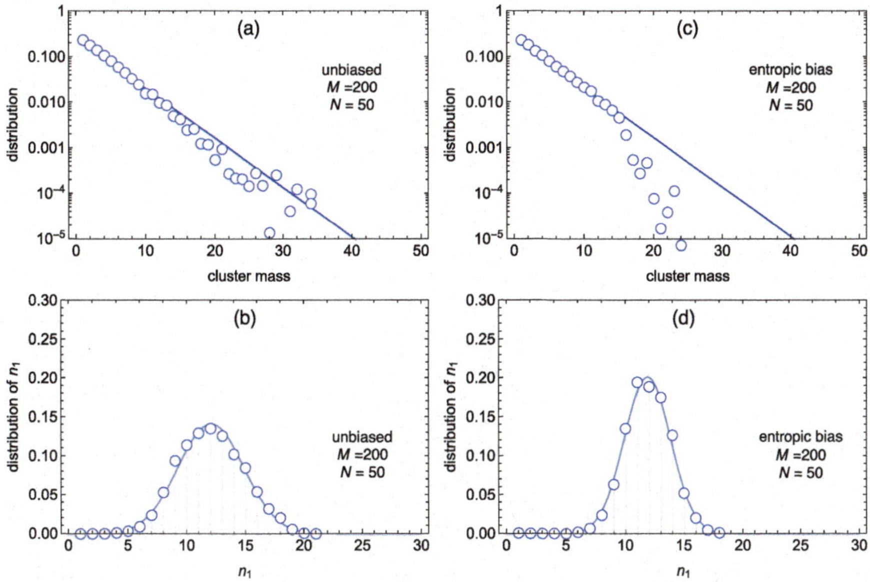

Fig. 4.4 Numerical simulation of unbiased and entropic ensembles by exchange reaction Monte Carlo using $M = 200$, $N = 50$ ($\bar{x} = 4$). (**a**) MPD of unbiased ensemble; (**b**) fluctuation in the number of monomers (unbiased); (**c**) MPD of entropic ensemble; (**d**) fluctuation in the number of monomers (entropic). The MPD is the same but fluctuations in the entropic ensemble are narrower. The solid lines in (**c**) and (**d**) are Gaussian distributions with mean and variance obtained from the simulation

4.4.1 The Laplace Relationship q-β-w

In the linear ensemble we obtained $\{q(\beta), w(x)\}$ as a Laplace-transform pair. In the nonlinear case this relationship is complicated by the fact that the cluster function also depends on $\bar{x}$. We begin by taking the Laplace transform of $w(x; \bar{x}) = xe^{x/\bar{x}}$:

$$\mathcal{L}\{w; x; \beta\} = \int_0^\infty \bar{x}e^{x/\bar{x}}e^{-\beta x}dx = \frac{\bar{x}^2}{\bar{x}\beta - 1}. \tag{4.55}$$

Using $\beta = 2/\bar{x}$ from Eq. (4.50) this becomes

$$\mathcal{L}\{w; x; \beta\} = \bar{x} = q, \tag{4.56}$$

and confirms that the forward Laplace relationship between q and w holds. Define the related function $\theta(\beta, y)$ as

$$\theta(\beta, y) = \frac{y^2}{y\beta - 1}. \tag{4.57}$$

This function appears on the right-hand side of Eq. (4.55) and satisfies the identity

$$\theta(\beta, \bar{x}) = q, \tag{4.58}$$

i.e., when y is replaced by $\bar{x}$ we obtain q. The inverse Laplace transform of $\theta(x, y)$ with respect to β and with transformed variable x is

$$\mathcal{L}^{-1}\{q^\star; \beta; x\} = ye^{x/y}, \tag{4.59}$$

and this at $y = \bar{x}$ is equal to $w(x; \bar{x})$. That is,[8]

$$\mathcal{L}^{-1}\{\theta(\beta, y); \beta; x\}\Big|_{y=\bar{x}} = w(x; \bar{x}), \tag{4.60}$$

which recovers w by inverse transform of θ. However, the function that is inverse-transformed is not q but the related function θ. The relationship between q and θ is asymmetric: if we know θ we obtain q from Eq. (4.58); but if we know q, either as a function of $\bar{x}$ or of β (the two are interrelated via $\beta = 2/\bar{x}$), we cannot construct the bivariate function θ if all we know is that Eq. (4.58) must be satisfied. Unlike the linear case, there is an irreversible loss of information when instead of the selection functional we are given the partition function.

4.4.2 Non-Uniqueness

We may eliminate $\bar{x}$ between Eqs. (4.50) and (4.51) to obtain a direct relationship between q and β:

$$q = 4/\beta^2. \tag{4.61}$$

We calculate the inverse Laplace transform $q(\beta)$ with respect to β using x as the transformed variable:

$$\mathcal{L}^{-1}\left\{\frac{4}{\beta^2}; \beta; x\right\} = 4x \equiv w^*(x). \tag{4.62}$$

[8] The notation on the right-hand side means that y is treated as a constant during the Laplace transform and is to be replaced by $\bar{x}$ only after the transform is completed. If y is replaced by $\bar{x}$ inside the transform and we use $\bar{x} = 2/\beta$, the result of the inverse transform is *not* $w(x; \bar{x})$ but $w^*(x)$ in Eq. (4.62).

The result represents a linear cluster function that depends on x but not on $\bar{x}$. Then the set

$$\{w^*(x) = 4x, q = \bar{x}^2, \beta = 2/\bar{x}\}$$

represents the parameters of a canonical MPD f^*,

$$f^*(x) = w^*(x)\frac{e^{-\beta x}}{q}$$

whose full form is

$$f^*(x) = \frac{4x}{\bar{x}}e^{-2x/\bar{x}}. \tag{4.63}$$

This distribution is normalized to unit area, its mean is $\bar{x}$ and has the same microcanonical partition function as the entropic MPD. The first two properties are confirmed by integrating the MPD. The last follows from the fact that this distribution has the same q and β as the entropic MPD, therefore the same microcanonical partition function $\log\omega = \beta\bar{x} + \log q$.

Table 4.3 summarizes the results for these three related ensembles: unbiased (a linear ensemble), entropic (nonlinear), and an associated linear ensemble that is associated with the nonlinear entropic ensemble. There is partial overlap between these ensembles. The unbiased and entropic ensembles have the same MPD but different $\{w, q, \beta, \omega\}$; the entropic and its associated linear ensemble have the same $\{q, \beta, \omega\}$ but different w, therefore different MPDs as well; the unbiased and the associated linear are both linear ensembles, while the entropic is not. The

Table 4.3 Summary of entropic ensemble in the ThL

	Ensemble		
Property	Unbiased	Entropic	Associated linear[a]
$\log W[Nf]$	0	$NS[f]$	$N\int\log(4x)f(x)dx$
$w(x;\bar{x})$	1	$\bar{x}e^{x/\bar{x}}$	$4x$
β	$1/\bar{x}$	$2/\bar{x}$	$2/\bar{x}$
q	$\bar{x}$	$\bar{x}^2$	$\bar{x}^2$
$\log\omega$	$1+\log\bar{x}$	$2+2\log\bar{x}$	$2+2\log\bar{x}$
$f(x)$	$\frac{e^{-x/\bar{x}}}{\bar{x}}$	$\frac{e^{-x/\bar{x}}}{\bar{x}}$	$\frac{4x}{\bar{x}}e^{-2x/\bar{x}}$

[a]Obtained by extracting the linear cluster function $w(x)$ from the entropic $q(\beta)$

entropic ensemble demonstrates that it is possible to associate the same MPD with more than one selection functional and the same microcanonical partition function with more than one MPD. No unique mapping exists between the MPD and the selection functional, nor between the selection functional and microcanonical partition function—unless the selection functional is linear. Nonlinearity breaks this one-to-one mapping. Knowing the MPD or the microcanonical partition function is not sufficient to recover the selection functional.

4.5 The Inverse Problem

The central property of the cluster ensemble is the microcanonical probability and this is completely specified by the selection functional. Once the selection functional is given, all properties of the ensemble are fixed and can be obtained, whether by derivation or by numerical simulation. This is the forward problem and describes the typical task in mathematical modeling: construct a physical model (W) and obtain its predictions for its observables (f).[9] As often we are faced with the *inverse* problem: we are in possession of observables and wish to *infer* the physical mechanism that produces them. In the context of the cluster ensemble the physical model is the selection functional and the observables are the MPD and associated variables. The transformation of information from the discrete domain to the thermodynamic limit can be represented schematically as

$$W \longrightarrow \{q, \beta; w\} \longrightarrow f.$$

The input in this transformation is the selection functional and the observable is the MPD whose canonical form is expressed via $\{\beta, q, w\}$. The microcanonical partition function does not appear here because it is fully reconstructed from the relationship $q = q(\beta)$.[10] If we know f, what can we infer about W? It is not possible to identify the selection functional from knowledge of the MPD for the same reason that we cannot identify a function $h(x)$ from its value at a single point x. As the ensemble reduces to a single distribution in the ThL, the best we may hope for is to obtain the value of the selection functional at that particular distribution. Neither can we tell whether the selection functional is linear or not. Fluctuations, assuming we have access to them, are confined to a linear neighborhood around the MPD and convey no information whatsoever regarding the existence or lack of nonlinear effects outside this neighborhood. In view of these limitations we pose a more modest question: is it possible to identify *a* functional that produces the observed MPD? We set out to answer this question.

---

[9] We will see specific examples on how to construct W in Chaps. 8 and 9 and 10.

[10] Recall that β and q are the slope and intercept of $\log \omega(\bar{x})$, therefore an equivalent representation of the ThL transformation is $W \longrightarrow \{\omega; w\} \Longleftrightarrow \{\beta, q; w\} \longrightarrow f$.

Suppose that the MPD that is obtained from the selection functional W is[11]

$$f(x;\bar{x}) = w(x;\bar{x})\frac{e^{-\beta x}}{q}. \tag{4.64}$$

Define the new functional W' to be

$$\log W'[Nf] = \log W[Nf] + a_1 M + a_0 N, \tag{4.65}$$

where a_0 and a_1 are arbitrary constants independent of x (they may depend on $\bar{x}$). Since the new term $a_0 N + a_1 M$ in W' has the same value in all distributions of the (M, N) ensemble, both W and W' assign the same relative weights to the distributions of the cluster ensemble, and both produce the same MPD. The cluster function of the new functional is

$$\log w'(x;\bar{x}) = \log w(x;\bar{x}) + a_1 x + a_0. \tag{4.66}$$

Let β' and q' be the parameters of the MPD that correspond to the new cluster weights w'; the MPD expressed in terms of (w', β', q') is identical to that in terms of (w, β, q), therefore,

$$f(x;\bar{x}) = w'(x;\bar{x})e^{a_1 x + a_0}\frac{e^{-\beta' x}}{q'} = w'(x;\bar{x})\frac{e^{-\beta x}}{q}.$$

From this we obtain the relationship between the two set of parameters,

$$\beta' = \beta + a_1, \quad \log q' = \log q + a_0. \tag{4.67}$$

The effect of shifting $\log W$ by $a_1 M + a_0 N$ is to shift β by $-a_1$ and $\log q$ by $-a_0$. The corresponding microcanonical partition function is

$$\log \omega' = \beta'\bar{x} + \log q' = \beta\bar{x} + \log q + a_1\bar{x} + a_0 = \log \omega + a_1\bar{x} + a_0,$$

and is shifted by $a_1\bar{x} + a_0$. It is easy to confirm the relationships

$$\frac{d \log \omega'}{d\beta'} = \bar{x};$$

$$\log q' = \log \omega' - \frac{d \log \omega'}{d\beta'}.$$

[11] In this section we notate the MPD as $f(x;\bar{x})$ to emphasize its dependence on $\bar{x}$.

The shifted variables satisfy the Legendre equations just as the original variables do, i.e., the set

$$\{\omega', \beta', q', w'\}$$

is an alternative canonical representation of the MPD. The constants a_0 and a_1 represent relative forms of $\log q$ and β, respectively, from a reference state set by the selection functional.[12] With the expressions in Eq. (4.67) the MPD becomes

$$f(x;\bar{x}) = w'(x;\bar{x})\frac{e^{-(\beta+a_1)x-a_0}}{q}. \tag{4.68}$$

The parameters a_0 and a_1 can be chosen entirely arbitrarily. We choose $a_1 = -\beta$, $a_0 = -\log q$, the above result becomes

$$w'(x;\bar{x}) = f(x;\bar{x}). \tag{4.69}$$

The parameters β' and q' are identified as the linear and constant terms, respectively, in $\log f - \log w'$ as a function of x. Having determined a feasible set of canonical parameters $\{w', \beta', q'\}$, we obtain their general form by solving Eqs. (4.66) and (4.67) for w, β, and q:

$$w(x;\bar{x}) = f(x;\bar{x})e^{-a_1x-a_0}, \tag{4.70}$$

$$\beta = \beta' - a_1, \tag{4.71}$$

$$q = q'e^{-a_0}, \tag{4.72}$$

These equations define an entire family of canonical sets $\{w, \beta, q\}$, all of which produce the same MPD, $f(x;\bar{x})$, regardless of the choice of a_1, a_2. This is because the transformation of the selection functional that produces these results is bias-neutral with respect to all distributions of the cluster ensemble.

Note 4.6 (Entropic Ensemble Revisited)
In our discussion of the entropic ensemble we found that the unbiased and entropic ensembles produce the same MPD. We will now show that their cluster functions satisfy Eq. (4.70). Substituting the exponential distribution

$$f(x) = \frac{e^{-x/\bar{x}}}{\bar{x}}$$

in Eq. (4.70) we obtain

$$w'(x;\bar{x}) = \frac{e^{-x/\bar{x}-a_1x-a_0}}{\bar{x}}. \tag{4.73}$$

[12]Once the selection functional is fixed, ω, β, and $\log q$ become immediately fixed. The (a_0, a_1) translation changes the functional and its associated parameters.

With $a_0 = \log \bar{x}$, $a_1 = -1/\bar{x}$, we obtain $w = 1$, the cluster function of the unbiased ensemble. With $a_0 = 2\log \bar{x}$, $a_1 = 2/\bar{x}$ the result is

$$w'(x; \bar{x}) = \bar{x}e^{x/\bar{x}}.$$

This is the cluster function of the entropic ensemble, which we previously obtained by Laplace transform of $q = q(\beta)$.

★

4.6 Canonical Representation of Some Common Distributions

The analysis of the previous section suggests that given any distribution $f(x; \bar{x})$ with $x \geq 0$ it is possible to construct a selection functional such that $f(x; \bar{x})$ is the MPD of the cluster ensemble under that functional. The process can be distilled into a recipe:

1. Obtain the cluster function as $\log w'(x; \bar{x}) = \log f(x; \bar{x})$; any linear or constant terms of x can be dropped because they may be absorbed into the parameters a_1 and a_2.
2. Obtain β' and $\log q'$ as the linear term and constant terms of $\log f - \log w'$ as a function of x.
3. Obtain the general form of w, β, and q from Eqs. (4.70)–(4.72)

The cluster functions obtained by this procedure may then be used in a Monte Carlo simulation of the cluster ensemble by exchange reactions. The equilibrium constant for the reaction between cluster masses i and j to produce masses k and l with $i + j = k + l$ is

$$K_{i+j \rightleftharpoons k+l} = \frac{w_k w_l}{w_i w_j}, \tag{4.74}$$

with cluster functions from Eq. (4.70). Notice that a_1 and a_0 cancel out by virtue of the condition $i + j = k + l$, which makes the equilibrium constant and the resulting MPD independent of a_1 and a_0. We demonstrate the method with four common distributions, exponential, Gaussian, Weibull, and uniform. The details are given below and the results are summarized in Table 4.4 and in Fig. 4.5, which shows the results of the Monte Carlo simulations.

Example 4.7 (Exponential Distribution)
Starting with the exponential distribution,

$$f(x) = \frac{e^{-x/\bar{x}}}{\bar{x}},$$

Table 4.4 Thermodynamic representation of some common distributions

Distribution	Canonical representation
Exponential $f(x) = \dfrac{e^{x/\bar{x}}}{\bar{x}}$	$w = e^{-a_1 x - a_0}$ $\beta = 1/\bar{x} - a_1$ $q = \bar{x}e^{-a_0}$
Gaussian $f(x) = \dfrac{e^{-(x-\bar{x})^2/2\sigma^2}}{\sqrt{2\pi\sigma^2}}$	$w = e^{-x^2/2\sigma^2 - a_1 x - a_0}$ $\beta = -\bar{x}^2/2\sigma^2 - a_1$ $q = \sqrt{2\pi\sigma^2}\, e^{\bar{x}^2/2\sigma^2 - a_0}$
Weibull $f(x) = \left(\dfrac{k}{\lambda}\right)\left(\dfrac{x}{\lambda}\right)^{k-1} e^{-(x/\lambda)^k}$	$w = (k-1)\log(x/\lambda) - (x/\lambda)^k e^{-a_1 x - a_0}$ $\beta = -a_1$ $q = \lambda/ke^{-a_0}$
Uniform $f(x) = \begin{cases} 1/(b-a), & a < x < b \\ 0 & \text{otherwise} \end{cases}$	$w = e^{-a_1 x - a_0}/(b-a) \quad (a \le x \le b)$ $\beta = -a_1$ $q = e^{-a_0}$

the cluster function is

$$\log w' = \log f = -\frac{x}{\bar{x}} - \log \bar{x}.$$

The linear term $x/\bar{x}$ and the constant term $\log \bar{x}$ can both be dropped from the right-hand side of this result to obtain the simpler cluster function

$$\log w' = 0 \quad \text{or} \quad w'(x) = 1.$$

We calculate $\log f - \log w'$,

$$\log f - \log w' = -\frac{x}{\bar{x}} - \log \bar{x},$$

and identify β as the coefficient of x and $\log q$ as the constant term:

$$\beta' = 1/\bar{x}; \quad q' = \bar{x}.$$

The general form for w, q, and b is

$$w = e^{-a_1 x - a_0},$$

$$\beta = \frac{1}{\bar{x}} - a_1,$$

$$q = \bar{x}e^{-a_0},$$

where a_0 and a_1 are arbitrary functions of $\bar{x}$.

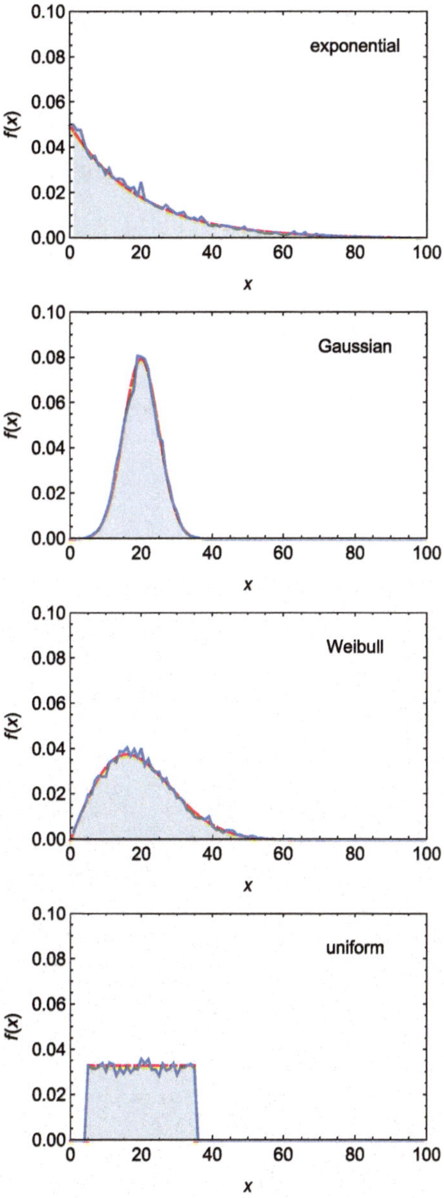

Fig. 4.5 Monte Carlo simulation of binary exchange reactions using the cluster functions in Table 4.4. Shaded curves are the results of the simulation and the dashed line shows the theoretical distribution. The parameters for these distributions are: *Gaussian:* $\sigma^2 = 4$; *Weibull:* $k = 22.5676, \lambda = 2$; *Uniform:* $a = 5, b = 35$. In all cases $\bar{x} = 20$

Example 4.8 (Gaussian Distribution)

The Gaussian distribution with mean $\bar{x}$ and variance σ^2 is

$$f(x) = \frac{e^{-(x-\bar{x})^2/2\sigma^2}}{\sqrt{2\pi\sigma^2}},$$

Applying Eq. (4.70) we obtain w'as

$$\log w' = \log f(x) = -\frac{x^2}{2\sigma^2} + \frac{x\bar{x}}{\sigma^2} - \frac{\bar{x}^2}{2\sigma^2} - \log\sqrt{2\pi\sigma^2}.$$

We simplify this result by dropping the constant and linear terms in x:

$$\log w' = -\frac{x^2}{2\sigma^2}$$

For $\log f - \log w'$ we have

$$\log f - \log w' = x\frac{\bar{x}}{\sigma^2} - \frac{\bar{x}^2}{2\sigma^2} - \log\sqrt{2\pi\sigma^2}$$

and make the identifications

$$\beta' = -\frac{\bar{x}}{\sigma^2}, \quad \log q' = \log\sqrt{2\pi\sigma^2} + \frac{\bar{x}^2}{2\sigma^2}.$$

The general expressions for the canonical parameters are

$$w = e^{-x^2/2\sigma^2 - a_1 x - a_0},$$
$$\beta = -\bar{x}/\sigma^2 - a_1,$$
$$q = \sqrt{2\pi\sigma^2} e^{\bar{x}^2/2\sigma^2 - a_0},$$

where a_0 and a_1 are arbitrary constants that may depend on $\bar{x}$ and σ. In general, a_0 and a_1 may depend on any parameters that appear in f (but they cannot depend on x).

Example 4.9 (Weibull Distribution)

The Weibull distribution with parameters k and λ is

$$f(x) = \left(\frac{k}{\lambda}\right)\left(\frac{x}{\lambda}\right)^{k-1} \exp\left[-\left(\frac{x}{\lambda}\right)^k\right],$$

and its mean is $\bar{x} = \lambda\Gamma(1 + 1/k)$. The cluster weight is

$$\log w' = \log(k/\lambda) + (k-1)\log(x/\lambda) - (x/\lambda)^k,$$

and after dropping the constant term $\log(k/\lambda)$ this becomes

$$\log w' = (k-1)\log(x/\lambda) - (x/\lambda)^k.$$

The corresponding parameters β', $\log q'$ are

$$\beta' = 0; \quad q' = \lambda/k.$$

(There is no linear term with respect to x in $\log f - \log w'$). The general form of the canonical parameters is

$$w = \left(\frac{x}{\lambda}\right)^{k-1} \exp\left[-\left(\frac{x}{\lambda}\right)^{k} - a_1 x - a_0\right],$$

with

$$\beta = -a_1; \quad q = \lambda e^{-a_0}/k,$$

where a_0 and a_1 may depend on λ and k.

Example 4.10 (Uniform Distribution)

The procedure is not limited to smooth distributions and for this last demonstration we use the uniform distribution in the interval (a, b),

$$f(x) = \begin{cases} 1/(b-a), & a < x < b \\ 0, & \text{otherwise} \end{cases}$$

The cluster weight is

$$\log w' = \log f(x) = \begin{cases} -\log(b-a), & a < x < b \\ -\infty, & \text{otherwise} \end{cases}$$

and the parameters β' and q' are

$$\beta' = 0; \quad q' = 1.$$

The general form of the canonical parameters is

$$w = \begin{cases} e^{-a_1 x - a_0}/(b-a) & a < x < b; \\ 0 & \text{otherwise,} \end{cases}$$

and

$$\beta = -a_1, \quad q = e^{-a_0},$$

with a_1, a_2, parameters that may depend on a and b. As Fig. 4.5 shows, the Monte Carlo simulation using these cluster functions is in very good agreement with the distribution.

★

Chapter 5
Phase Transitions: The Giant Cluster

A problem of great theoretical interest in thermodynamics is the possibility of phase transitions. Many dynamical systems exhibit behavior that is remarkably similar to phase transitions. Polymer gelation, shattering in fragmentation, the spread of epidemics and the emergence of long-range connectivity in artificial and neural networks are examples of the emergence of a giant coherent structure, a behavior that is often discussed qualitatively in the language of phase transitions. If generic population obey thermodynamics, do they also undergo phase transitions? The answer is, *yes*. As in molecular thermodynamics, phase splitting in the cluster ensemble is associated with the violation of the stability conditions that guarantee the existence of a maximum in the microcanonical weight that defines the most probable distribution. In this chapter we formalize the stability conditions that ensure the existence of the most probable distribution and identify the *giant cluster* as a phase that is distinct from the *sol*, a stable population of dispersed clusters. We discuss two mathematical models that give rise to a giant cluster and solve them analytically. A kinetic model of gelation with a closer connection to a physical system of reacting polymers will be discussed in Chap. 9.

5.1 Variational Properties of the Partition Function

In the thermodynamic limit the partition function converges to the microcanonical weight evaluated at the most probable distribution,

$$\log \Omega_{M,N} = \log \tilde{\mathbf{n}}! W(\tilde{\mathbf{n}}).$$

The logarithm of the microcanonical weight, $\log \mathbf{n}! W(\mathbf{n})$, is a homogeneous concave functional of $\mathbf{n}$ and these properties transfer to $\log \Omega$ as a function of M and N, as we see next. We start with two microcanonical ensembles, (M_A, N_A) and (M_B, N_B)

T. Matsoukas, *Generalized Statistical Thermodynamics*,
Understanding Complex Systems, https://doi.org/10.1007/978-3-030-04149-6_5

whose corresponding MPDs are $\tilde{\mathbf{n}}_A$ and $\tilde{\mathbf{n}}_B$, respectively. We add them to form a new distribution $\mathbf{n}_{AB} = \tilde{\mathbf{n}}_A + \tilde{\mathbf{n}}_B$. As a member of the ensemble $(M_A + M_B, N_A + N_B)$, the composite distribution satisfies the inequality

$$\log \tilde{\mathbf{n}}_{AB}! W(\mathbf{n}_{AB}) \geq \log(\tilde{\mathbf{n}}_A + \tilde{\mathbf{n}}_B)! W(\tilde{\mathbf{n}}_A + \tilde{\mathbf{n}}_B)$$

where $\tilde{\mathbf{n}}_{AB}$ is the MPD of the composite ensemble; and since the right-hand side is the $\log \Omega_{M_A+M_B,N_A+N_B}$, the above inequality becomes

$$\log \Omega_{M_A+M_B,N_A+N_B} \geq \log(\tilde{\mathbf{n}}_A + \tilde{\mathbf{n}}_B)! W(\tilde{\mathbf{n}}_A + \tilde{\mathbf{n}}_B).$$

By the concave/homogeneous inequality, Eq. (2.19), we have

$$\begin{aligned}\log(\tilde{\mathbf{n}}_A + \tilde{\mathbf{n}}_B)! W(\tilde{\mathbf{n}}_A + \tilde{\mathbf{n}}_B) \geq \\ \log \tilde{\mathbf{n}}_A! W(\tilde{\mathbf{n}}_A) + \log \tilde{\mathbf{n}}_B! W(\tilde{\mathbf{n}}_B) = \\ \log \Omega_{M_A,N_A} + \log \Omega_{M_B,N_B}.\end{aligned}$$

Combining the last two inequalities we obtain the relationship between the partition functions of the parts and of the whole:

$$\log \Omega_{M_A+M_B,N_A+N_B} \geq \log \Omega_{M_A,N_A} + \log \Omega_{M_B,N_B}. \tag{5.1}$$

In the special case that the two distributions are homogeneous copies of each other, i.e., $\tilde{\mathbf{n}}_A = \lambda \tilde{\mathbf{n}}_B$, Eq. (5.1) applies as an exact equality and the sum of the MPDs of the parts is the MPD of the whole:

$$\tilde{\mathbf{n}}_{A+B} = \tilde{\mathbf{n}}_A + \tilde{\mathbf{n}}_B. \tag{5.2}$$

As homogeneous copies of each other, distributions $\mathbf{n}_A$ and $\mathbf{n}_B$ have the same normalized distribution and share the same intensive properties:

$$\bar{x}_A = \bar{x}_B = \bar{x}, \tag{5.3}$$

$$\beta_A = \beta_B = \beta, \tag{5.4}$$

$$q_A = q_B = q. \tag{5.5}$$

The inequality of Eq. (5.1) implies that $\log \Omega(M, N)/N$ is a concave function of M/N. To see this we use the homogeneity property of $\log \Omega$ to express the inequality in the form

$$\log \Omega_{(M_A+M_B)/(N_A+N_B),1} \geq \alpha \log \Omega_{M_A/N_A,1} + (1-\alpha) \log \Omega_{M_B/N_B,1}$$

with $\alpha = N_A/(N_A + N_B)$.[1] This can be expressed in the equivalent form,

$$\log \omega(\alpha \bar{x}_A + (1-\alpha)\bar{x}_B) \geq \alpha \log \omega(\bar{x}_A) + (1-\alpha) \log \omega(\bar{x}_B).$$

The result states that $\log \Omega_{M,N}$ is concave in M/N and this further implies that $\log \Omega_{M,N}$ is concave in both M, N.[2] Concavity requires that both second derivatives of $\log \Omega_{M,N}$ be negative:

$$\left(\frac{\partial^2 \log \Omega}{\partial^2 M}\right)_N \leq 0; \quad \left(\frac{\partial^2 \log \Omega}{\partial^2 N}\right)_N \leq 0. \tag{5.6}$$

Using Eqs. (3.23) and (3.24) we also have

$$\left(\frac{\partial \beta}{\partial M}\right)_N \leq 0, \tag{5.7}$$

$$\left(\frac{\partial \log q}{\partial N}\right)_M \leq 0. \tag{5.8}$$

These can be simplified by noting that β and q are intensive and therefore functions of $\bar{x} = M/N$ rather than of M or N individually. We have

$$\left(\frac{\partial \beta}{\partial M}\right)_N = \left(\frac{\partial \beta}{\partial M}\right)_N = N \frac{d\beta}{d\bar{x}},$$

$$\left(\frac{\partial \log q}{\partial N}\right)_M = \left(\frac{\partial \log q}{\partial N}\right)_M = \frac{1}{N}\left(\frac{\partial \log q}{\partial (1/\bar{x})}\right)_N = -\frac{\bar{x}^2}{N}\frac{d \log q}{d\bar{x}}.$$

[1]Apply the homogeneity condition in the form $\log \Omega_{M,N} = N \log \Omega_{M/N,1}$ and notice that

$$\log \Omega_{M/N,1} = \frac{\log \Omega_{M,N}}{N} = \log \omega(\bar{x}).$$

[2]Let $F(x, y) = x f(x/y)$, which makes F homogeneous in x and y with degree 1. The second derivatives of F are related to those of f as follows:

$$F_{xx} = \frac{y^2}{x^3} f''(y/x), \quad F_{yy} = \frac{1}{x} f''(y/x).$$

If x and y are both positive, F has the same curvature (convex or concave) with respect to both variables as f. The relationship between f and F is the same as between $\log \omega$ and $\log \Omega$ and since $\log \omega$ is concave, so is $\log \Omega$.

With these results and the fact that M and N are always positive, Eqs. (5.7) and (5.8) become

$$\frac{d\beta}{d\bar{x}} \le 0; \qquad \frac{d\log q}{d\bar{x}} \ge 0. \tag{5.9}$$

The first of these restates the fact that $\log\omega$ is concave in $\bar{x}$ (recall that $\beta = d\log\omega/d\bar{x}$). These inequalities require β and $\bar{x}$ to vary in opposite directions, while q and $\bar{x}$ vary in the same direction. We also conclude that β and q must vary in opposite directions to each other,

$$\frac{d\beta}{dq} \le 0, \tag{5.10}$$

but the result is not new, it is implied in the relationship[3]

$$\frac{d\log q}{d\beta} = -\bar{x}, \tag{5.11}$$

which was obtained already in Chap. 3 as Eq. (3.18).

We refer to the various inequalities that follow from the concave property of the partition function as *stability* conditions. They ensure that the extremum that defines the MPD is indeed a maximum. We may also view stability in the context of the binary exchange reaction. We have used the binary reaction to construct a Markov chain of configurations whose limiting distribution is the MPD. The predicted MPD will be reached only if it satisfies the stability criteria. If it does not, the Markov chain will transition to a stable solution that is different from the MPD that is calculated from the partition function and its derivatives. Monte Carlo examples that demonstrate this behavior are discussed later in this chapter.

5.2 The Equilibrium Phase

We are now set to address the main question of this chapter: *is it possible for a population to undergo the equivalent of a phase transition?* Before we can answer this question must first define *phase*. To facilitate the discussion we adopt terms from molecular thermodynamics but redefine them to fit our purpose.[4] The

[3]Both $\bar{x}$ and q are positive.

[4]The advantage in recycling terminology is that the theory may be more transparent and accessible to those familiar with standard thermodynamics. The potential downside is that this familiarity may carry with it preconceived notions about these terms that conflict with our definitions.

system is an extensive population of M members distributed among N groups with probabilities assigned to all distributions via a fixed selection functional. The *state* is the distribution $\mathbf{n}$ of this population. We will say that *the system is in equilibrium* to indicate that the distribution in question is the MPD of the ensemble in the thermodynamic limit, therefore it maximizes its partition function.[5] The *intensive state* of the system is the normalized MPD, $\tilde{\mathbf{p}} = \tilde{\mathbf{n}}/N$. A *partitioning* of the system is the splitting of distribution $\mathbf{n}$ into two distributions $\mathbf{n}_A$ and $\mathbf{n}_B$ that satisfy the conservation condition

$$\mathbf{n} = \mathbf{n}_A + \mathbf{n}_B.$$

It is an *equilibrium partitioning* if the distribution in each part is the MPD of the corresponding ensemble, i.e., it maximizes $\Omega_{M',N'}$, where M' and N' are the mass and number of clusters in the partition. We can now give a definition of *phase*:

> *A system in the thermodynamic limit forms a single phase if every equilibrium partitioning consists of homogeneous copies of the whole, i.e., all parts have the same intensive distribution.*

It follows from this definition that all intensive properties are the same in every equilibrium partitioning of a single phase system, and all extensive properties are proportional to a single factor that defines the scale of the system.

5.3 The Giant Cluster

The maximum possible cluster in a population of M members distributed into N groups is found when the remaining $N - 1$ clusters contain one member each. The size of the maximum cluster is equal to the remaining number of members:

$$i_{\max} = M - N + 1. \tag{5.12}$$

The discrete MPD is

$$\frac{\tilde{n}_i}{N} = \frac{\tilde{w}_i e^{-\beta i}}{\sum_{i=1}^{i_{\max}} \tilde{w}_i e^{-\beta i}}, \tag{5.13}$$

[5]The statement *"the MPD maximizes the partition function"* is technically incorrect. The partition function is a function of M, N, not of the MPD, and cannot be maximized by the MPD. What we mean is that the MPD maximizes the microcanonical weight $\mathbf{n}!W(\mathbf{n})$, whose maximum value is equal to the partition function. This should be understood as implied when we use the shorthand *"MPD maximizes Ω."*

where q has been replaced by the summation that defines it in terms of $\tilde{w}_i$ and β. We can use this expression to calculate the discrete MPD numerically if the cluster functions are known. The system is *stable* if the MPD becomes invariant as M and N go to infinity at constant $M/N = \bar{x}$. This ensures that $\tilde{n}_i/N$ satisfies homogeneity with degree 0. The maximum cluster in this limit scales as

$$i_{\max} = M\left(1 - \frac{N}{M} + \frac{1}{M}\right) \sim M\left(1 - \frac{1}{\bar{x}}\right),$$

i.e., it increases in proportion to M. To remain invariant as the domain increases indefinitely, the normalized distribution must decay sufficiently fast with cluster size i so that the new elements of the MPD that appear in the summation make infinitesimal contribution to the total. If for some cluster function this condition is not met, it is possible to reach a state in which the homogeneity of the MPD is not satisfied. In this case the state cannot be represented by the MPD in Eq. (5.13), even though this MPD is an extremum of the partition function. When this situation arises the extremum is a minimum, not a maximum of the microcanonical weight. To explore this further we consider clusters at the high end of the distribution.

Only one distribution of the (M, N) ensemble contains a cluster with size $i_{\max}$. In general, the largest cluster in a distribution is smaller than $i_{\max}$ and larger than M/N. Within this region the sub-range defined by the condition

$$\frac{i_{\max} + 1}{2} < i < i_{\max}, \tag{5.14}$$

has a special property: *it may contain no more than one cluster*. Here is why. Suppose there are two clusters in this region and assume that they both have the smallest size possible in this range, i.e., they are both equal to $(i_{\max} + 1)/2 + 1$. The combined size of these clusters is

$$2\left(\frac{i_{\max} + 1}{2} + 1\right) = M - N + 4,$$

leaving $M - (M - N + 4) = N - 4$ members to be distributed over the remaining $N - 2$ clusters. But this is not possible because there are more clusters than the total mass. Therefore the region defined in Eq. (5.14) may contain at most one cluster. A cluster in this range, say of size i^*, represents a fraction $1/N \to 0$ of the total number of clusters, and a fraction of mass of the order

$$\frac{i^*}{M} \sim \left(\frac{i_{\max} + 1}{2M}\right) \sim \frac{\bar{x} - 1}{2}.$$

In the thermodynamic limit this cluster represents a vanishing fraction of the population in terms of number, but a finite fraction in terms of mass. We call the region $1 \le (i_{\max} + 1)/2$ *sol* and the region $(i_{\max} + 1)/2 \le i_{\max}$ *gel*:

$$\begin{aligned} \text{sol:} \quad & 1 \le i < \frac{i_{\max}+1}{2} \\ \text{gel:} \quad & \frac{i_{\max}+1}{2} \le i \le i_{\max} \end{aligned}$$

We will say that the system contains a *gel phase* if a finite fraction of distributions in the vicinity of the MPD contains a gel cluster. A system that contains a gel phase must necessarily contain a *sol phase*. This is forced by the microcanonical condition that fixes the number of clusters to be constant. In the microcanonical ensemble therefore, the state can be either a single phase system (sol) or a two-phase system (sol/gel). The cluster size distribution of a sol-gel state is discontinuous. Unless the gel cluster sits exactly on the sol-gel boundary, there is a gap between the largest sol cluster and the gel cluster that contains no clusters.

Example 5.1 (Convergence of Stable MPD)

The behavior of a stable MPD is demonstrated in Fig. 5.1. The MPD is calculated with cluster function $w_i = e^{-\gamma i^2}$, which produces in ThL a Gaussian distribution

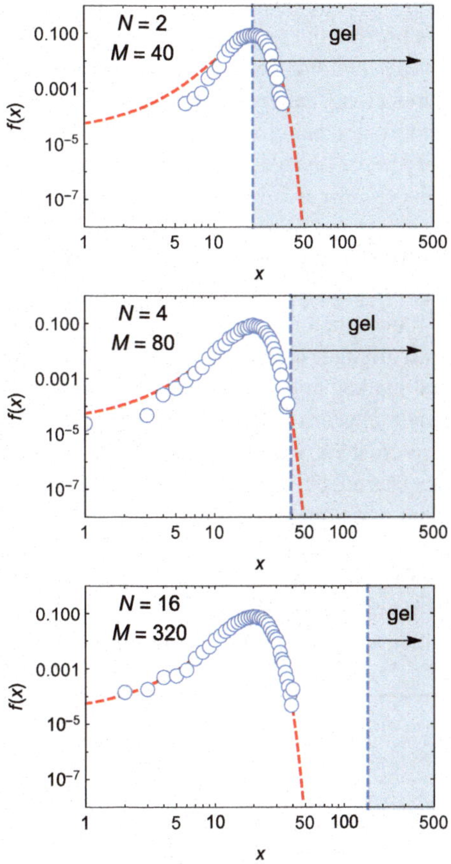

Fig. 5.1 Convergence of stable MPD in the thermodynamic limit at fixed mean cluster size. With increasing M, N, at fixed $M/N = 20$, the distribution converges to a stable MPD that is located entirely within the sol region. Circles are MC simulations with linear weights $w_i = e^{-0.02i^2}$ and lines are the MPD in the ThL (Gaussian distribution with $\bar{x} = 20, \sigma^2 = 25$). The gel region is shaded and moves to the right as M and N increase and the dashed line marks the sol/gel boundary

with variance $\sigma^2 = 1/2\gamma$ (in this example $\gamma = 0.02$). The distribution is simulated by binary exchange reaction at various values of M, N, keeping the average cluster mass constant at $M/N = 20$. Results are shown for $N = 2, 4$, and 16. These values are intentionally small to demonstrate the approach to the thermodynamic limit. In all cases the theoretical MPD is calculated as a Gaussian distribution with mean $\bar{x} = 20$ and variance $\sigma^2 = 25$. At very small N the distribution extends into the gel region. This does *not* mean that a gel phase exists, phases are defined in the thermodynamic limit only. The simulated distribution at $N = 2$ may leave the impression that the gel region contains more than one cluster. This is not so. The simulated distributions are the average of several distributions; each contains a single gel cluster but its position varies, resulting in a spread of average distribution over a range of gel clusters. The crossover into the gel region at small N is unimportant and disappears as N increases. Even at the relatively small value of $N = 16$ the features of stable convergence are clearly seen: the MPD converges to the predicted MPD and most importantly, it is fully contained in the sol region. The probability to find a cluster in the gel region vanishes as the size of the system increases.

Note 5.2 (Sol, Gel, Giant Component)

We borrow the terms *sol* and *gel* from colloidal science. Sol refers to a suspension of free colloidal clusters or polymers, gel to an infinite network of macroscopic dimensions that emerges from the sol and coexists with it. The sol-gel process is an important technology in ceramic synthesis. Polymer gelation is equally important in polymer science and has been the subject of intense theoretical investigations. We use the term *sol* to refer to a population whose clusters are distributed continuously over a range of sizes and whose normalized first and second moments are finite and converge with increasing size. We use *gel* to refer to a single cluster in a very large population that contains a finite fraction of the total mass. The emergence of a giant structure, a single member of a population that competes in terms of mass with all other clusters taken together, is a feature that appears frequently in physical and mathematical systems. In network theory it is known as the *giant component*, a single cluster of interconnected nodes within a quasi-continuous medium of finite-size clusters. Its existence and uniqueness are two of its primary features.[6] Both are shared by the gel cluster as defined here. We have chosen gel, rather than giant component in part because we will treat later (Chap. 9) an explicit model for polymer gelation, but more from the practical advantage that the twin terms sol/gel provide short descriptive names for *both* phases present. *Coexistence* of phases is a defining feature of phase equilibrium in molecular systems. The same is true, as we will see, in generic probability distributions.

★

[6]See, for example, Spencer (2010).

5.4 Sol-Gel Equilibrium

The picture we construct to study the emergence of the gel phase and its interaction with the sol is shown in Fig. 5.2. It is a representation of the average distribution of a population that contains a sol and a gel phase and illustrates the qualitative features of a two-phase population. We fix M and N and distribute the mass into the specified number of clusters according to the selection functional using binary exchange reactions. We view the exchange reaction as an algorithm that samples distributions from the ensemble, not as a real process that actually takes place in the population. A distribution either contains one gel cluster or none. The variable N_{gel}, defined as the sum of n_i in the gel region is a binary variable:

$$N_{\text{gel}} \equiv \sum_{i=(i_{\max}+1)/2}^{i_{\max}} n_i = \begin{cases} 1 & \text{if } \mathbf{n} \text{ contains a gel cluster} \\ 0 & \text{otherwise.} \end{cases} \tag{5.15}$$

The mean value of N_{gel} over the ensemble of equilibrium distributions is the probability to find a gel cluster in the system (Fig. 5.2). In the scaling limit this variable is also binary but in finite systems we expect it to exhibit a continuous change from 0 (sol only) to 1 (sol/gel equilibrium).

We now consider a system in which a single gel cluster of mass m_{gel} coexists alongside with an equilibrium sol that contains $N - 1$ clusters that carry the remaining mass $M - m_{\text{gel}}$. This is a partitioning of the microcanonical system $\mu C(M, N)$ into two microcanonical systems, $\mu C(M - m_{\text{gel}}, N - 1)$ (the sol) and $\mu C(m_{\text{gel}}, 1)$ (the gel). The partition function of the two-phase system is

$$\log \Omega_{M,N} = \log \Omega^{\text{sol}} + \log \Omega^{\text{gel}}.$$

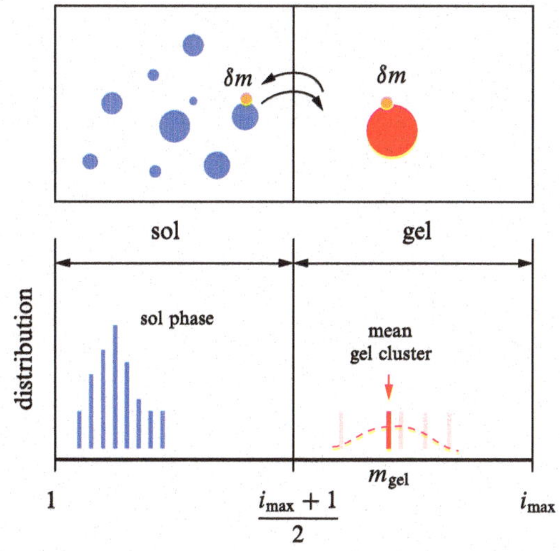

Fig. 5.2 The gel phase contains a single cluster whose position fluctuates as a result of mass exchange with the sol. The gel cluster disappears if its mass drops below $(i_{\max} + 1)/2$. The distribution in the gel region represents fluctuations of the mass of the gel cluster due to exchange with the sol

The partition function of the sol is the equilibrium partition function of the ensemble $M_{\text{sol}}, N-1$; the partition function of gel is a function of m_{gel} only, since this phase consists of a single cluster. Then we have

$$\log \Omega_{M,N} = \log \Omega_{M_{\text{sol}},N-1} + \log \Omega^{\text{gel}}, \tag{5.16}$$

with $M_{\text{sol}} + m_{\text{gel}} = M$. At equilibrium this is maximized with respect to all possible partitions of the total mass. If we transfer mass δm between the two phases, the transfer leaves the total partition function unchanged:

$$\delta \log \Omega_{M,N} = \underbrace{\delta \log \Omega_{M_{\text{sol}},N-1}}_{\beta^{\text{sol}}\delta m} + \delta \log \Omega^{\text{gel}} = 0,$$

The change in the partition function of the sol is

$$\delta \log \Omega_{M_{\text{sol}},N-1} = \beta^{\text{sol}} \delta m, \tag{5.17}$$

where β^{sol} is the derivative of $\log \Omega^{\text{sol}}$ with respect to M_{sol}. This leads to the equilibrium condition

$$\beta^{\text{sol}} = \frac{\delta \log \Omega^{\text{gel}}}{\delta m} \equiv \beta^{\text{gel}}. \tag{5.18}$$

We identify the derivative $\delta\Omega_{\text{gel}}/\delta m$ as the parameter β of the gel phase. We conclude that the condition for sol-gel equilibrium requires the generalized temperatures in each phase to be the same. There is no corresponding condition on q. The two phases exchange mass but not *clusters*. This is a constrained equilibrium analogous to two bodies separated by a conducting wall (temperatures equalize but not the pressures or chemical potentials), or osmotic equilibrium (chemical potentials equalize but not the pressures).[7] The gel phase is a single cluster with mass m_{gel} and its MPD is

$$\tilde{n}_i^{\text{gel}} = \delta_{i,m_{\text{gel}}}, \tag{5.19}$$

[7]In the multivariate ensemble $(M_1, M_2 \cdots ; N)$ Eq. (5.18) applies to all β_i. If we make the associations

$$M_1 \to \text{energy},\ M_2 \to \text{volume},\ M_{2+i} \to \text{number of particles of component } i,$$

then

$$\beta_1 = 1/k_B T,\ \beta_2 = p/k_B T,\ \beta_{2+i} = \mu_i/k_B T,$$

and we recover the equality of temperature, pressure, and all chemical potentials in every part as the condition of equilibrium. We defer the discussion of the connection between the cluster ensemble and statistical mechanics until Chap. 7.

where $\delta_{i,j}$ is the Kronecker delta. The partition function of the gel is equal to the microcanonical weight of this distribution and since the multiplicity of $\mathbf{n}^{\text{gel}}$ is 1 (the distribution contains a single cluster),

$$\Omega_{\text{gel}} = W(\mathbf{n}^{\text{gel}}). \tag{5.20}$$

If the functional is known, this condition leads to the calculation of the gel mass.

5.4.1 *Special Case: Linear Functional*

If the selection functional is linear, then

$$\log W(\mathbf{n}^{\text{gel}}) = \sum \tilde{n}_i^{\text{gel}} \log w_i = w(m_{\text{gel}}). \tag{5.21}$$

In this case the partition function of the gel phase is known explicitly as a function of m_{gel} and this allows us to calculate the phase diagram by direct application of the equilibrium condition in Eq. (5.18). This condition can now be expressed as

$$\beta^* = \left(\frac{d \log w(m)}{dm}\right)_{m=m_{\text{gel}}}, \tag{5.22}$$

where β^* is the common value of β in the two phases at equilibrium. The MPD of the sol phase is

$$\frac{\tilde{n}_i^{\text{sol}}}{N-1} = w_i \frac{e^{-\beta^* i}}{q}, \tag{5.23}$$

and satisfies the normalization conditions

$$\sum_i \tilde{n}_i^{\text{sol}} = N - 1 \tag{5.24}$$

$$\sum_i i\tilde{n}_i^{\text{sol}} = M - m_{\text{gel}}. \tag{5.25}$$

Equations (5.22)–(5.25) provide all the necessary information to calculate the phase diagram of the system.

5.5 A Case Study: Linear Ensemble with $w_i = i^{-3}$

As a concrete example we consider the linear selection functional

$$W(\mathbf{n}) = \prod_i i^{-3n_i} \tag{5.26}$$

with power-law cluster function

$$w_i = 1/i^3. \tag{5.27}$$

In Sect. 4.3 we studied power-law functions with positive exponents. In that case the MPD is a stable gamma function whose stability can be easily confirmed by showing that both conditions in Eq. (5.9) are satisfied. Here the exponent is $a = -3 < 0$, and this, as we will see, alters stability.

We consider a population of mass M distributed into N clusters. Both M and N are finite but large enough for the thermodynamic limit to apply. The MPD is

$$\frac{\tilde{n}_i}{N} = i^{-3}\frac{e^{-\beta i}}{q} \tag{5.28}$$

with β and q given by

$$q = \sum_{i=1}^{i_{\max}} i^{-3} e^{-\beta i}, \tag{5.29}$$

$$\bar{x} = \frac{1}{q}\sum_{i=1}^{i_{\max}} i^{-2} e^{-\beta i}, \tag{5.30}$$

where

$$i_{\max} = M - N + 1 \tag{5.31}$$

is the maximum cluster mass that can exist in a distribution of the (M, N) ensemble. The next example demonstrates the calculation of the parameters β and q.

Example 5.3 (Graphical Demonstration of the Calculation of β, q)

We demonstrate the calculation of the β, q, parameters of the MPD by a graphical construction. We choose $M = 200$ and calculate $\bar{x}$ from Eq. (5.30) along with (5.29). Figure 5.3 shows $\bar{x}$ as a function of β. The graph transitions sharply from $\bar{x} = i_{\max}$ at $\beta \to \infty$ to $\bar{x} = 1$ at $\beta \to \infty$. To obtain (β, q) for given $\bar{x}$ we first solve Eq. (5.30) for β, which amounts to reading the corresponding β from Fig. 5.3, then use this value in (5.29) to obtain q. For example, at $\bar{x} = 4$ we find

$$\beta = -0.05249, \quad q = 1.3229. \tag{5.32}$$

The corresponding MPD is

$$\frac{\tilde{n}_i}{N} = i^{-3}\frac{e^{-0.05249i}}{1.3229},$$

and we can confirm that it is properly normalized and its mean is $\bar{x} = 4$.

★

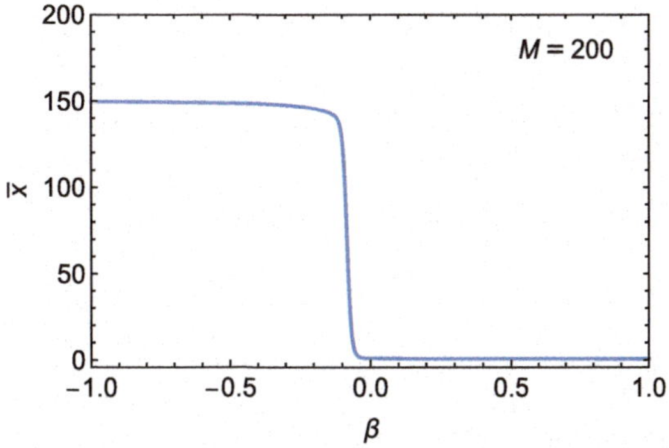

Fig. 5.3 The mean cluster size as a function of β from Eq. (5.30). The mean cluster size ranges from $i_{\max}$ at $\beta = -\infty$ to 1 at $\beta = \infty$. To obtain (β, q) for given $\bar{x}$, obtain β from this graph, then use Eq. (5.29) to find q

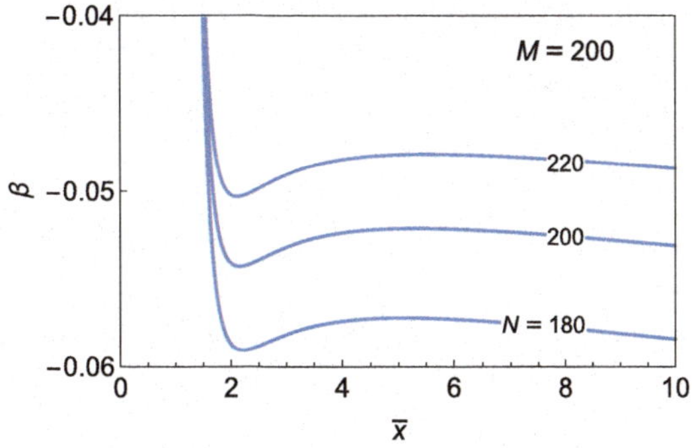

Fig. 5.4 The parameter β of discrete ensemble with $w_i = i^{-3}$ has a minimum that marks the limit of stability. The system is stable to the left of the minimum and unstable to the right

The procedure in the previous example can be used to obtain the set of (β, q) values in the entire range $\bar{x} = 1$ to $\bar{x} = M - N + 1$ for any M. Figure 5.4 shows the results of this calculation for three values of N in the vicinity of $N = 200$. The stability criteria in Eq. (5.9) require β to be a monotonically decreasing function of $\bar{x}$. The presence of a minimum in Fig. 5.4 indicates the onset of instability. To the left of the minimum $\beta(\bar{x})$ is a monotonically decreasing function of $\bar{x}$ and satisfies the first stability criterion in Eq. (5.9). To the right of the minimum the system is unstable. The calculations in this figure are based on the assumption that the system forms a single phase and when we say "the system is unstable" we mean that the system cannot exist as the single phase given by Eqs. (5.28)–(5.30). Figure 5.4 reveals an additional feature. In the stable region all β collapse to a single curve,[8] indicating that β is an intensive function of $\bar{x}$ This is not true in the unstable region, where β depends on both $\bar{x}$ and N. The breakdown of homogeneity is a sign of phase splitting: the system cannot satisfy homogeneity as a single phase, instead it splits into multiple phases, each internally homogeneous.

[8]This can be confirmed by expanding the scale in Fig. 5.4. We choose a narrow range in the vertical axis to highlight the minimum.

5.5.1 Tie Line

In the unstable region a system with overall mean cluster size $\bar{x} = M/N$ forms two phases, a dispersed phase (sol) whose MPD is given in terms of the parameters $(\bar{x}_{\text{sol}}, \beta, q)$, and a gel phase that contains a fraction ϕ_{gel} of the total mass. The set

$$\{\bar{x}_{\text{sol}}, \beta, q, \phi_{\text{gel}}\}$$

that describes the intensive properties of the two phases is the *tie line* of the two-phase system, a term we borrow from chemical thermodynamics. At equilibrium the system maximizes the microcanonical partition function with respect to all equilibrium partitioning of the total mass between a sol with $N-1$ clusters and a gel cluster with mass m_{gel}. We will use this condition to determine the tie line of a finite population (M, N).

Suppose that the gel cluster has mass m_{gel}. This mass must be in the gel range,

$$\frac{i_{\max}+1}{2} < m_{\text{gel}} \le i_{\max}, \tag{5.33}$$

with $i_{\max} = M - N + 1$. The equilibrium sol phase has $N-1$ clusters with total mass $M_{\text{sol}} = M - m_{\text{gel}}$. Therefore we have

$$M_{\text{sol}} = M - m_{\text{gel}}, \quad N_{\text{sol}} = N - 1, \qquad N_{\text{gel}} = 1, \tag{5.34}$$

which satisfy the conservation conditions $M_{\text{sol}} + m_{\text{gel}} = M$ and $N_{\text{sol}} + N_{\text{gel}} = N$. The maximum cluster mass in the sol phase is

$$i_{\text{sol}} = M_{\text{sol}} - N_{\text{sol}} + 1 = M - m_{\text{gel}} - N + 2. \tag{5.35}$$

The MPD of the sol phase is given by Eqs. (5.28), (5.30), and (5.29) with M, N, and $i_{\max}$ replaced by $M_{\text{sol}} = M - m_{\text{gel}}$, $N_{\text{sol}} = N - 1$, and i_{sol}, respectively:

$$\left(\frac{\tilde{n}_i}{N}\right)^{\text{sol}} = w_i \frac{e^{-\beta i}}{q}, \tag{5.36}$$

$$q = \sum_{i=1}^{i_{\text{sol}}} w_i e^{-\beta i}. \tag{5.37}$$

$$\frac{M_{\text{sol}}}{N_{\text{sol}}} = \sum_{i=1}^{i_{\text{sol}}} i w_i e^{-\beta i}. \tag{5.38}$$

The partition function of the two-phase system is

$$\log \Omega_{12} = \log \Omega_{\text{sol}} + \log \Omega_{\text{gel}}. \tag{5.39}$$

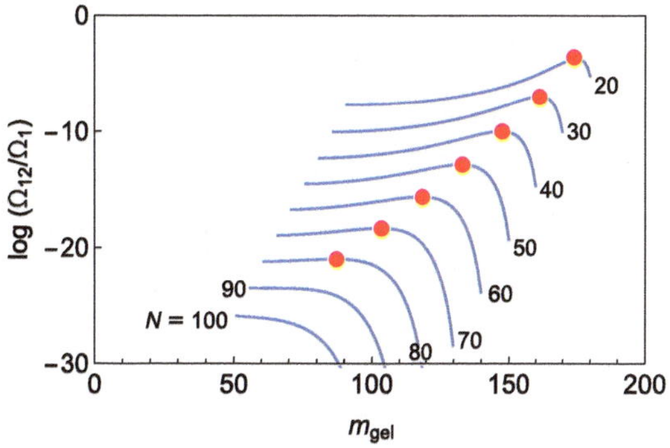

Fig. 5.5 We hypothesize that the system contains a gel phase and calculate the partition function of the two-phase system as a function of the hypothesized m_{gel}. If there is a maximum, it corresponds to the equilibrium mass of the gel. The calculation is for $M = 200$ and various values of N that correspond to $\bar{x}$ from 2 to 10

Using $\log \Omega_{\text{sol}} = \beta M_{\text{sol}} + N_{\text{sol}} \log q$, $\log \Omega_{\text{gel}} = \log w(m_{\text{gel}})$ this becomes

$$\log \Omega_{12} = \beta M_{\text{sol}} + N_{\text{sol}} \log q + \log w(m_{\text{gel}}). \tag{5.40}$$

We can calculate the combined partition function as a function of m_{gel} as follows: pick a value of m_{gel} in the range defined by Eq. (5.33), calculate the mass and number of clusters in the sol from the conservation equations in Eq. (5.34), obtain β and q of the sol from Eqs. (5.37) and (5.38), and finally the partition function from Eq. (5.40).

Results of this calculation are shown in Fig. 5.5 for $M = 200$ and for N ranging from 100 to 20, corresponding to $\bar{x}$ from 2 to 10. For large N (small $\bar{x}$) the two-phase partition function has no maximum in the gel region, which means that a gel phase is not stable. Below some critical value of N, somewhere between 90 and 80, the partition function begins to develop a maximum that indicates the equilibrium mass of the gel phase. This mass moves to larger values of m_{gel} as N increases. We demonstrate this calculation with the example below.

Example 5.4 (Numerical Calculation of the Tie Line)

Here we show the calculation of the tie line for $M = 200$, $N = 40$ ($\bar{x} = 4$). By direct tabulation of $\log \Omega_{12}$ we find that the maximum is at

$$m_{\text{gel}} = 133.5, \quad M_{\text{sol}} = 66.5, \quad x_{\text{sol}} = 1.357, \quad \beta = -0.0239, \quad q = 1.240.$$

The mass fraction of the gel is

$$\phi_{\text{gel}} = \frac{m_{\text{gel}}}{M} = 0.668.$$

With these results we construct the distribution in both phases:

$$F_{\text{sol}}(x) = (N-1)x^{-3}\frac{e^{-0.0239x}}{1.240}; \qquad F_{\text{gel}}(x) = \delta(x - 133.5), \tag{5.41}$$

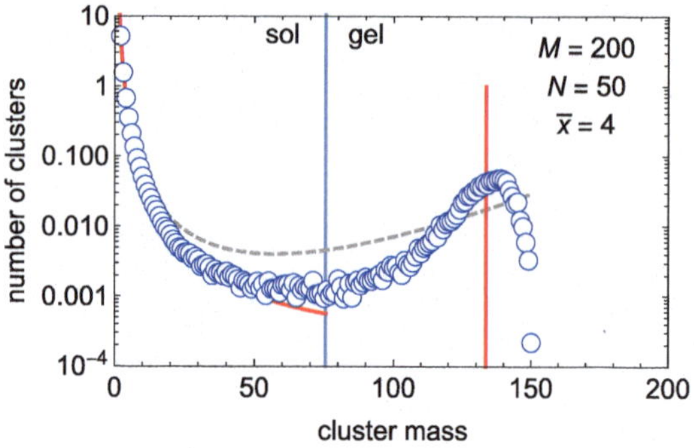

Fig. 5.6 Sol and gel equilibrium phases calculated by maximizing the partition function and plotted against data from Monte Carlo simulation

where $F(x)$ stands for the extensive size distribution and $\delta(x)$ is the Dirac delta. Figure 5.6 shows the calculated tie line from Eq. (5.41) and the result from Monte Carlo simulation. The equilibrium sol distribution, shown by the solid line, is in very good agreement with the sol region of the data. The delta function places the mass of the gel ($m_{\text{gel}} = 133.5$) fairly close to the observed maximum ($m'_{\text{gel}} \approx 138$). The gel distribution in the simulation is not a delta function but broader peak that connects smoothly to the sol branch. In a finite system the gel mass fluctuates due to exchange with the sol. These fluctuations refer to position of the gel mass, not the number of gel clusters, which is always 1. The construction of the tie line is based on the equilibrium (most probable) gel mass and does not reflect these fluctuations. For comparison, the unstable MPD of the single-phase system is also shown as a dashed line. This distribution has on average the expected behavior: it is a fairly good representation of the distribution in the sol region and has a diverging tail that suggests the presence of a gel phase. Quantitatively, however, it does not describe the gel phase correctly and fails to scale homogeneously with the size of the system, a condition for its derivation, because of the diverging tail.

5.5.2 *Scaling Limit*

The numerical maximization of the partition function is not practical except for the smallest systems, and it was only shown as a demonstration of the construction of the tie line. For large M we pass in the scaling limit and it becomes possible to obtain simple analytic expressions for the pre- and post-gel state of the system. Since the ensemble is linear we apply the equilibrium condition in Eq. (5.22) with $w(x) = x^{-3}$ to obtain

$$\beta^* = \left(\frac{d \log w(m)}{dm}\right)_{m=m_{\text{gel}}} = -\frac{3}{m_{\text{gel}}^4}. \tag{5.42}$$

In the scaling limit $m_{\text{gel}} \to \infty$ (recall that m_{gel} is always larger than $(M - N + 1)/2 + 1$), therefore

$$\beta^* = 0. \tag{5.43}$$

The corresponding value of q^* is obtained from Eq. (5.29) with $\beta = \beta^*$ and $i_{\max} \to \infty$:

$$q^* = \zeta(3) = 1.20206. \tag{5.44}$$

The MPD of the equilibrium sol is given by Eq. (5.29) with $\beta = \beta^*$ and $q = q^*$:

$$\left(\frac{\tilde{n}_i}{N}\right)^* = \frac{i^{-3}}{\zeta(3)} = \frac{i^{-3}}{1.20206}, \tag{5.45}$$

and its mean is

$$\bar{x}^* = \frac{\pi^2}{6\zeta(3)} = 1.36843, \tag{5.46}$$

where $\zeta(x)$ is the zeta function,

$$\zeta(x) = \sum_{k=0}^{\infty} k^{-x}.$$

Equations (5.43)–(5.46) define the state of the sol in the presence of a gel phase. We may now obtain the pre- and post-gel behavior of the system (the results are summarized in Table 5.1):

Pre Gel: This region is defined by the condition

$$0 \le \bar{x} \le \bar{x}_* \tag{5.47}$$

The system consists of a stable sol with MPD given by

$$\frac{\tilde{n}_i}{N} = i^{-3} \frac{e^{-\beta i}}{q} \tag{5.48}$$

with

$$q = \mathrm{Li}_3\left(e^{-\beta}\right), \quad \bar{x} = \frac{\mathrm{Li}_2\left(e^{-\beta}\right)}{\mathrm{Li}_3\left(e^{-\beta}\right)}. \tag{5.49}$$

Table 5.1 Results in the scaling limit

	pre gel ($\bar{x} < 1.36843$)	**post gel** ($\bar{x} > 1.36843$)
sol MPD	$\dfrac{\tilde{n}_i}{N} = i^{-3}\dfrac{e^{-\beta i}}{q}$	$\dfrac{\tilde{n}_i}{N} = \dfrac{i^{-3}}{1.20206}$
$\bar{x}_{\text{sol}}$	$\bar{x}$	1.36843
β	$\bar{x} = \dfrac{\text{Li}_2\left(e^{-\beta}\right)}{\text{Li}_3\left(e^{-\beta}\right)}$	0
q	$\text{Li}_3\left(e^{-\beta}\right)$	1.20206
$\log\omega$	$\beta\bar{x} + \log q$	0.184034
ϕ_{gel}	0	$\dfrac{\bar{x} - 1.36843}{\bar{x}}$

where $\text{Li}_a(x)$ is the polylogarithm function

$$\text{Li}_a(x) = \sum_{k=1}^{\infty} \frac{x^k}{k^a}.$$

Equations (5.49) follow directly from (5.29) and (5.30) with $i_{\max} \to \infty$. The partition function is

$$\log\omega = \beta\bar{x} + \log q \tag{5.50}$$

with β and q from Eq. (5.49).

Post Gel: In the region

$$\bar{x} > \bar{x}^* \tag{5.51}$$

the system contains a sol phase in equilibrium with a gel. The state of the sol is given by Eqs. (5.43)–(5.46). The gel fraction $\phi_{\text{gel}} = m_{\text{gel}}/M$ is obtained from mass conservation, which we write as

$$N\bar{x} = (N-1)\bar{x}_{\text{sol}} + M\phi_{\text{gel}},$$

where $\bar{x} = M/N$ is the overall mean cluster mass and $\bar{x}_{\text{sol}} = M_{\text{sol}}/(N-1)$ is the mean cluster within the sol phase. For the sol we have $\bar{x}_{\text{sol}} = \bar{x}^*$, then solving for the gel fraction we find

$$\phi_{\text{gel}} = \frac{\bar{x} - \bar{x}^*}{\bar{x}} = \frac{\bar{x} - 1.36843}{\bar{x}}. \tag{5.52}$$

To obtain the partition function in the post gel region we begin with the extensive form

$$N \log \omega = (N-1)\log \omega_{\text{sol}} + \log \Omega_{\text{gel}}. \tag{5.53}$$

The partition function of the sol is given by Eq. (5.50) with $\beta = \beta^*$ and $q = q^*$,

$$\log \omega_{\text{sol}} = \log q^* \equiv \log \omega^* \tag{5.54}$$

The partition function of the gel is

$$\log \Omega_{\text{gel}} = \log w(m_{\text{gel}}) = -3 \log m_{\text{gel}}. \tag{5.55}$$

Making these substitutions into Eq. (5.53) we obtain

$$\log \omega = \log \omega^* - \frac{3 \log m_{\text{gel}}}{N}, \tag{5.56}$$

and since $\log m_{\text{gel}}$ increases logarithmically with size and vanishes when divided by N in the scaling limit the final result is

$$\log \omega = \log \omega^*. \tag{5.57}$$

The partition function in the post-gel region contains no contribution from the gel phase and is equal to the partition function of the equilibrium sol. We then have

$$\omega = \omega^*, \quad \beta = \beta^*,$$

in the entire post gel region.

Gel Point: The gel phase appears at $\bar{x} = \bar{x}^*$. At this point the sol phase has just reached its equilibrium state and the gel fraction from Eq. (5.52) is 0.

5.5.3 *Alternative Description of the Tie Line*

The phase behavior in the scaling limit can be expressed more compactly in terms of the progress variable θ, defined as

$$\theta = 1 - 1/\bar{x}. \tag{5.58}$$

It represents a transformed size that ranges from $\theta = 0$ at $\bar{x} = 1$ to $\theta = 1$ at $\bar{x} = \infty$, and conveniently maps the entire trajectory in phase space onto the interval [0, 1]. The gel point is at

$$\theta^* = 1 - 1/\bar{x}^* = 1 - \frac{6\zeta(3)}{\pi^2} = 0.26923. \tag{5.59}$$

For the mean size in the sol and gel fraction we then have:

$$\bar{x}_{\text{sol}} = \begin{cases} \dfrac{1}{1-\theta} & 0 \le \theta \le \theta^* \\ \dfrac{1}{1-\theta^*} & \theta^* < \theta \end{cases} \tag{5.60}$$

$$\phi_{\text{gel}} = \begin{cases} 0 & 0 \le \theta \le \theta^* \\ \dfrac{\theta^* - \theta}{\theta^* - 1} & \theta^* < \theta \end{cases} \tag{5.61}$$

All other variables ($\log\omega, \beta, q$) can similarly be expressed in terms of θ, as shown in Fig. 5.7. In the two-phase region, $\bar{x}_{\text{sol}}$, $\log\omega$, β, and q remain constant. What varies is the overall mean cluster size $\bar{x} = 1/(1-\theta)$ and the fraction of the gel phase ϕ_{gel}. In this two-phase region the properties of the sol phase are those of point S in Fig. 5.7, which refers to the pure sol at the sol-gel boundary. As a singular phase, the gel does not have a proper MPD, its distribution is a single cluster with mass $m_{\text{gel}} \to \infty$. In the scaling limit this is the same state as $\theta = 1$, at which point the entire mass of the population is gathered into a single cluster. Therefore point G, located at $\theta = 1$ may be understood to represent the state of the equilibrium gel phase. This phase has the same β as the equilibrium sol at point S. The straight line SG has the properties of a tie line that connects the states of two coexisting phases. A point A on this line is a mixture of two phases, a sol with the properties of point S, namely,

$$\theta_{\text{sol}} = \theta^*, \quad \beta = \beta,^* \quad q = q^*;$$

and a gel phase with the properties of point G with

$$\theta_{\text{gel}} = 1, \quad \beta = \beta^*.$$

The corresponding mean cluster sizes in each phase are

$$\bar{x}_{\text{sol}} = \frac{1}{1-\theta^*} = \bar{x}^*; \quad \bar{x}_{\text{gel}} = \lim_{\theta\to 1} \frac{1}{1-\theta} = \infty.$$

We now rewrite Eq. (5.61) as

$$\theta = \phi_{\text{sol}}\theta_{\text{sol}} + \phi_{\text{gel}}\theta_{\text{gel}}, \tag{5.62}$$

with

$$\theta_{\text{sol}} = \theta^*, \quad \theta_{\text{gel}} = 1, \quad \phi_{\text{sol}} + \phi_{\text{gel}} = 1.$$

Equation (5.62) is in the form of the lever rule, a relationship in thermodynamics that gives property θ of a two-phase system as an average of the corresponding properties of the pure phases, θ_{sol} and θ_{gel}, in proportion to the mass fraction in each phase. This identifies θ as the intensive variant of extensive property $\theta M = M - N$ that is partitioned between phases in the same manner that volume, energy, and other extensive properties are partitioned between phases in vapor/liquid equilibrium. If we imagine point A in Fig. 5.7 to move from left to right, we have the equivalent of isothermal condensation, as more mass is transferred from the sol to the gel.

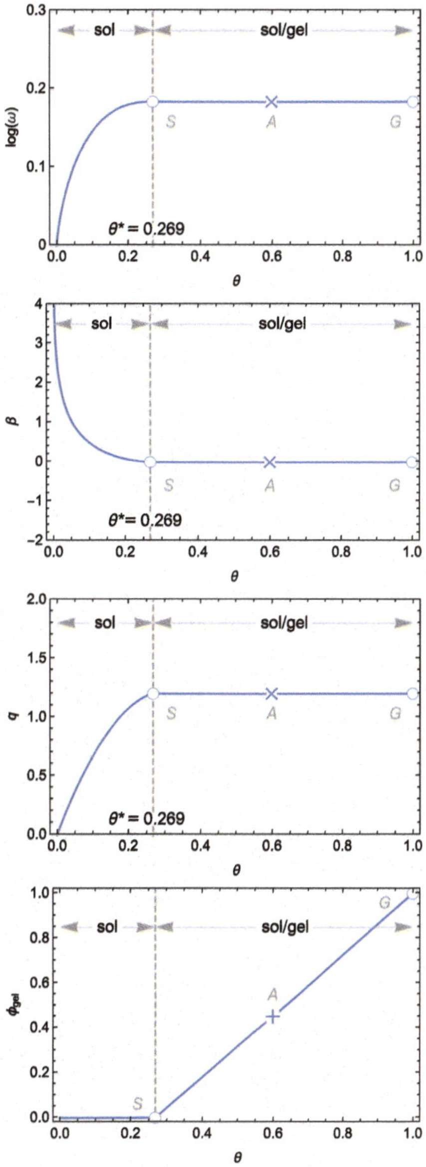

Fig. 5.7 Phase diagram for the model $w_i = i^{-3}$ as a function of the rescaled size $\theta = 1 - 1/\bar{x}$. The region $0 \le \theta < \theta^*$ is a single phase sol; the region $\theta^* < \theta < 1$ contains a sol in equilibrium with a gel. A point A in the two-phase region is a mixture of the two phases. The sol in the mixture has the properties of point S, which represents a pure sol at the boundary of the sol/gel transition. The gel phase is represented by point G at $\theta = 1$. The parameters β and q in the two-phase region remain constant. If point A moves to the right, the process represents conversion of sol into gel ("condensation"). Conversely, if point A moves to the left, the process represents conversion of the gel into sol ("evaporation"). During the conversion the gel fraction varies linearly with θ between $\theta = \theta^*$ (all sol) and $\theta = 1$ (all gel). If we interpret β as "temperature," these processes are isothermal in both directions

In the reverse direction we are dealing with isothermal evaporation. In both cases the isothermal condition is implied by the constant value of β. The equilibrium phases are separated by a finite gap, which can be expressed as $\theta_{\text{gel}} - \theta_{\text{sol}}$, or equivalently as $\bar{x}_{\text{gel}} - \bar{x}_{\text{sol}}$.

As a final exercise we calculate the entropy in each phase. In the sol we have

$$\frac{S_{\text{sol}}}{N_{\text{sol}}} = -\sum_i \left(\frac{\tilde{n}_i}{N}\right)_{\text{sol}} \log \left(\frac{\tilde{n}_i}{N}\right)_{\text{sol}} = \log \omega_{\text{sol}} - \frac{\log \tilde{W}}{N_{\text{sol}}}$$

where $\log \omega_{\text{sol}}$ is the partition function of the sol and $\log \tilde{W}_{\text{sol}}$ is the sum of $\log w_i$ over the MPD of the sol,

$$\log \tilde{W}_{\text{sol}} = \sum_i \tilde{n}_i \log w_i = -3M_{\text{sol}}.$$

At equilibrium $\omega = \omega^*$, $M_{\text{sol}} = \bar{x}^* N_{\text{sol}}$, and the entropy of the sol phase is

$$\left(\frac{S_{\text{sol}}}{N_{\text{sol}}}\right)^* = \log \omega^* + 3\bar{x}^* = 3\bar{x}^* + \log q^* = 4.289. \tag{5.63}$$

The entropy of the gel phase is zero because its MPD is a delta function:

$$\frac{S_{\text{gel}}}{N_{\text{gel}}} = 0. \tag{5.64}$$

Therefore the entropy change for the sol-gel transition is

$$\Delta s_{\text{sol}\to\text{gel}} \equiv \left(\frac{S_{\text{gel}}}{N_{\text{gel}}}\right)^* - \left(\frac{S_{\text{sol}}}{N_{\text{sol}}}\right)^* = -4.289.$$

The entropy change is negative because as a delta function in the gel phase has the absolute minimum entropy (zero) of any distribution in the cluster ensemble.

Example 5.5 (Scaling Theory Versus Monte Carlo Simulation)

Here we check the scaling theory against simulation results. In this example the simulations are done by binary exchange reaction in a population with $M = 200$ and with cluster function $w_i = i^{-3}$. The average cluster size in the sol is calculated as the average of cluster masses in the region $1 \le i < (i_{\max} + 1)/2$ with $i_{\max} = M + N - 1$; the gel fraction is calculated as the mean mass fraction in the range $(i_{\max} + 1)/2 \le i \le i_{\max}$. The results are shown in Fig. 5.8. The good agreement in the pre-gel region shows that no gel phase is detected in the simulations before the predicted gel point. Past the gel point there are discrepancies. The onset of gelation is delayed, as reflected by the overshoot in $\bar{x}_{\text{sol}}$ and undershoot in ϕ_{gel}. Once the gel phase is established, both $\bar{x}_{\text{sol}}$ and ϕ_{gel} converge to the scaling predictions. As the state approaches $\theta = 1$, the scaling theory begins to fail as the number of clusters becomes of the order of 1. The scatter in $\bar{x}_{\text{sol}}$ is significantly larger than that in ϕ_{gel}. The scatter in both phases is due to exchange of mass between sol and gel

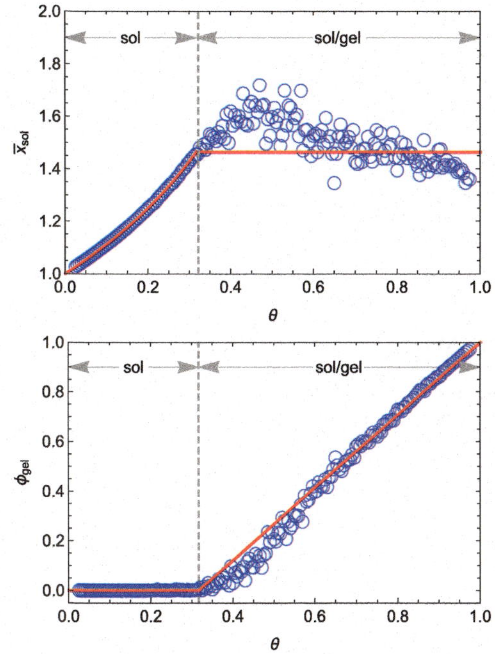

Fig. 5.8 Comparison between scaling solution and Monte Carlo simulation with $w_i = i^{-3}$. The simulation is run with $M = 200$ and N from 195 to 5. (Top) Mean size in sol; (bottom) mass fraction of gel. The horizontal axis is $\theta = 1 - 1/\bar{x}$

but since the mean cluster size in the sol is much smaller than the gel mass, these exchanges represent a much larger fluctuation in $\bar{x}_{sol}$. Also notice that the scaled size θ effectively amplifies the region of small $\bar{x}$ at the expense of large $\bar{x}$ and exaggerates the magnitude of the discrepancy. The region above $\theta \approx 0.7$ where the scaling limit is in good agreement with the simulation covers the size range $3.3 < \bar{x} < 200$, a very wide range indeed.

5.5.4 Evolution of the Distribution

We gain additional insight by examining the evolution of the distribution as the system passes through the gel point (Fig. 5.9). The MPD in the scaling theory is obtained from Table 5.1 and the location of the gel phase is calculated as $m_{gel} = M\phi_{gel}$, with ϕ_{gel} from the same table. We arrange the results in the order of decreasing N, which represents a system of constant mass whose mean cluster size increases by aggregation, or by "compression," if we imagine that particles are being forced to occupy fewer clusters. We may arrange them in reverse order to represent fragmentation or expansion, but ordering in increasing size has the advantage that the system begins as a single phase and then crosses into the two-phase region (in the reverse direction we always start with two phases). Starting at $N = 160$ ($\bar{x} = 1.25$) we are well below the gel point and the distribution is entirely contained in the sol region. At $N = 106$ ($\bar{x} = 1.88$), scaling theory

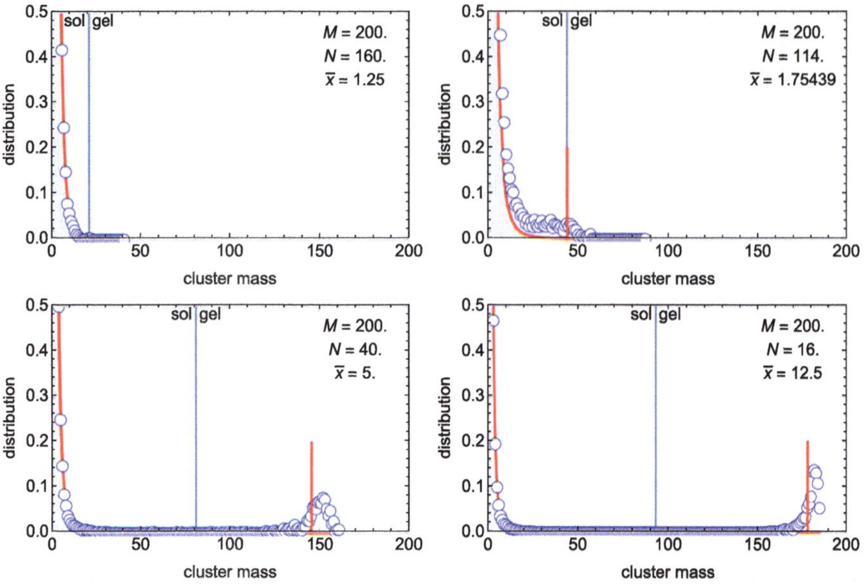

Fig. 5.9 Distributions for the model $w_i = i^{-3}$ with $M = 200$. Lines are the predictions of the theory in the scaling limit

places the system just above the gel point ($\bar{x}^* = 1.36843$) and predicts a gel phase at the sol/gel boundary. The simulated distribution lacks a well-defined peak but shows clear signs of a shoulder near the sol/gel boundary. As N decreases further the system advances into the post gel region and the formation of a gel phase is clearly manifested by a peak that is separated from the sol. The vertical bar that marks the gel phase is located at the value of m_{gel} that is predicted by the scaling theory (the height of the bar is not of any significance). The theory represents the gel phase fairly well. As with the construction of the tie line in Sect. 5.5.1, the scaling theory does not predict the shape of the gel peak, only its most probable position. Past the gel point the normalized distribution of the sol remains unchanged, only the number of clusters changes (decreases with increasing $\bar{x}$). The difference represents mass that is transferred to the gel, whose growth is indicated by its advancing to the right. The gel phase is always accompanied by a sol. It is obtained as a single pure phase only in the limit $\bar{x} = M \to \infty$, when all mass accumulates into a single cluster.

5.6 Nucleation, Fluctuations, and the Order Parameter

The mass of the gel cluster satisfies the condition

$$\frac{M - N + 2}{2} \leq m_{\text{gel}} \leq M - N + 1, \tag{5.14}$$

which defines the gel region. In the scaling limit the fraction of mass that carried by the gel cluster satisfies

$$\frac{1}{2} < \frac{m_{\text{gel}}}{M} < 1. \tag{5.65}$$

The gel fraction within a distribution is either zero, if no gel cluster is present, or some number between 1/2 and 1, if the distribution contains a gel cluster. One might think then that the gel fraction ought to jump abruptly from zero to a value around 0.5 as the system passes through the gel point. This is not so. Both theory and simulations agree that ϕ_{gel} grows continuously from 0 at the gel point to 1 and the fully gel state. This continuity is due to fluctuations near the gel point. Fluctuations arise from the exchange of mass between the gel and the sol and cause the position of the gel cluster to fluctuate. If the mass transferred to the sol is large enough such that what is left of the gel cluster is smaller than $(i_{\text{max}} + 1)/2$, the gel cluster disappears and the resulting distribution contains only a sol phase. This means that not all distributions of the ensemble contain a gel component. Within the subset of distributions that do, the mean size of the gel cluster is of the order of M, as Eq. (5.14) requires. Within the entire ensemble of distributions the mean gel cluster is smaller because distributions with no gel cluster do not contribute to this average. What we call ϕ_{gel} is the average size of the gel cluster weighted by the fraction of distributions that contain such cluster, and this could be less than the minimum gel mass. Figure 5.10 explains the situation. In this case,

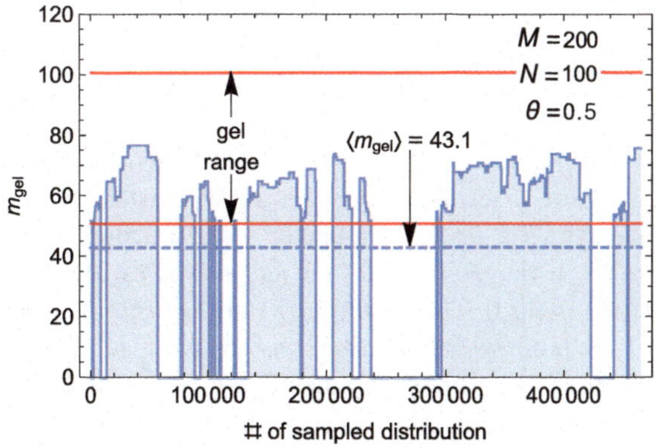

Fig. 5.10 Fluctuations of the gel mass in Monte Carlo simulation with $M = 200$, $N = 100$ in a large sample of distributions. The gel region is $51 \le m_{\text{gel}} \le 101$ and is marked by the solid lines. The vertical axis is the mass of the largest cluster in the region $i \ge i^* = 56$ and is set to zero if no such cluster exists. About 65% of the sampled distributions contain a gel cluster and 35% consist only of a sol

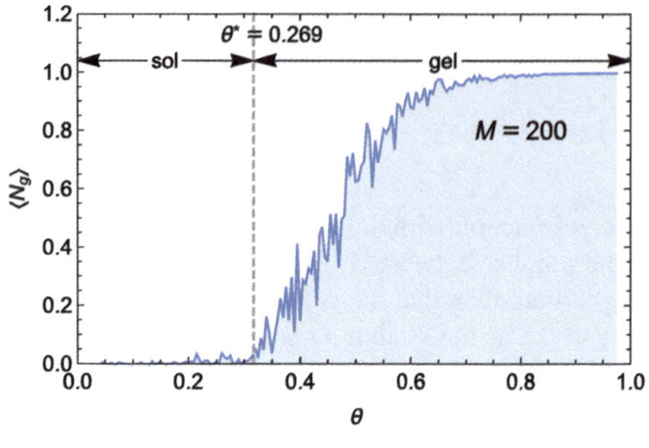

Fig. 5.11 Mean fraction of distributions with a gel cluster as a function of θ for $M = 200$. Before the gel point θ^* the fraction of distributions with a gel cluster is essentially zero. Past the gel point this fraction increases continuously at approaches 1 as $\theta \to 1$

$M = 200$, $N = 100$, which places the system at $\theta = 0.5$, above the gel point $\theta^* = 0.26923$. The gel region is $56 \le i \le 101$. If a distribution contains a cluster in this region, its mass is plotted on the vertical axis, otherwise the mass of the gel cluster is shown as zero. Approximately 65% of the sampled distributions contain a gel cluster; within this subset the mean cluster size is $m'_{\text{gel}} = 65.5$. The overall mean gel clusters is $m_{\text{gel}} = 43.1$, roughly 65% of m'_{gel}, and in this particular case it is smaller than $i^* = 56$. The corresponding gel fraction is $\phi_{\text{gel}} = 43.1/200 = 0.215$.

We can quantify the fraction of distributions that carry a gel cluster and track it by simulation. We define the gel number N_g in a distribution by the summation of all n_i in the gel region,

$$N_g = \sum_{i=i^*}^{i_{\max}} n_i, \tag{5.66}$$

with $i^* = (i_{\max} + 1)/2$ and $i_{\max} = M - N + 1$. This is a binary variable that takes the value 1 if the distribution contains a gel cluster, and 0 if it does not. The ensemble average $\langle N_g \rangle$ of this parameter is the fraction of distributions that contain a gel cluster. It ranges from 0 to 1 and serves as an order parameter for the sol/gel transition, with 0 representing the pure sol (no distribution contains a gel cluster) and 1 the pure gel (all distributions contain a gel cluster). Its evolution has the typical features of an order parameter (Fig. 5.11). At $\theta < \theta^*$ the fraction $\langle N_g \rangle$ is practically zero. In this region a gel cluster appears only as a short-lived fluctuations. At the gel point predicted by the scaling theory $\langle N_g \rangle$ begins to grow. From $\theta = \theta^*$ to $\theta \approx 7$ the gel cluster appears as an increasingly long-lived fluctuation, but fluctuations without a gel cluster are also long-lived. As $\theta \to 1$, almost all distributions contain a gel fraction and those with none represent short-lived fluctuations. The evolution of $\langle N_g \rangle$ is S-shaped but continuous. It remains so in the scaling limit because any discontinuities would be reflected in ϕ_{gel} as well.

The nucleation of the gel phase is closely related to the fluctuations in Fig. 5.10. The life-time of the gel cluster is the fraction of successive distributions that contain a gel cluster, while the remaining fraction represents the life-time of fluctuations in which the gel cluster has evaporated. We mention this to draw the connection between kinetics of nucleation and the cluster ensemble, but we will not pursue this here any further.

5.7 An Abrupt Phase Transition: The i^2 Model

The nucleation of the gel phase generally leads to continuous growth of the gel fraction but in certain cases it is possible for the gel phase to appear abruptly. We will consider one such example here that produces an abrupt transition. The selection function is again linear and the cluster function in this case is

$$w_i = e^{\gamma i^2}. \tag{5.67}$$

If $\gamma < 0$, this functional produces a stable Gaussian MPD with variance $\sigma^2 = 1/2\gamma$, provided that $\bar{x} \gg 1/\sqrt{2|\gamma|}$.[9] When γ is negative it produces a potentially unstable MPD. To see why, we express the MPD in canonical form,

$$\frac{\tilde{n}_i}{N} = w_i \frac{e^{-\beta i}}{q} = \frac{e^{\gamma i^2 - \beta i}}{q}. \tag{5.68}$$

with

$$q = \sum_{i=1}^{M-N+1} e^{\gamma i^2 - \beta i}, \tag{5.69}$$

$$\frac{M}{N} = \frac{1}{q} \sum_{i=1}^{M-N+1} i e^{\gamma i^2 - \beta i} \tag{5.70}$$

In the scaling limit ($M > N \to \infty$) these summations diverge because of the square term in the exponent, but when M and N are both finite it is possible to obtain stable or unstable solutions, depending on the values of M and N. We call this the i^2 model. To study the stability of this model we assume M to be finite but large enough for the system to be in the ThL.

The condition for sol-gel equilibrium is

$$\beta = \left(\frac{d \log w(x)}{dx}\right)_{m_{\text{gel}}} = 2\gamma m_{\text{gel}}, \tag{5.71}$$

[9] If $\bar{x}$ is of the order $1/\sqrt{2|\gamma|}$ the distribution is a truncated Gaussian function and can be calculated numerically.

where m_{gel} is the equilibrium size of the gel phase. The parameter β^* is that of the equilibrium sol phase and satisfies

$$\frac{M_{\text{sol}}}{N_{\text{sol}}} = \sum_{i}^{i^*_{\text{max}}} i e^{\gamma i^2 - \beta i}, \tag{5.72}$$

$$q = \sum_{i=1}^{i^*_{\text{max}}} e^{\gamma i^2 - \beta i}, \tag{5.73}$$

with $M_{\text{sol}} = M - m_{\text{gel}}$, $N_{\text{sol}} = N - 1$ and $i^*_{\text{max}} = M_{\text{sol}} - N_{\text{sol}} + 1$. Form Eq. (5.71) we obtain

$$\phi_{\text{gel}} = \frac{\beta}{2\gamma M}. \tag{5.74}$$

Using $M_{\text{sol}} = M(1 - \phi_{\text{gel}})$ Eq. (5.72) can be similarly solved for the gel fraction,

$$\phi_{\text{gel}} = 1 - \frac{N-1}{M} \frac{\sum_i i e^{\gamma i^2 - \beta i}}{\sum_i e^{\gamma i^2 - \beta i}}. \tag{5.75}$$

Equating Eqs. (5.74) and (5.75) we obtain a single equation for β. If an acceptable solution exists, the system forms sol-gel equilibrium with parameter β. The values of q and $\bar{x}_{\text{sol}}$ are obtained by back substitution from Eq. (5.73) and (5.71), respectively, once β is known.

This procedure can be demonstrated graphically: if we plot ϕ_{gel} from Eqs. (5.74) and (5.75) as a function of β, the solution is given by their intersection, as shown in Fig. 5.12. We use $\gamma = 0.02$ and $M = 100$, which will serve as the standard case from here on. Equation (5.74) is a straight line and does not depend on N. Equation (5.75) is an S-shaped function of β (it contains a flat segment at negative β that is not shown) that moves up as N is decreased. For large N (small $\bar{x}$), as in Fig. 5.12a, there is no intersection between the two lines at positive β. The system exists as a single phase sol whose parameters are obtained by solving Eqs. (5.68)–(5.70) numerically. As N is decreased, the line representing Eq. (5.75) moves up and past a critical value $N^* = 48.44$. The two lines intersect for all $N < N^*$. In fact, there are three intersections but only the one with the largest β represents a stable phase. The calculation is demonstrated in the example below.

Example 5.6 (Numerical Calculation of the Tie Line)

We calculate the tie line at $M = 100$, $N = 20$, with $\gamma = 0.02$. We define

$$F_k(\beta) = \sum_{i=1}^{i_{\text{max}}} i^k e^{-\beta i + \gamma i^2}, \tag{5.76}$$

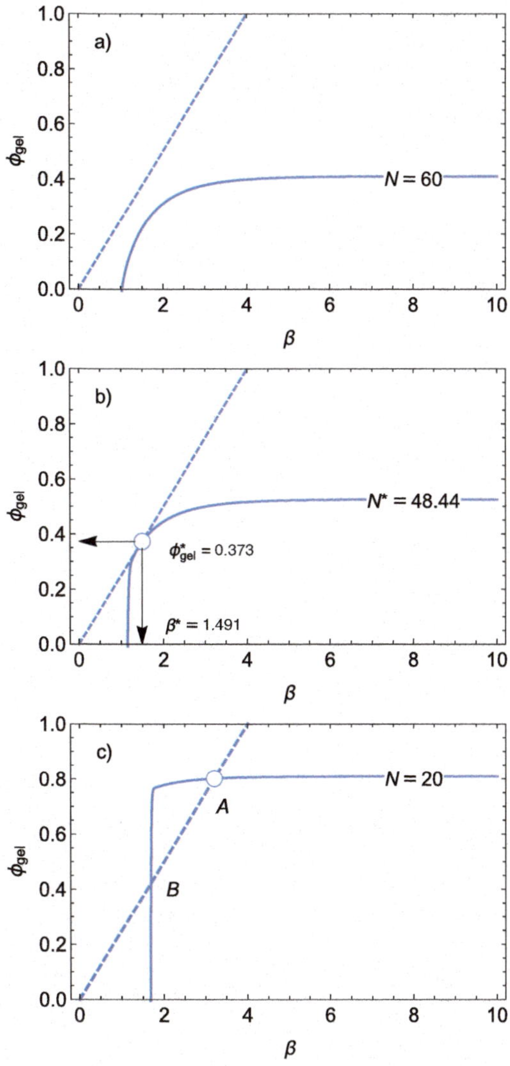

Fig. 5.12 Graphical construction of the tie line of the i^2 model with $M = 100$, $\gamma = 0.02$. The dashed line is ϕ_{gel} from Eq. (5.74) and the solid line is ϕ_{gel} from Eq. (5.75). Equation (5.75) is an S-shaped curve that becomes sharper and moves upward as N decreases. (**a**) For $N > N^* = 48.44$ there is no intersection and the system exists as a single sol. (**b**) At $N = N^*$ the two lines intersect at $\beta^* = 1.491$, $\phi^*_{gel}0.373$, and this defines the gel point. (**c**) For $N < N^*$ there are two intersections at A and B. Only A represents a stable sol-gel mixture (see Fig. 5.13). A third intersection at negative β is rejected as unphysical. If β is negative, the MPD of the sol increases monotonically with size and extends into the gel region

and write the equation of the tie line as

$$\frac{\beta}{2\gamma M} = 1 - \frac{N-1}{M}\frac{F_1(\beta)}{F_0(\beta)}, \tag{5.77}$$

which is obtained by equating (5.74) with (5.75). The upper limit in the summations in Eq. (5.76) is the maximum possible cluster size in the sol,

$$(i_{max})^{sol} = M - m_{gel} - N + 2,$$

and depends on the mass of the gel cluster, which is not known. To simplify the calculation we replace the upper limit with the maximum possible cluster in the population, $i_{\max} = M - N + 1$. This is acceptable, provided that the range of cluster masses in the sol satisfy the condition $i \ll \beta/\gamma$ because then the exponent in the summation is dominated by the linear term. We will check this condition afterwards. Solving Eq. (5.77) numerically we find

$$\beta = 3.20569.$$

The value of q is

$$q = F_0(\beta) = 0.043213,$$

and the gel fraction is

$$\phi_{\text{gel}} = \frac{\beta}{2\gamma M} = 0.801422.$$

The corresponding mass of the gel cluster is

$$m_{\text{gel}} = \phi_{\text{gel}} M = 80.14,$$

which is indeed in the gel region, which for $M = 100$, $N = 20$, is defined by the condition $41 \le m_{\text{gel}} \le 81$. In fact, the gel cluster is only slightly less than the largest possible mass. Finally, the mean size in the sol is

$$\bar{x}_{\text{sol}} = \frac{M - m_{\text{gel}}}{N - 1} = 1.045.$$

The mean cluster size is barely larger than 1, which means that the equilibrium sol consists almost entirely of monomers. This is consistent with the fact that the gel cluster is nearly the maximum possible mass for the given M and N. Since the sol consists almost of monomers, the condition $i \ll \beta/2\gamma = 80$ is satisfied, therefore the replacement of the upper limit in the summations of Eq. (5.76) is justified.

There is a second positive solution to Eq. (5.77). It corresponds to the second intersection in Fig. 5.12 and gives the following results:

$$\beta' = 1.6874, \quad q' = 0.2409, \quad \phi'_{\text{gel}} = 0.4219, \quad m'_{\text{gel}} = 42.186, \quad \bar{x}'_{\text{sol}} = 3.043.$$

To identify the stable solution we calculate the tie line by maximizing the partition function following the procedure discussed in Sect. 5.5.1. This calculation is shown in Fig. 5.13. The partition function has a maximum at $m_{\text{gel}} = 80.51$, very close to $m_{\text{gel}} = 80.14$ that was obtained by solving the equilibrium condition. There is also a minimum around $m'_{\text{gel}} = 38.2$ that corresponds to the second solution of Eq. (5.77). The stable solution is point A at the maximum of the partition function; point B represents an unstable solution and is rejected. In serial calculations of the tie line it is not necessary to perform this analysis at each point. Instead, we always select the root with the largest β.

★

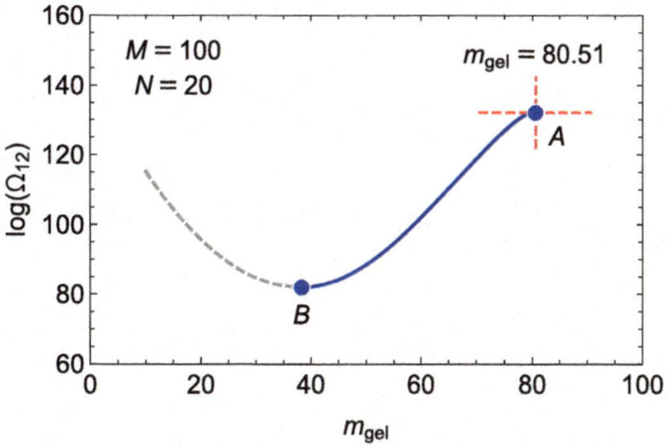

Fig. 5.13 The partition function of a sol-gel mixture as a function of the mass of the gel cluster. The gel region is defined by the condition $41 \le m_{\text{gel}} \le 81$ and the dashed line is a continuation of the calculation into the unphysical region of gel masses below this range. The stable solution is point A at the maximum

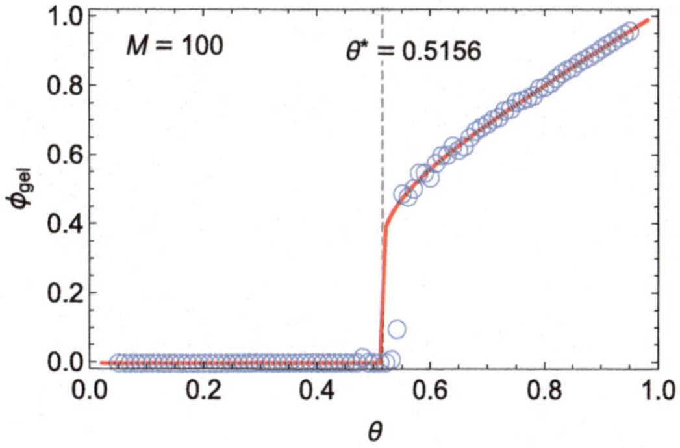

Fig. 5.14 Phase behavior of the i^2 system ($\gamma = 0.02$, $M = 100$) as a function of $\theta = 1 - 1/\bar{x}$. The lines are obtained from the numerical calculation of the tie line and the symbols are Monte Carlo simulations at various N. The dashed line marks the sol-gel point predicted by the numerical tie line ($\theta^* = 0.5156$)

5.7.1 The Sol-Gel Point

The gel point for $M = 100$, $\gamma = 0.02$, is at $N = N^*$ and produces the tie line

$$\beta^* = 1.491,\ q^* = 0.3032,\ \bar{x}^*_{\text{sol}} = 1.3222,\ \phi^*_{\text{gel}} = 0.3728,\ \theta^* = 0.5156.$$

At the critical value $\theta^* = 0.5156$ the corresponding gel fraction is $\phi^*_{\text{gel}} = 0.3728$; for $\theta < \theta^*$ $\phi_{\text{gel}} = 0$. The gel phase emerges abruptly when approached from the sol side; it disappears as abruptly when the critical point is crossed from the gel side. We confirm this behavior by simulation (see Fig. 5.12). The simulation shows remarkable agreement with the results of the calculated tie line. At the sol-gel transition the gel fraction jumps from zero just below θ^* to $\phi_{\text{gel}} = 0.3728$ right after, then continues to grow smoothly (Fig. 5.14).

To understand this behavior we examine $\bar{x}_{\text{sol}}$, β and q in Fig. 5.15. The jump in ϕ_{gel} is accompanied by a similar jump in the mean cluster in the sol phase. Before the transition the system consists of a single sol and the mean cluster is trivially related to θ by

$$\bar{x} = \frac{1}{1-\theta},$$

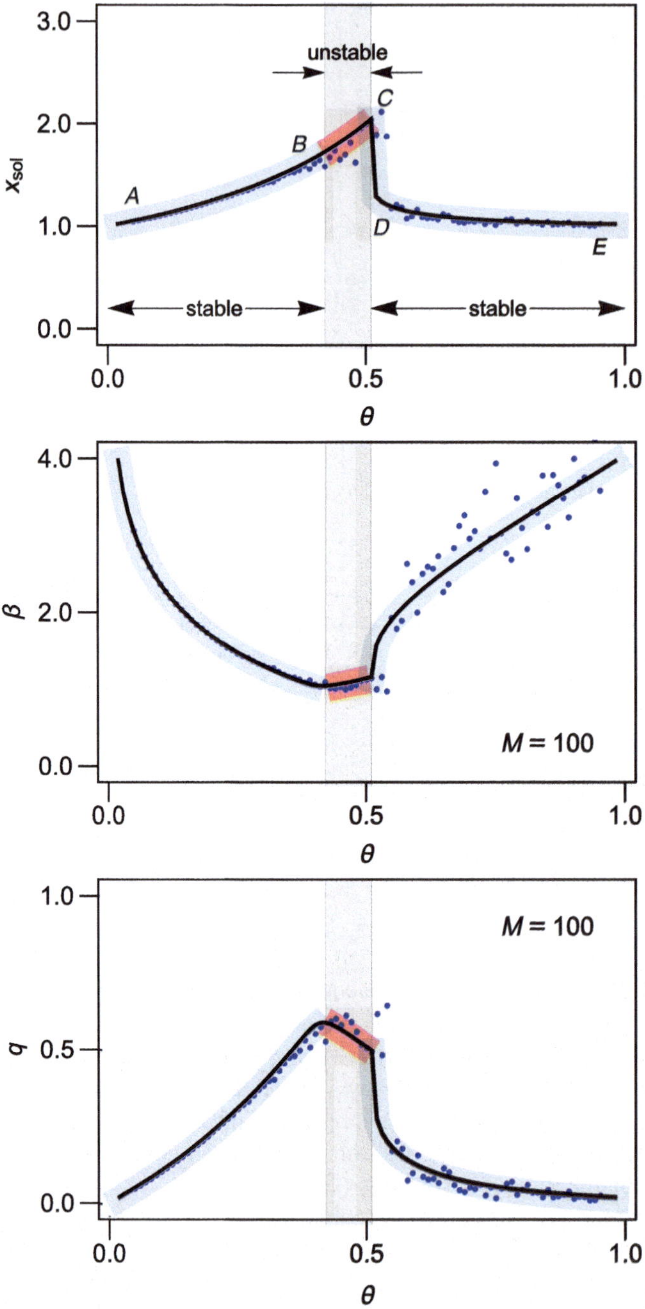

Fig. 5.15 In the region AB, defined by the condition $0 \le \theta < 0.41$ the system satisfies the stability criteria: as $\bar{x}$ increases, β decreases, q increases. In the region BS ($0.41 < \theta < \theta^*$) $\bar{x}$ continues to increase while both β and q change direction. Here the system is unstable but continues to form a single sol phase. At θ^* the state transitions abruptly from the unstable branch to the stable one and forms the gel phase. At $\theta > \theta^*$ the system exists as a sol-gel mixture with an increasing fraction of the mass transferred to the sol as $\theta \to 1$

which follows from the definition of θ. Right before the transition the cluster size reaches a maximum value $\bar{x} = 1/(\theta^* - 1) \approx 2$ but then drops to $\bar{x}^*_{\text{sol}} = 1.3222$ right after. From there on it decreases and approaches 1 as $\theta \to 1$. A similar jump can be seen in β and q. In Sect. 5.1 we obtained the stability conditions,

$$\frac{d\beta}{d\bar{x}} \leq 0; \qquad \frac{d \log q}{d\bar{x}} \geq 0. \qquad [5.9]$$

Both conditions are violated in the region marked as BC in Fig. 5.15. This region of instability divides the behavior of the system into four distinct regions. Starting from $\theta = 0$, the region AB is stable: the system forms a single sol phase whose mean cluster size increases as it advances to the right. At B the system enters the unstable region, and yet, it continues to form a single sol phase whose mean size increases. At C the instability can no longer be sustained and the state snaps to the stable branch at D. This marks the sol-gel point. Past C the system is a stable mixture of a sol phase in equilibrium with the gel.

Why does the state persist as a single sol phase even though the system has entered the unstable region? To answer this question we look at the evolution of the cluster distribution in Fig. 5.16. The distributions are arranged from (a) to (j) in the direction of increasing θ. The unstable region is $0.41 < \theta < 0.5156$, or equivalently $59 > N > 48.44$. In (a) the system is still in the stable one-phase region and the MPD obtained by simulation is in excellent agreement with the simulation. This MPD is a rather rapidly decaying function of cluster size and remains contained in the sol region. Panels (b)–(g) show the system at various stages as it moves through the unstable region. Here the cluster distribution increasing deviates from the theoretical MPD and the deviation is manifested as a significant growth of large clusters that nonetheless remain contained in the sol region. As the system approaches the gel point the population of these clusters continues to grow until at (h) the instability is relieved abruptly through the formation of the gel phase. At the same time the MPD of the sol relaxes to its theoretical value. In the simulation, the sol/gel transition is observed at $N = 46$, slightly delayed compared to theory which places the gel point at $N^* = 48.44$.

In the unstable region the deviation between the distribution obtained by mean distribution obtained simulation and the theoretical MPD is manifested as excess mass accumulated at the tail of the distribution. This mass may be calculated as

$$\delta M = \sum_{i=1}^{i^{\text{sol}}} i \left(\langle n_i \rangle - \tilde{n}_i \right) \tag{5.78}$$

with the summation going up to the maximum cluster size i^{sol} in the sol region,

$$i^* = \frac{M - N + 2}{2}. \tag{5.79}$$

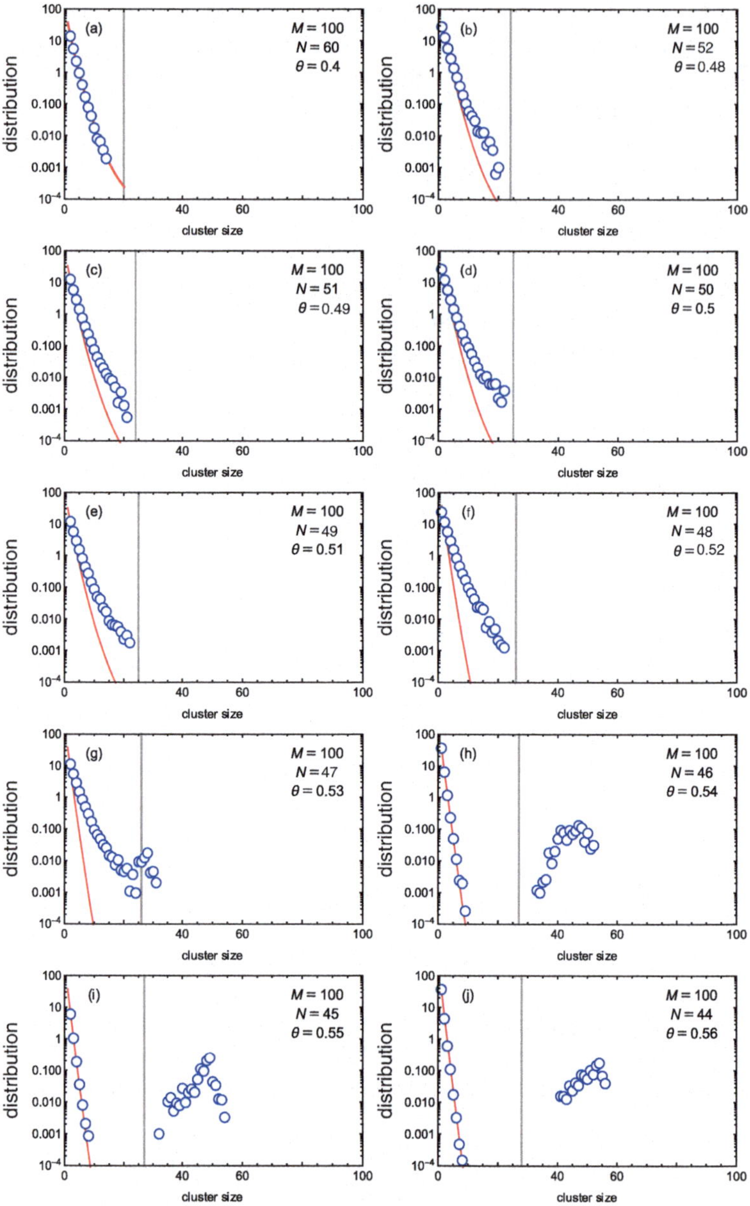

Fig. 5.16 Evolution of the cluster distribution as it passes through the unstable region. Lines are from theory, symbols are results from Monte Carlo simulation. The unstable region is $59 > N > 48.44$, or $0.41 < \theta < 0.5156$. The system enters the unstable region in (**b**) and exits in (**h**). The calculations are for $M = 100$, $\gamma = 0.02$

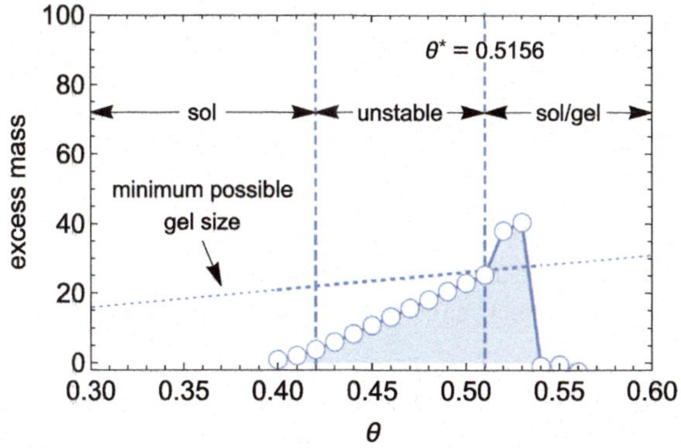

Fig. 5.17 Excess mass in sol relative to the equilibrium MPD ($M = 100$, $\gamma = 0.02$). The dotted line shows the minimum possible gel size as a function of θ. As long as the excess mass is less than the minimum gel cluster, the system is unable to produce a gel phase and remains as an unstable sol. The instability is relieved when the excess mass reaches the minimum gel size at θ^*. (In the simulation the formation of the gel is delayed somewhat until $\theta \approx 0.54$)

At $N = 50$ we find $\delta M = 22.86$. On the other hand, the minimum possible gel cluster at $M = 100$, $N = 50$, is

$$\frac{M - N + 2}{2} = 26.$$

This, however, is more than the available excess mass in the sol. The excess mass is not sufficient to nucleate the gel phase and as a result the state remains in the sol phase. Upon decreasing N, more excess mass accumulates until enough is available to nucleate the gel phase. Figure 5.17 shows the growth of the excess mass as the system approaches the gel point at θ^*. If the excess mass, shown by the symbols, is less than the minimum size of the gel cluster at given θ the system remains as a single phase sol.[10] At θ^* the accumulated excess mass is just enough to form a gel and opens up a pathway to relieve the instability. Once the gel forms the excess mass drops to zero as the sol distribution relaxes to its equilibrium MPD. The excess mass grows smoothly up to the gel point, where it drops to zero, while an almost equal[11] fraction of mass appears at the same point as ϕ_{gel}. The abrupt jump in ϕ_{gel} can now be seen to correspond to the passage of the excess mass from the sol region to the gel.

[10]The minimum gel size is

$$i_{\text{gel}} = \frac{M - N + 2}{2} = \frac{M\theta}{2} + \frac{2}{M}.$$

This is plotted as dotted line in Fig. 5.17.

[11]There is a discontinuity between the fraction of the excess mass right before the gel point and the gel fraction right after it, though of smaller magnitude than the discontinuity in ϕ_{gel}. It arises from the fact that β and q of the equilibrium sol vary abruptly in the region marked CD in Fig. 5.15.

Note 5.7 (On the Excess Mass)
We write Eq. (5.78) as

$$\delta M = \sum_{i=1}^{i^{\text{sol}}} i \langle n_i \rangle - \sum_{i=1}^{i^{\text{sol}}} i \tilde{n}_i .$$

Both summations cover the sol region. Since the actual distribution in the simulation is entirely contained in the sol range (there is no gel cluster), the first summation is the total mass M. Then

$$\delta M = M - \sum_{i=1}^{i^{\text{sol}}} i \tilde{n}_i , \tag{5.80}$$

which expresses the excess mass entirely in terms of the equilibrium MPD. For a stable sol the excess mass goes to zero. Accordingly, the condition

$$\sum_{i=1}^{i^{\text{sol}}} i \tilde{n}_i \to M, \tag{5.81}$$

is an equivalent statement of stability of the sol phase and states that the mass of the MPD is contained entirely in the sol region.

★

5.7.2 Scaling Limit

From a mathematical standpoint the instability is caused by the presence of the term γi^2 in the exponent of the MPD. The logarithm of the MPD is a parabola with a minimum at cluster size $\beta/2\gamma$. This is the same as the size of the gel cluster at the gel point in Eq. (5.74):

$$m^*_{\text{gel}} = M \phi^*_{\text{gel}} = \frac{\beta^*}{2\gamma} . \tag{5.82}$$

The formation of the gel phase is how a finite population responds when the single-phase MPD extends past the minimum of the parabola: instead of forming a distribution with a diverging tail, it forms two distributions, one in the sol region that is properly decaying, and one in the gel at a single cluster such that mass balance is satisfied. In the scaling limit, $M \to \infty$, Eq. (5.82) gives $m^*_{\text{gel}} \to \infty$, $\phi^*_{\text{gel}} \to 0$, $\beta^* \to \infty$, and the corresponding MPD of the sol is a delta function at $\bar{x}_{\text{sol}} = 1$. The transition to the scaling limit is shown in Fig. 5.18. As M increases the gel point

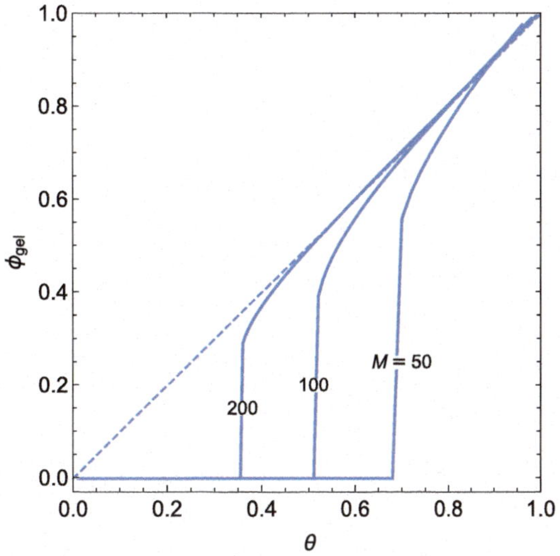

Fig. 5.18 Approach to scaling limit in the i^2 model with $\gamma = 0.02$. When $M \to \infty$ the sol-gel transition point approaches $\theta = 0$

moves to the right and the magnitude of the discontinuity in ϕ^*_{gel} decreases. In the scaling limit the discontinuity disappears and the gel phase is present at all θ and the gel fraction is given by the diagonal $\phi_{gel} \to \theta$, which also gives the asymptotic gel fraction in finite populations in the limit $\theta \to 1$. In both limits, $M \to \infty$ and $\theta \to 1$, the population forms a fully segregated state that contains one gel particle at size $m_{gel} = M - N + 1$, and a sol population that entirely consists of monomers. In the scaling limit this is represented by two delta functions, one at $x = 1$, and one at $x = \infty$.

The i^2 model exhibits several features that are specific to that model, however two conclusions may be drawn that are more general than the specific example: First, that abrupt discontinuous changes of the gel fraction at the gel point appear to require extreme conditions in order to appear; and second, the theory of the cluster ensemble is capable of describing discontinuous responses to differential changes in large finite systems, even though such discontinuities evaporate when the size of the system becomes infinite.

Chapter 6
The Bicomponent Ensemble

The simplest cluster ensemble is formed by partitioning an extensive variable M, which we have called "mass," into N clusters. In Chap. 3 we generalized this to the partitioning of a set of extensive variables, $X_1, X_2, \cdots$ into N clusters. X_i may represent energy, volume, or any other extensive attribute that is distributed. A special case is when this attribute refers to a *distinct species* that we recognize as a *component*. A population that contains two or more components forms a *mixture* and its behavior is quite different from that of the generic multivariate ensemble. All extensive properties of a multicomponent population may be sub-partitioned with respect to components. In this chapter we formulate the bicomponent cluster ensemble, derive its thermodynamics, and study the mixing of components for certain classes of selection functionals.

6.1 Bicomponent Cluster Ensemble

The simplest multicomponent cluster ensemble consists of one extensive attribute, *mass* M, distributed over fixed number N of clusters of two components, A and B, under fixed total mass of each component. This ensemble is defined by the extensive set

$$M_A,\ M_B\ , N$$

where M_A and M_B are the total masses of components A and B. The total mass in the ensemble,

$$M = M_A + M_B, \tag{6.1}$$

T. Matsoukas, *Generalized Statistical Thermodynamics*,
Understanding Complex Systems, https://doi.org/10.1007/978-3-030-04149-6_6

is also fixed but is not an independent variable. We must have $M_A + M_B \geq N$ because clusters must contain at least one unit of mass. The mean cluster size is

$$\bar{x} = \frac{M_A + M_B}{N}. \tag{6.2}$$

The mean amounts of A and B per cluster are

$$\bar{a} = \frac{M_A}{N} = \bar{x}\phi_A, \quad \bar{b} = \frac{M_B}{N} = \bar{x}\phi_B \tag{6.3}$$

where ϕ_A and ϕ_B are the mass fractions of components in the population,

$$\phi_A = \frac{M_A}{M_A + M_B}, \quad \phi_B = \frac{M_B}{M_A + M_B}. \tag{6.4}$$

The mean amounts of A and B per cluster satisfy the balance equation

$$\bar{a} + \bar{b} = \bar{x}. \tag{6.5}$$

These results are an extension of the one component cluster ensemble (M, N) to one in which mass is sub-partitioned with respect to components A and B.

6.1.1 Cluster and Configuration

The bicomponent cluster is an ordered list of A's and B's. It is characterized by the number a of A's and the number b of B's it contains, and by its total mass (size) $k = a + b$. The clusters below ($\bullet = A$, $\circ = B$)

$$(\circ\,\circ\,\bullet\,\bullet\,\circ), \quad (\bullet\,\circ\,\circ\,\circ\,\bullet), \quad (\circ\,\circ\,\circ\,\circ\,\circ),$$

are examples of distinguishable clusters with size $k = 5$. A bicomponent *configuration* is an ordered list of clusters, as in

$$\mathbf{m} = \Big((\circ\bullet), (\circ\,\circ\,\bullet\,\bullet\,\circ), (\bullet\circ), (\bullet)\Big) \equiv (m_1, m_2, m_3, m_4). \tag{6.6}$$

A configuration is distinguished by the order of the clusters it contains and by the order of components within a cluster. Therefore, the following configurations are distinctly different from each other:

$$\Big((\circ\bullet),(\circ\circ\bullet\bullet\circ),(\bullet\circ),(\bullet)\Big)$$

$$\Big((\circ\bullet),(\circ\circ\bullet\bullet\circ),(\bullet),(\bullet\circ)\Big)$$

$$\Big((\bullet\circ),(\circ\bullet\circ\bullet\circ),(\bullet),(\bullet\circ)\Big)$$

The order of species in the cluster and of clusters in the configuration are unobservable internal variables that determine the proper enumeration of states. The observable is *composition*, the number of A's and B's independently of the order in which they appear.

6.1.2 Bicomponent Distribution

The bicomponent distribution $\mathbf{n}_{A,B}$ is a two-dimensional vector,

$$\mathbf{n}_{A,B} = \begin{pmatrix} n_{0,0} & n_{0,1} & \cdots \\ n_{1,0} & n_{1,1} & \cdots \\ \vdots & \vdots & \vdots \end{pmatrix} \tag{6.7}$$

whose element $n_{a,b}$ is the number of clusters that contain a units of A and b units of B regardless of the order of the units inside the cluster. This distribution satisfies the conditions,

$$\sum_{a=0}^{\infty}\sum_{b=0}^{\infty} n_{a,b} = N, \tag{6.8}$$

$$\sum_{a=0}^{\infty}\sum_{b=0}^{\infty} a n_{a,b} = M_A, \tag{6.9}$$

$$\sum_{a=0}^{\infty}\sum_{b=0}^{\infty} b n_{a,b} = M_B, \tag{6.10}$$

Either a or b may be zero but the case $a = b = 0$ is not meaningful because a cluster must contain at least one unit of mass. We set $n_{0,0} = 0$ and allow both a and b to range from 0 to ∞. As an example, the distribution for the configuration shown in Eq. (6.6) is

$$n_{1,0} = 1, n_{1,1} = 2, n_{2,3} = 1,$$

with all other elements equal to zero. By definition the bicomponent distribution is oblivious to the internal order of components in the cluster. Accordingly, clusters

(∘•) and (•∘) are both counted as "clusters with one A and one B." The primary moments of the distribution are

$$\begin{aligned} N &= n_{1,0} + n_{1,1} + n_{2,3} = 4 \\ M_A &= (1)n_{1,0} + (1)n_{1,1} + (2)n_{2,3} = 5 \\ M_B &= (0)n_{1,0} + (1)n_{1,1} + (3)n_{2,3} = 5. \end{aligned}$$

and are obtained from Eqs. (6.8) to (6.10).

6.1.3 Color-Blind Size Distribution

The *color-blind size distribution* (or simply *size distribution*) is a vector $\mathbf{n} = (n_1, n_2, \cdots)$, whose element n_k is the number of clusters with mass $a + b = k$:

$$n_k = \sum_{a=0}^{k} n_{a,k-a}. \tag{6.11}$$

It satisfies the conditions

$$\sum_{k=1}^{\infty} n_k = N, \tag{6.12}$$

$$\sum_{k=1}^{\infty} k n_k = M_A + M_B, \tag{6.13}$$

which are the familiar constraints of the one-component ensemble with the same total mass and number of clusters as the bicomponent ensemble. The color-blind distribution is the same as that of this associated one-component population and for this reason we use the same symbol n_k to notate it. In vector form the color blind distribution will be notated as $\mathbf{n}_{A+B}$, i.e.,

$$\mathbf{n}_{A+B} = (n_1, n_2, \cdots) \tag{6.14}$$

As an example, the color-blind distribution for the configuration in Eq. (6.6) is

$$n_1 = 1, n_2 = 2, n_5 = 1,$$

with all other elements equal to zero. We then have

$$N = n_1 + n_2 + n_5 = 4,$$

$$M_A + M_B = (1)n_1 + (2)n_2 + (5)n_5 = 10.$$

These are in agreement with the results obtained from $n_{a,b}$.

6.1.4 Sieve-Cut Distribution

Distribution $\mathbf{n}_{A,B}$ treats both components symmetrically: $n_{a,b}$ is the number of clusters with a units of A and b units of B, and both a and b range from 0 to ∞. It is convenient, however, to express the distribution in a form that is not symmetric with respect to components. We take A to be the primary component and express the distribution as a function of the mass of A in the cluster, and of the total cluster mass $k = a + b$. We represent this distribution by the two-dimensional vector

$$\mathbf{n}_{A|k} = (n_{0|k}, n_{1|k}, \cdots n_{k|k})$$

where $n_{a|k}$ is the number of clusters of size k that contain a units of A ($a = 0 \cdots k$). Its relationship to distribution $\mathbf{n}_{A,B}$ is given by

$$n_{a|k} = n_{a,k-a}. \tag{6.15}$$

The sieve-cut distribution satisfies the conditions

$$\sum_{a=0}^{k} n_{a|k} = n_k, \tag{6.16}$$

$$\sum_{a=0}^{k} a n_{a|k} = n_k \bar{a}, \tag{6.17}$$

where n_k is the number of clusters with size k and $\bar{a}_k$ is the mean number of A within clusters of size k. Summing over all k we obtain the global conditions

$$\sum_{k=1}^{\infty} \sum_{a=0}^{k} n_{a|k} = N, \tag{6.18}$$

$$\sum_{k=1}^{\infty} \sum_{a=0}^{k} a n_{a|k} = M_A, \tag{6.19}$$

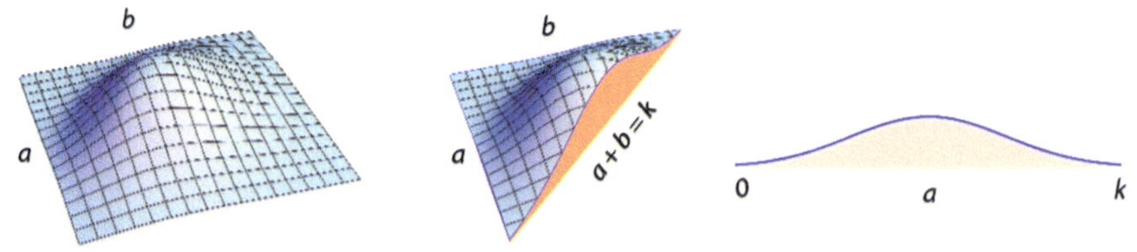

Fig. 6.1 Views of a two-component distribution. (Top) Bicomponent distribution on the (a, b) plane. (Middle) Intersection at $a = b = k$. (Bottom) Sieve-cut distribution at fixed $a + b = k$

$$\sum_{k=1}^{\infty}\sum_{a=0}^{k}(k-a)n_{a|k} = M_B, \tag{6.20}$$

The term *sieve cut* alludes to the extraction of size cut k from the distribution $\mathbf{n}$. We will also refer to $\mathbf{n}_{A|k}$ as the *compositional* distribution because it gives the distribution of components within fixed size k (Fig. 6.1).

As an example, the sieve-cut distributions corresponding to the distribution in Eq. (6.6) are

$$\begin{aligned}
\mathbf{n}_{A|1} &= (n_{0|1}, n_{1|1}) = (0, 1)\\
\mathbf{n}_{A|2} &= (n_{0|2}, n_{1|2}, n_{2|2}) = (0, 2, 0)\\
\mathbf{n}_{A|5} &= (n_{0|5}, n_{1|5}, n_{2|5}, n_{3|5}, n_{4|5}, n_{5|5}) = (0, 0, 1, 0, 0)
\end{aligned}$$

and all other $\mathbf{n}_{A|k}$ are zero. The sum of elements in each $\mathbf{n}_{A|k}$ is equal to n_k and we may further confirm that Eqs. (6.18)–(6.20) are satisfied.

6.1.5 *Probability Distributions*

To appreciate the usefulness of expressing the distribution in terms of sieve cuts we will obtain its relationship to the bicomponent distribution. We start with $n_{a,k-a} = n_{a|k}$ in Eq. (6.15) and express it in the form

$$\frac{n_{a,k-a}}{N} = \left(\frac{n_k}{N}\right)\left(\frac{n_{a|k}}{n_k}\right). \tag{6.21}$$

The left-hand side is the probability within distribution $\mathbf{n}_{A,B}$ to find a cluster with a units of A and b units of B; on the right-hand side we have the product of the probability to find a cluster of size k, and the probability to find a units of A in a cluster given that the size of the cluster is k. Equation (6.21) may be expressed as the joint probability to find a cluster with a units of A and $k - a$ units of b,

$$p_{a,k-a} = p_k \, p_{k|a}, \tag{6.22}$$

where

$$p_{a,b} = \frac{n_{a,b}}{N}, \tag{6.23}$$

$$p_k = \frac{n_k}{N}, \tag{6.24}$$

$$p_{a|k} = \frac{n_{a|k}}{n_k}. \tag{6.25}$$

It should be understood that these probabilities refer to the specific distribution $\mathbf{n}_{A,B}$ from which they were derived. To avoid cluttering the notation we will not indicate $\mathbf{n}$ explicitly in the notation of these probabilities.

Note 6.1 (On Nomenclature)

We have introduced several different distributions, all of which are represented by the symbol $\mathbf{n}$. We summarize them below:

$\mathbf{n}_{A,B}$: Notated with two subscripts, $\mathbf{n}$ refers to the bicomponent distribution whose element $n_{a,b}$ is the number of clusters with a units of A and b units of B.

$\mathbf{n}_{A+B}$: Notated with one subscript, $\mathbf{n}$ refers to the color-blind distribution $(n_1, n_2 \cdots)$ whose element n_k is the number of clusters with size k.

$\mathbf{n}_{A|k}$: Notated with a conditional expression as subscript, $\mathbf{n}$ refers to the sieve-cut distribution at size k whose element $n_{a|k}$ is the number of clusters of size k that contain a units of A.

Probability distributions $\mathbf{p}_{A,B}$, $\mathbf{p}_{A+B}$, and $\mathbf{p}_{A|k}$ follow the same naming conventions and are scaled forms of the corresponding extensive distributions normalized to unit area:

$$\sum_{a=0}^{\infty}\sum_{a=0}^{\infty} p_{a,b} = \sum_{k=1}^{\infty} p_k = \sum_{a=0}^{\infty} p_{a|k} = 1.$$

The summations also indicate the stochastic variables in each case: $p_{a,b}$ is the distribution of a and b, p_k is the distribution of k, and $p_{a|k}$ is the distribution of a. Notice that even though $p_{a|k}$ is a function of two variables, a and k, it is a probability distribution only with respect to a.

In double summations involving $n_{a,b}$ or $p_{a,b}$ the limits run from 0 to ∞ for both indices:

$$\sum_a\sum_b = \sum_{a=0}^{\infty}\sum_{b=0}^{\infty}.$$

The unphysical case $a = b = 0$ is handled by setting $n_{0,0} = p_{0,0} = 0$. In summations involving n_k or p_k, k runs from 1 to ∞:

$$\sum_k = \sum_{k=1}^{\infty}$$

In summations involving $n_{a|k}$ or $p_{a|k}$, a runs from 1 to k, and k from 1 to ∞:

$$\sum_k \sum_a = \sum_{k=1}^{\infty} \sum_{a=0}^{k}$$

We will use the simpler notation without showing the limits, with the understanding that they are set according to the conventions described here.

★

6.2 Microcanonical Probability

The microcanonical probability of distribution $\mathbf{n}_{A,B}$ is

$$P\left(\mathbf{n}_{A,B}\right) = \mathscr{M}(\mathbf{n}_{A,B}) \frac{W(\mathbf{n}_{A,B})}{\Omega} \tag{6.26}$$

where $\mathscr{M}[\mathbf{n}_{A,B}]$ is multiplicity of $\mathbf{n}_{A,B}$, $W(\mathbf{n}_{A,B})$ is the selection functional and $\Omega = \Omega_{M_A,M_B,N}$ is the partition function. The multiplicity of distribution is the number of distinct configurations with the same distribution of components and is equal to the number of permutations of the internal coordinates in a configuration that produce the same distribution. In a bicomponent configuration there are two levels of permutations: permutations in the order of clusters, and permutations in the ordering of components within a cluster. The multiplicity of the distribution is the product of the two. The part that arises from permutations in the order of clusters is given by the multinomial coefficient of the distribution:

$$\mathbf{n}_{A,B}! = \frac{N!}{n_{0,1}! n_{0,2}! \cdots n_{1,0}! \cdots} = N! \prod_a \prod_b \frac{1}{n_{a,b}!}.$$

The multiplicity of a cluster that contains a units of A and a units of B is given by the binomial coefficient, and since there are $n_{a,b}$ clusters with the same number of A's and B's, this multiplicity is

$$\prod_a \prod_b \left(\frac{(a+b)!}{a!b!} \right)^{n_{a,b}}.$$

The multiplicity of distribution $\mathbf{n}_{A,B}$ is obtained as the product of both terms:

$$\mathscr{M}[\mathbf{n}_{A,B}] = \mathbf{n}_{A,B}! \prod_a \prod_b \left\{ \frac{(a+b)!}{a!b!} \right\}^{n_{a,b}}. \tag{6.27}$$

We may express multiplicity in terms of the sieve-cut distribution. First we rewrite the multinomial factor $\mathbf{n}_{A,B}!$ by collecting terms of fixed $a+b$:

$$\begin{aligned} \mathbf{n}_{A,B}! &= \frac{N!}{(n_{01}!n_{10}!)(n_{02}!n_{11}!n_{20}!)\cdots} \\ &= \underbrace{\left(\frac{N!}{n_1!n_2\cdots} \right)}_{\mathbf{n}_{A+B}!} \underbrace{\left(\frac{n_1!}{n_{01}!n_{10}!} \right)}_{\mathbf{n}_{A|1}!} \underbrace{\left(\frac{n_2!}{n_{02}!n_{11}!n_{20}!} \right)}_{\mathbf{n}_{A|2}!} \cdots . \end{aligned}$$

The first term on the far right is the multinomial factors of the color-blind distribution and is followed by the product of the multinomial factors of all sieve cuts. The final result is

$$\mathbf{n}_{A,B}! = \mathbf{n}_{A+B}! \mathbf{n}_{A|1}! \mathbf{n}_{A|2}! \cdots \tag{6.28}$$

Here the multinomial factor of the bicomponent distribution is decomposed into a product of multinomial factors, one for the color-blind size distribution $\mathbf{n}$ and one for every sieve-cut distribution $\mathbf{n}_{A|k}$. Substituting into Eq. (6.27) the multiplicity is expressed as:

$$\mathscr{M}[\mathbf{n}_{A,B}] = \mathbf{n}_{A+B}! \mathbf{n}_{A|1}! \mathbf{n}_{A|2}! \cdots \prod_k \prod_a \binom{k}{a}^{n_{a|k}}. \tag{6.29}$$

In this form the multiplicity shows explicitly the contribution from all degrees of freedom: ordering of color-blind clusters, ordering of bicomponent clusters within each sieve cut, and ordering of components within individual clusters.

Bicomponent Entropy

We define *bicomponent entropy* as the logarithm of the multiplicity factor of the bicomponent distribution. Using Eq. (6.27) for the multiplicity, Stirling's formula for the factorials and dropping terms that grow logarithmically with size we obtain

$$S^{\text{bi}}(\mathbf{n}_{A,B}) = -\sum_a \sum_b n_{a,b} \log \frac{n_{a,b}}{N} + N \sum_a \sum_b n_{a,b} \log \binom{a+b}{a}. \tag{6.30}$$

Dividing by N this becomes

$$\frac{S^{\text{bi}}(\mathbf{n}_{A,B})}{N} = -\sum_a \sum_b \frac{n_{a,b}}{N} \log \frac{n_{a,b}}{N} + \sum_a \sum_b \frac{n_{a,b}}{N} \log \binom{a+b}{a}. \tag{6.31}$$

Using $p_{a,b} = n_{a,b}/N$ from Eq. (6.23) we obtain the bicomponent entropy of probability distribution $\mathbf{p}_{A,B}$:

$$S^{\text{bi}}(\mathbf{p}_{A,B}) = -\sum_a \sum_b p_{a,b} \log \left\{ p_{a,b} \Big/ \binom{a+b}{a} \right\}, \tag{6.32}$$

This is the intensive bicomponent entropy $S^{\text{bi}}(\mathbf{p}_{A,B}) = S^{\text{bi}}(\mathbf{n}_{A,B})/N$. We may also express it in terms of the sieve-cut distribution. Using Eq. (6.29) for the multiplicity we have

$$S^{\text{bi}}(\mathbf{n}_{A,B}) = -\sum_k n_k \log \frac{n_k}{N} - \sum_k \sum_a n_{a|k} \log \frac{n_{a|k}}{n_k} + \sum_k \sum_a n_{a|k} \log \binom{k}{a}.$$

Finally, expressing all extensive distributions in terms of probability distributions we obtain

$$S^{\text{bi}}(\mathbf{p}_{A,B}) = -\sum_k p_k \log p_k - \sum_k p_k \sum_a p_{a|k} \log \left\{ p_{a|k} \Big/ \binom{k}{a} \right\}. \tag{6.33}$$

The result expresses entropy as a sum of the entropy of the color-blind size distribution and the entropy of all sieve cuts.

Note 6.2 (Bicomponent Entropy Versus One-Component Entropy)

We have defined bicomponent entropy as the logarithm of the natural multiplicity of distribution, and took "natural multiplicity" to include the order of components within a mixed cluster. As a result the entropy of the bicomponent population includes a binomial term that is not part of the entropy in one-component systems. Let us explore this further. The first summation on the right-hand side of Eq. (6.33) is the entropy of the color-blind probability and is given by the familiar entropy functional,

$$S(\mathbf{p}_{A+B}) = -\sum_k p_k \log p_k. \tag{6.34}$$

The inner summation on the second term of the right-hand side is the bicomponent entropy of sieve cut k:

$$S^{\mathrm{bi}}(p_{a|k}) = -\sum_a p_{a|k} \log \left\{ p_{a|k} \bigg/ \binom{k}{a} \right\}. \tag{6.35}$$

Indeed, if $\mathbf{p}_{A+B}$ is a Kronecker delta at $a+b=k$, Eq. (6.33) reduces to (6.35). This entropy is not the same as the entropy functional in the one-component ensemble because of the binomial term. But the two are closely related. We expand the logarithm in (6.35) to write

$$S^{\mathrm{bi}}(p_{a|k})] = -\sum_a p_{a|k} \log p_{a|k} + \sum_a p_{a|k} \log \binom{k}{a} \tag{6.36}$$

$$= S(\mathbf{p}_{A,B}) + \log W^{\mathrm{bi}}(\mathbf{p}_{A|k}), \tag{6.37}$$

where $\log W^{\mathrm{bi}}$ is the linear functional

$$\log W^{\mathrm{bi}}(\mathbf{p}_{A|k}) = \sum_{a=0}^{k} p_{a|k} \log \binom{k}{a}. \tag{6.38}$$

Substituting into Eq. (6.33) we obtain

$$S^{\mathrm{bi}}(\mathbf{p}_{A,B}) = S(\mathbf{p}_{A+B}) + \sum_k p_k S(\mathbf{p}_{a|k}) + \log W^{\mathrm{bi}}(\mathbf{p}_{A,B}). \tag{6.39}$$

Here $S^{\mathrm{bi}}(\mathbf{p}_{A,B})$ is the bicomponent entropy, $S(\mathbf{p}_{A,B})$ is the one-component entropy functional applied to the color-blind distribution, $S(\mathbf{p}_{a|k})$ is the one-component entropy applied to sieve cut k, and $\log W^{\text{bi}}[\mathbf{p}_{A,B}]$ is the bicomponent functional applied to $\mathbf{p}_{A,B}$,

$$\log W^{\text{bi}}(\mathbf{p}_{A,B}) = \sum_a \sum_b p_{a,k-a} \log \binom{a+b}{a} = \sum_k p_k \sum_a p_{a|k} \log \binom{k}{a}. \tag{6.40}$$

The microcanonical weight of probability distribution $\mathbf{p}_{A,B}$ is

$$\begin{aligned} & S^{\text{bi}}(\mathbf{p}_{A,B}) + \log W(\mathbf{p}_{A,B}) \\ &= \left\{ S(\mathbf{p}_{A+B}) + \sum_k p_k S(\mathbf{p}_{a|k}) \right\} + \left\{ \log W^{\text{bi}}(\mathbf{p}_{A,B}) + \log W(\mathbf{p}_{A,B}) \right\} \\ &= S(\mathbf{p}_{A,B}) + \left\{ \log W^{\text{bi}}(\mathbf{p}_{A,B}) + \log W(\mathbf{p}_{A,B}) \right\}. \end{aligned}$$

The right-hand side is expressed in terms of one-component entropies. If we incorporate W^{bi} into the selection functional by defining

$$W'(\mathbf{p}_{A,B}) = W(\mathbf{p}_{A,B})\, W^{\text{bi}}(\mathbf{p}_{A,B}),$$

we obtain an equivalent description of the ensemble that does not require the notion of bicomponent entropy and applies the one-component entropy functional to all distributions regardless of number of components. Mathematically the two formulations are equivalent but the notion of the *unbiased* selection functional is different in each case. In the formulation that uses one-component entropies, unbiased means $W' = 1$, which amounts to ignoring the ordering of components within a cluster. If that ordering must be included, then the ensemble is *biased*. This convention is appropriate for attributes other than components. In the formulation that utilizes the notion of *bicomponent entropy*, the unbiased ensemble accounts for the order of components in the cluster. Here *unbiased* means that the multiplicity of inter-cluster ordering is equal to the number of permutations. As we will find out, this convention is indeed the proper one for the attribute we call *component*. The two conventions represent the adoption of different reference states.[1] When we use *bicomponent entropy*, the reference state, which is represented by the condition $W = 1$, is the *random mixture*.

---

[1]The reference state is a rule that allows us to factor the microcanonical weight into two factors, one that we call *multiplicity*, and one that we call *selection functional*. When we use the term "natural multiplicity," we are in fact referring to the multiplicity given the chosen reference state. "Natural" in this context should not be taken to imply a universal quality.

6.3 Selection Functional

The selection bias is a functional of $\mathbf{n}_{A,B}$ that satisfies the homogeneity condition,

$$\log W(\mathbf{n}_{A,B}) = \sum_a \sum_b n_{a,b} \log w_{ab;\mathbf{n}_{A,B}}, \tag{6.41}$$

where $\log w_{a,b}$ is the cluster bias, defined as

$$\log w_{ab;\mathbf{n}_{A,B}} = \frac{\partial \log W(\mathbf{n}_{A,B})}{\partial n_{a,b}}. \tag{6.42}$$

We may classify selection functionals according to their special properties. Three major types are given below.

- *Linear functional* The cluster bias generally depends on the distribution $\mathbf{n}_{A,B}$ to which the functional is applied. In the special case that $w_{a,b}$ is independent of distribution (i.e., it is an intrinsic function of a and b) the selection functional is linear. The cluster bias of linear functional is of the form

$$w_{a,b;\mathbf{n}_{A,B}} = c_{a,b} \tag{6.43}$$

where $c_{a,b}$ is some function of a and b. The corresponding selection bias is

$$W(\mathbf{n}_{A,B}) = \prod_a \prod_b \left(c_{a,b}\right)^{n_{a,b}} \tag{6.44}$$

and this is the general form of linear bias.

- *Composition-independent (color blind) functional* If the selection functional reduces to a functional on the color-blind cluster size distribution, it becomes independent of composition. In this case all distributions with the same cluster distribution have the same bias regardless of how components distribute among clusters. The *unbiased functional*, $W(\mathbf{n}_{A,B}) = 1$, is an example of a linear functional that is also independent of composition.
- *Separable homogeneous functional* This class contains selection functionals whose cluster function can be factored into two terms, one that depends on size only and one that depends on composition, i.e., it is of the form

$$w_{a,k-a} = k\, w'_{a+b}\, w''\left(\frac{a}{k}, \frac{k-a}{k}\right),$$

where w'' is a function of the mass fractions $\xi_A = a/k$ and $\xi_B = 1 - \xi_A$ of the two components in the cluster.

This classification does not encompass all possible bicomponent functionals but includes several important and fairly general classes of functionals that we will discuss in more detail.

6.4 The Bicomponent MPD

The microcanonical bicomponent ensemble (M_A, M_B, N) consists of all distributions $\mathbf{n}_{AB}$ that satisfy the conditions:

$$\sum_{a=0}^{\infty}\sum_{b=0}^{\infty} n_{a,b} = N, \qquad [6.8]$$

$$\sum_{a=0}^{\infty}\sum_{b=0}^{\infty} a n_{a,b} = M_A, \qquad [6.9]$$

$$\sum_{a=0}^{\infty}\sum_{b=0}^{\infty} b n_{a,b} = M_B. \qquad [6.10]$$

The probability of distribution $\mathbf{n}_{A,B}$ is the ratio of the microcanonical weight over the partition function:

$$P(\mathbf{n}_{A,B}) = \frac{(\mu C \text{ weight})}{\Omega_{A,B,N}}, \qquad (6.45)$$

and the microcanonical weight of distribution $\mathbf{n}_{A,B}$ is

$$(\mu C \text{ weight}) = \mathbf{n}_{A,B}! \prod_a \prod_b \binom{a+b}{a}^{n_{a,b}} W(\mathbf{n}_{A,B}).$$

The most probable distribution maximizes the microcanonical weight under the constraints that define the ensemble, Eqs. (6.8)–(6.10). Equivalently, we maximize the log of the microcanonical weight, which we write in terms of the probability distribution $p_{a,b}$ as follows:

$$\frac{\log(\mu C \text{ weight})}{N} = -\sum_a \sum_b p_{a,b} \log\left\{ p_{a,b} \Big/ \binom{a+b}{a} \right\} + \sum_a \sum_b p_{a,b} \log w_{a,b}.$$

The most probable distribution is obtained by standard Lagrange maximization and the result is easily found to be

$$\tilde{p}_{a,b} \equiv \frac{\tilde{n}_{a,b}}{N} = w_{a,b}\binom{a+b}{a}\frac{e^{-\beta_A a-\beta_B b}}{q}, \tag{6.46}$$

where q, β_A, and β_B are Lagrange multipliers associated with the constraints in Eqs. (6.8)–(6.10). By back substitution of the MPD into the constraints we obtain the following relationships for the parameters q, β_A, and β_B:

$$\frac{\partial \log q}{\partial \beta_A} = -\frac{M_A}{N} = -\bar{x}\phi_A, \tag{6.47}$$

$$\frac{\partial \log q}{\partial \beta_B} = -\frac{M_B}{N} = -\bar{x}\phi_B. \tag{6.48}$$

The bicomponent entropy of the MPD is

$$S(\tilde{\mathbf{n}}_{A,B}) = -N\sum_{a,b}\tilde{p}_{a,b}\log\left\{\tilde{p}_{a,b}\Big/\binom{a+b}{a}\right\} = -\log\tilde{W}+\beta_A M_A+\beta_B M_B+Nq, \tag{6.49}$$

where $\tilde{W}$ is the selection functional evaluated at the MPD. Since $\tilde{S}+\tilde{W}\to\log\Omega$, the final result is

$$\log\Omega_{A,B,N} = \beta_A M_A + \beta_B M_B + (\log q)N. \tag{6.50}$$

By the homogeneous property of the ensemble we then have

$$\beta_A = \left(\frac{\partial \log \Omega_{A,B,N}}{\partial M_A}\right)_{M_B,N}, \tag{6.51}$$

$$\beta_B = \left(\frac{\partial \log \Omega_{A,B,N}}{\partial M_B}\right)_{M_A,N}, \tag{6.52}$$

$$q = \left(\frac{\partial \log \Omega_{A,B,N}}{\partial N}\right)_{M_A,M_B}. \tag{6.53}$$

By combining Eq. (6.49) with (6.50) we obtain

$$\log\Omega = S^{\mathrm{bi}}(\mathbf{n}_{A,B}) + \log W(\mathbf{n}_{A,B}). \tag{6.54}$$

Thus we have recovered the canonical relationships of the multivariate ensemble derived in Chap. 3.

6.4.1 Special Case: Unbiased Ensemble

As an immediate application of the above results we obtain the solution to the unbiased ensemble. With $w_{a,b} = 1$ Eq. (6.46) becomes

$$\tilde{p}_{a,b} = \binom{a+b}{a} \frac{e^{-\beta_A a - \beta_B b}}{q}. \tag{6.55}$$

To determine the canonical parameters β_A, β_B, and q we first write the MPD as

$$\tilde{p}_{a,k-a} = \binom{k}{a} c_0 c_1^a c_2^{k-a}, \tag{6.56}$$

with $c_0 = 1/q$, $c_1 = e^{-\beta_A}$, $c_2 = e^{-\beta_B}$, and express the constraints in the form

$$1 = \sum_{k=1}^{\infty} \sum_{a=0}^{k} \binom{k}{a} c_0 c_1^a c_2^{k-a} = \frac{c_0(c_1 + c_2)}{1 - c_1 - c_2}, \tag{6.57}$$

$$\bar{x}\phi_A = \sum_{k=1}^{\infty} \sum_{a=0}^{k} a \binom{k}{a} c_0 c_1^a c_2^{k-a} = \frac{c_0 c_1}{(1 - c_1 + c_2)^2}, \tag{6.58}$$

$$\bar{x}\phi_B = \sum_{k=1}^{\infty} \sum_{a=0}^{k} (k-a) \binom{k}{a} c_0 c_1^a c_2^{k-a} = \frac{c_0 c_2}{(1 - c_1 + c_2)^2}. \tag{6.59}$$

Solving for c_0, c_1, and c_2 we obtain

$$c_0 = \frac{1}{\bar{x} - 1}, \quad c_1 = \frac{\bar{x} - 1}{\bar{x}} \phi_A, \quad c_2 = \frac{\bar{x} - 1}{\bar{x}} \phi_B. \tag{6.60}$$

Substituting into Eqs. (6.51)–(6.53) we obtain the canonical parameters β_A, β_B, and q,

$$q = \bar{x} - 1, \tag{6.61}$$

$$\beta_A = \log \frac{\bar{x}}{(\bar{x} - 1)\phi_A}, \tag{6.62}$$

$$\beta_B = \log \frac{\bar{x}}{(\bar{x} - 1)\phi_B}, \tag{6.63}$$

and finally the bicomponent MPD:

$$\tilde{p}_{a,k-a} = \frac{1}{\bar{x}-1}\left(\frac{\bar{x}}{\bar{x}-1}\right)^k \binom{k}{a} \phi_A^a \phi_B^{k-a}. \tag{6.64}$$

When $\bar{x} \gg 1$ the canonical variables take the simpler forms,

$$q \to \bar{x} \tag{6.65}$$

$$\beta_A \to \frac{1}{\bar{x}} - \log \phi_A, \tag{6.66}$$

$$\beta_B \to \frac{1}{\bar{x}} - \log \phi_B, \tag{6.67}$$

and the MPD goes over to

$$\tilde{p}_{a,k-a} = \underbrace{\frac{e^{-k/\bar{x}}}{\bar{x}}}_{\tilde{p}_k} \cdot \underbrace{\binom{k}{a} \phi_A^a \phi_B^{k-a}}_{\tilde{p}_{a|k}}. \tag{6.68}$$

In this form the MPD is the product of an exponential distribution in the color-blind size k and a binomial distribution in A and B in sieve cut k. The exponential distribution is the unbiased MPD of the one-component problem. Since the left-hand side of this product is the MPD of the bicomponent population, the binomial distribution must also be an MPD of some kind that maximizes its own microcanonical functional. In Sect. 6.5 we will see that this is indeed the case and we will identify the functional that is maximized by the binomial distribution.

6.5 The Sieve-Cut Ensemble

Consider the following special case of bicomponent ensemble: M_A particles of type A and M_B particles of type B are distributed over N_k clusters of fixed mass k. We may view this ensemble to be a sieve cut from a larger bicomponent ensemble that contains $N_{k'}$ clusters of size $k' = 1, 2, \cdots$. By mass balance we have the following relationships between M_A, M_B, and n_k:

$$kN_k = M_A + M_B, \tag{6.69}$$

$$\bar{a}_k N_k = M_A, \tag{6.70}$$

$$\bar{b}_k N_k = M_B, \tag{6.71}$$

where $\bar{a}_k$ and $\bar{b}_k = k - \bar{a}_k$ are the mean number of A's and B's, respectively, in clusters of size k. We will calculate the MPD in the unbiased case $W = 1$. The ensemble is defined by the variables (M_A, N_k, k). The mass of component B does not need to be specified because it can be obtained as $M_B = kN_k - M_A$. All distributions $p_{a|k}$ in the ensemble satisfy the conditions

$$\sum_a p_{a|k} = 1 \tag{6.72}$$

$$\sum_a a p_{a|k} = \frac{M_A}{N_k} = \bar{a}_k. \tag{6.73}$$

We also have

$$\sum_a k p_{a|k} = k$$

and

$$\sum_a (k-a) p_{a|k} = k - \bar{a}_k,$$

but these are not independent, because they follow from (6.72) and (6.72) along with Eqs. (6.69)–(6.71). The microcanonical weight, including the Lagrange multipliers for the two constraints, is

$$-\sum_a p_{a|k} \log\left\{p_{a|k} \Big/ \binom{k}{a}\right\} - \lambda_0 \left(\sum_a p_{a|k} - 1\right) - \lambda_1 \left(\sum_a a p_{a|k} - \bar{a}_k\right) \tag{6.74}$$

and the distribution that maximizes it is

$$\tilde{p}_{a|k} = \binom{k}{a} e^{-\lambda_1 a - \lambda_0}. \tag{6.75}$$

To evaluate the Lagrange multipliers we first define a new set of parameters c_1 and c_2,

$$e^{-\lambda_1} = \frac{c_1}{1-c_1}, \quad e^{-\lambda_0} = c_0 (1-c_1)^k \tag{6.76}$$

and write the MPD in the form

$$\tilde{p}_{a|k} = c_0 \binom{k}{a} c_1^a (1-c_1)^{k-a}. \tag{6.77}$$

Substituting into the first constraint gives $c_0 = 1$, then from the second constraint we find $c_1 = \phi_A$.[2] The MPD therefore is the binomial distribution

$$\tilde{p}_{a|k} = \binom{k}{a} \phi_A^a \phi_B^{k-a}. \tag{6.78}$$

This result provides the entropic interpretation of the binomial distribution: it maximizes the bicomponent entropy functional in sieve cut k.

6.5.1 Entropy and the Partition Function

The one-component entropy of the binomial distribution follows by application of the entropy functional

$$S(\tilde{p}_{a|k}) = -\sum_a \tilde{p}_{a|k} \log \tilde{p}_{a|k} \tag{6.79}$$

and the result is

$$S(\tilde{p}_{a|k}) = -k\,(\phi_A \log \phi_A + \phi_B \log \phi_B) - \log W^{\text{bi}}(\tilde{p}_{a|k}). \tag{6.80}$$

with $\log W^{\text{bi}}(\tilde{p}_{a|k})$ from Eq. (6.38). The bicomponent entropy of the binomial distribution follows from Eq. (6.35),

$$S^{\text{bi}}(\tilde{p}_{a|k}) = -k\,(\phi_A \log \phi_A + \phi_B \log \phi_B)\,. \tag{6.81}$$

To obtain the partition function we first write the MPD in canonical form

---

[2] We make use of the identities,

$$\sum_{a=0}^{k} \binom{k}{a} c_1^a (1-c_1)^{k-a} = 1, \quad \sum_{a=0}^{k} \binom{k}{a} a c_1^a (1-c_1)^{k-a} = k c_1.$$

From the latter identity and Eqs. (6.72) and (6.69) we obtain

$$k c_1 = \frac{M_A}{N_k} = \frac{k M_A}{M_A + M_B} = k\phi_A,$$

and finally $c_1 = \phi_A$.

$$\tilde{p}_{a|k} = \binom{k}{a} \frac{e^{-\beta'' a}}{q''}, \tag{6.82}$$

with $\beta'' = \lambda_1$, $q'' = \log \lambda_0$. Comparing with the binomial distribution in Eq. (6.78) we obtain the canonical parameters β'' and q'' in terms of the mass fractions and the size of the cluster:

$$\beta'' = -\log \frac{\phi_A}{1-\phi_A} = -\log \frac{\phi_A}{\phi_B}, \tag{6.83}$$

$$q'' = (1-\phi_A)^{-k} = \phi_B^{-k}. \tag{6.84}$$

The microcanonical partition function is

$$\frac{\log \Omega''}{N_k} = \frac{1}{N_k}\left(\beta'' M_A + N_k \log q''\right) = \beta'' \bar{a}_k + \log q'', \tag{6.85}$$

where $\bar{a}_k$ is the mean amount of A per cluster. Substituting Eqs. (6.83) and (6.84) we obtain the equivalent form

$$\left(\log \omega''\right)_{\text{rnd mix}} = -k(\phi_A \log \phi_A + \phi_B \log \phi_B). \tag{6.86}$$

The right-hand side is known in solution thermodynamics as the ideal entropy of mixing and will be discussed in Sect. 6.6.

Note 6.3 (Monte Carlo Simulation of the Unbiased Sieve-Cut Ensemble)
Here we will demonstrate the unbiased sieve-cut ensemble by exchange-reaction Monte Carlo. This will help clarify concepts behind this fundamental ensemble and also prepare the ground for simulations in the presence of bias, as we will see in Sect. 6.7. The exchange reaction in the sieve-cut ensemble involves the transfer of components between two clusters while maintaining the size of the clusters unchanged. Schematically,

$$(\bullet\circ\circ\bullet) + (\circ\bullet\bullet\bullet) \rightarrow (\bullet\bullet\circ\circ) + (\bullet\circ\bullet\circ)$$

The exchange produces two new clusters whose composition and internal ordering of components are generally different from that of the reactant clusters. In the simulation each cluster is represented by a list of As and Bs with k elements. The size of the clusters k, their number N_k, and the total masses of A and B are fixed. The exchange reaction is implemented as follows: we pick two clusters at random, we combine them into a single list, randomize their order, and then divide the list into two equal parts to produce two new clusters. The new clusters replace the old ones and the process is repeated for the desired number of iterations. At each step we calculate the compositional distribution and average the results over the number of iterations. This procedure produces the mean compositional distribution, which in the

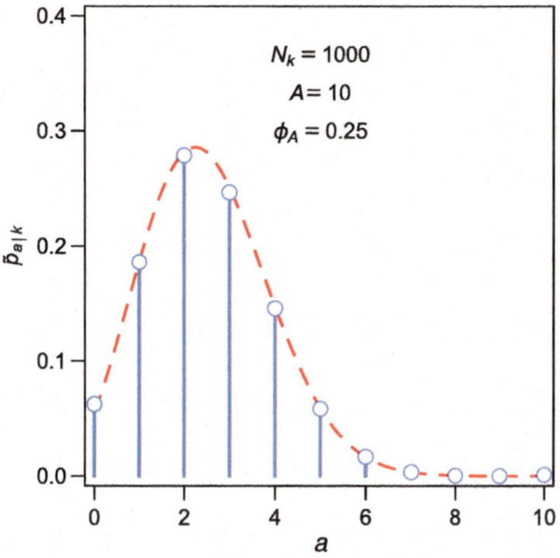

Fig. 6.2 Monte Carlo simulation of the unbiased sieve-cut ensemble with $k = 10$, $N_k = 1000$, and $\phi_A = 0.25$. Symbols with sticks are the result of Monte Carlo and the dashed line is the binomial distribution

thermodynamic limit converges to the MPD. By combining two clusters, randomizing their order, and then separating into two new clusters we simulate the random exchange of mass between clusters. By accepting the new clusters as replacement of the old pair we are simulating unbiased exchange. As Fig. 6.2 shows, the algorithm performs as intended and the distribution is in excellent agreement with the binomial distribution.The point of this demonstration is to show that the bicomponent entropy, as defined via the binomial coefficient, represents *random mixing*, which we simulate via the randomization of the merged clusters before separation.

⋆

6.6 Random Mixing

A special case of a biased ensemble is when the selection functional depends on the distribution of clusters but not on their composition. This means that cluster function $\log w_{a,b}$ is of the form

$$\log w_{a,b} = \log w'_{a+b}, \tag{6.87}$$

where w' is a function of the cluster mass $a + b$ but not of a and b individually. The selection functional is

$$\begin{aligned}
&\log W(p_{a,k-a}) \\
&\quad = \log W(p_k p_{k_{a|b}}) = \sum_k \sum_a p_k \, p_{a|k} \log w'_k \\
&\quad = \sum_k p_k \log w'_k = \log W'(p_k).
\end{aligned} \tag{6.88}$$

This result expresses the fact that the bias is color-blind: the selection bias depends entirely on the color-blind distribution and is independent of the distribution of components.

The microcanonical weight of the bicomponent distribution $p_{a,k-a}$ is

$$S^{\mathrm{bi}}(p_{a,k-a}) + \log W(p_{a,k-a}) = \sum_k p_k \left(-\log p_k + \log w'_k\right) + \sum_k p_k \sum_a p_{a|k} \left(-\log p_{a|k} + \binom{k}{a}\right), \tag{6.89}$$

to be maximized with respect to the product $p_k p_{a|k}$ under the constraints

$$\sum_k p_k = 1, \tag{6.90}$$

$$\sum_k k p_k = \frac{M_A + M_B}{N} = \bar{x}, \tag{6.91}$$

$$\sum_a p_{a|k} = 1, \quad k = 1, 2 \cdots, \tag{6.92}$$

$$\sum_k p_k \sum_a a\, p_{a|k} = \frac{M_A}{N} = \phi_A \bar{x}. \tag{6.93}$$

The functional to be maximized is of the form

$$\mathscr{A}[p_k] + \sum_k p_k \mathscr{B}[p_a|k], \tag{6.94}$$

with

$$\mathscr{A}[p_k] = \sum_k p_k \left(-\log p_k + \log W'(p'_k)\right), \tag{6.95}$$

$$\mathscr{B}[p_{a|k}] = \sum_a p_{a|k} \left(-\log p_{a|k} + \binom{k}{a}\right). \tag{6.96}$$

Since $\mathscr{A}[p_k]$ depends only on p_k and $\mathscr{B}$ only on $p_{a|k}$, Eq. (6.89) is maximized when $\mathscr{A}$ and $\mathscr{B}$ are individually at maximum with respect to their argument. The variational problem therefore is equivalent to two independent maximizations, one with respect to p_k and one with respect to $p_{a|k}$. The cluster size MPD is obtained by maximizing $\mathscr{A}[p_k]$ with respect to p_k under constraints (6.90) and (6.91). The result is the familiar canonical MPD,

$$\tilde{p}_k = w'_k \frac{e^{-\beta' k}}{q'}, \tag{6.97}$$

Next we turn to the compositional MPD but first we must address a point that at first glance would seem to complicate the maximization: $p_{a|k}$ is not a single distribution but a family of compositional distributions in sieve cuts $k = 1, 2 \cdots$. Equation (6.92), which is a set of constraints, fixes the normalization of every distribution in this family. Equation (6.93) on the other hand is a single constraint that involves a summation over the entire family of $p_{a|k}$. It might appear there are no enough constraints to solve for the set of $p_{a|k}$. This, however, is not the case. To see why, we multiply Eq. (6.91) by ϕ_A and subtract from (6.93). The result is

$$\sum_k p_k \left(-k\phi_A + \sum_a a\, p_{a|k} \right) = 0$$

and must be satisfied by all distributions p_k with mean $\bar{x}$. Since p_k is nonnegative, the quantity in parenthesis must be identically equal to zero:

$$\sum_a a\, p_{a|k} = k\phi_A, \quad k = 1, 2, \cdots. \tag{6.98}$$

This is a set of constraints that apply to each $p_{a|k}$. The compositional MPD then is obtained by maximizing $p_{a|k}$ in every sieve cut k under Eqs. (6.90), which fixes the normalization, and (6.98), which fixes the mean amount of A in all sieve cuts k. Equation (6.98) also says something else that is important: the mean amount of A in sieve cut k is equal to the expected amount $k\phi_A$, given that the overall mass fraction of A in the ensemble is ϕ_A (the same applies to component B).

We are now ready to conduct the maximization. The functional to be maximized is

$$\sum_a p_{a|k} \left(-\log p_{a|k} + \binom{k}{a} \right) - \lambda_0 \left(\sum_a p_{a|k} - 1 \right) - \lambda_1 \left(\sum_a a p_{a|k} - k\phi_A \right). \tag{6.99}$$

where λ_0 and λ_1 are Lagrange multipliers for the constraints in Eq. (6.92) and (6.98), respectively. The distribution that maximizes this functional is

$$\tilde{p}_{a|k} = \binom{k}{a} c_A^a c_B^{k-a}. \tag{6.100}$$

with $c_A = e^{-\lambda_1 - \lambda_0/k}$, $c_B = e^{-\lambda_0/k}$. Substituting this result into constraint (6.92) we obtain

$$1 = \sum_a \binom{k}{a} c_A^a c_B^{k-a} = c_A + c_B. \tag{6.101}$$

From constraint (6.98) we find

$$k\phi_A = \sum_a a\binom{k}{a} c_A^a (1-c_A)^{k-a} = k c_A \tag{6.102}$$

from which we obtain $c_A = \phi_A$. Then $c_B = 1 - \phi_A = \phi_B$ and the compositional MPD is given by the binomial distribution

$$\tilde{p}_{a|k} = \binom{k}{a} \phi_A^a \phi_B^{k-a}. \tag{6.103}$$

Finally we assemble the bicomponent MPD by forming the product of $\tilde{p}_k$ and $\tilde{p}_{a|k}$:

$$\tilde{p}_{a,k-a} = \tilde{p}_k \tilde{p}_{a|k} = \left(w'_k \frac{e^{-\beta' k}}{q'} \right) \binom{k}{a} \phi_A^a \phi_B^{k-a}. \tag{6.104}$$

The binomial distribution is the most probable compositional distribution when the selection functional is independent of composition. The precise form of the selection functional is immaterial; it affects only the color-blind distribution but not the distribution of components within clusters of fixed size. As a corollary of this result we have the following: Suppose that M_A particles of type A and M_B of type B are distributed over N clusters with fixed color-blind distribution p_k with mean $\bar{x}$, such that mass balance is satisfied,

$$N\bar{x} = M_A + M_B. \tag{6.105}$$

If the selection functional is independent of composition, then the most probable compositional distribution is the binomial. This follows from the fact the binomial distribution maximizes the microcanonical function in Eq. (6.94) for *any* p_k. In the special case that p_k is monodisperse at size $k = k^*$ we obtain the entropic interpretation of the binomial experiment: If we pick k times randomly from an infinite urn that contains A's and B's at ratio $\phi_A : \phi_B$, the binomial distribution is the most likely distribution among all possible distributions.

6.6.1 Entropy of Mixing

The entropy of the bicomponent MPD is obtained from (6.33) with $\tilde{p}_k$ for p_k and $\tilde{p}_{k|a}$ for $p_{a|k}$. Combining with Eq. (6.81) for the bicomponent entropy of the binomial distribution we find

$$S^{\mathrm{bi}}(p_{a,k-a}) = S(\tilde{p}_k) - \bar{x}(\phi_A \log \phi_A + \phi_B \log \phi_B), \tag{6.106}$$

where $\bar{x}$ is the mean cluster size,

$$\bar{x} = \sum_k k p_k = \frac{M_A + M_B}{N_A + N_B}. \tag{6.107}$$

To interpret Eq. (6.106), suppose we mix two ensembles, one of pure A and one of pure B, to form a mixed ensemble:

$$\mu C(M_A, N_A) + \mu C(M_B, N_B) \to \mu C(M_A + M_B, N_A + N_B).$$

The probabilities in all ensembles are governed by the same composition-independent functional. Additionally we require that they satisfy the condition

$$\frac{M_A}{N_A} = \frac{M_B}{N_B} = \frac{M_A + M_B}{N_A + N_B} = \bar{x}, \tag{6.108}$$

which fixes the mean cluster size to be the same in all ensembles. The selection bias is color blind and the mean cluster size is the same in all ensembles, therefore all ensembles have the same color-blind MPD, $\tilde{p}_k$. In the initial state we have two one-component ensembles whose entropy is

$$S(\mathbf{n}_A) = -N_A \sum_k \tilde{p}_k \log \tilde{p}_k,$$

$$S(\mathbf{n}_B) = -N_B \sum_k \tilde{p}_k \log \tilde{p}_k.$$

The entropy of the mixed state is given by Eq. (6.106) multiplied by the number of clusters,

$$\begin{aligned} S^{\mathrm{bi}}(\mathbf{n}_{A,B}) &= -(N_A + N_B) S^{\mathrm{bi}}(\tilde{p}_{a,k-a}) \\ &= (N_A + N_B) S(\tilde{p}_k) - (M_A + M_B)(\phi_A \log \phi_A + \phi_B \log \phi_B). \end{aligned}$$

The entropy change for this mixing process is

$$\frac{\Delta S}{M_A + M_B} = -\phi_A \log \phi_A - \phi_B \log \phi_B. \tag{6.109}$$

This is the same as the *ideal entropy of mixing* in solution thermodynamics.[3] The requirement that $\bar{x}$ be the same in the pure and mixed states is equivalent to the condition of constant temperature and pressure in the definition of mixing properties. In solution thermodynamics the ideal entropy of mixing arises either when there are no interactions between components (ideal-gas mixture), or all interactions are of identical strength (ideal solution). In the context of the cluster ensemble this is equivalent to the condition that the selection functional is oblivious to composition.

6.7 Nonrandom Mixing

We have seen that if the selection functional is independent of composition the MPD is expressed as a product of two MPDs: the MPD of the one-component problem and the compositional MPD for random mixing. The factorization of the MPD can be generalized to composition-dependent bias provided that the cluster functions can be factored into a size-dependent term and a composition-dependent term:

$$w_{a,b} = w'_{a+b}\, w''_{a,b}. \tag{6.110}$$

Moreover, the compositional term satisfies the homogeneity condition

$$\log w''(a, b) = (a + b) \log w''(\xi_A, \xi_B), \tag{6.111}$$

where $\xi_A = a/k$ and ξ_B are the mass fractions of the components in the cluster. Accordingly, the compositional term is homogeneous in a and b with degree 1, a property that is shared asymptotically by the logarithm of the binomial coefficient. The selection functional for this cluster function is

$$\begin{aligned}
\log W(p_{a,k-a}) &= \sum_k \sum_a p_k p_{a|k} \left(\log w'_k + \log w''_{a,k-a}\right) \\
&= \sum_k p_k \log w'_k + \sum_k p_k \sum_a p_{a|k} \log w''_{a,k-a} \\
&= \log W'(p_k) + \sum_k p_k \log W''(p_{a|k}).
\end{aligned}$$

[3]"Ideality" is a dated term still used in thermodynamics to refer to a convenient reference state (several such states are in common use, including ideal gas, ideal solution, and ideal solute). The preferred term is *entropy of random mixing*, which states the reference state explicitly.

Here $W'(p_k)$ depends only on the color-blind distribution p_k and $W'(p_{a|k})$ only on the compositional distribution $p_{a|k}$. We will obtain the compositional MPD by maximizing the microcanonical weight

$$S^{\text{bi}}(p_{a,k-a}) + \log W(p_{a,k-a}) = -\sum_k p_k(\log p_k - \log w'_k)$$
$$-\sum_k p_k \sum_a p_{a|k}\left(\log\left\{p_{a|k}\Big/\binom{k}{a}\right\} + \log w''_{a,k-a}\right) \quad (6.112)$$

subject to the constraints in Eqs. (6.92) and (6.98). This microcanonical weight differs from that in Eq. (6.89) only in the presence of the summation of $\log w''_{a,k-a}$ over the cluster p_k. Since $w''_{a,k-a}$ is independent of p_k, the variational problem is equivalent to two independent maximizations for precisely the same reasons as in the case of the composition-independent bias. Thus we will obtain $\tilde{p}_k$ by maximizing

$$\sum_k p_k(\log p_k - \log w'_k) \quad (6.113)$$

with respect to p_k under (6.90)–(6.91), and $\tilde{p}_{a|k}$ by maximizing

$$\sum_a p_{a|k}\left(\log\left\{p_{a|k}\Big/\binom{k}{a}\right\} + \log w''_{a,k-a}\right) \quad (6.114)$$

with respect to $\tilde{p}_{a|k}$ under the constraints in Eqs. (6.92) and (6.98). Following the same steps as in the case of composition-independent kernels, the MPD of the cluster size distribution is

$$\tilde{p}_k = w'_k \frac{e^{-\beta' k}}{q'}, \quad (6.115)$$

with canonical variables β' and q' that satisfy the familiar relationships

$$\beta' = \frac{d\log\omega'}{d\bar{x}}, \quad \frac{d\log q'}{d\beta'} = -\bar{x}, \quad \log\omega' = \beta'\bar{x} + \log q', \quad (6.116)$$

where $\omega' = \omega'(\bar{x})$ is the partition function of the one-component ensemble and $\bar{x} = (M_A + M_B)/N$ is the mean cluster size. The compositional MPD is

$$\tilde{p}_{a|k} = w''_{a,k-a}\binom{k}{a} e^{-\lambda_0 - a\lambda_1}, \quad (6.117)$$

where λ_0 and λ_1 are Lagrange multipliers for the constraints in Eqs. (6.92) and (6.98) and may depend on k. We define the new set of constants c_A and c_B,

$$c_A = \frac{e^{-\lambda_1+\lambda_0/k}}{(1+e^{-\lambda_i})^2}, \quad c_B = \frac{e^{\lambda_0/k}}{(1+e^{-\lambda_i})^2}, \tag{6.118}$$

and express the compositional MPD as

$$\tilde{p}_{a|k} = w''_{a,k-a} \binom{k}{a} c_A^a c_B^{k-a}. \tag{6.119}$$

The homogeneous property of $\log w''$, Eq. (6.111), allows us to express it as

$$\log w''(a, b) = a \log u_A + b \log u_B, \tag{6.120}$$

where $\log u_A$ and $\log u_B$ are the partial derivatives

$$\log u_A = \frac{\partial \log w''_{a,b}}{\partial a}, \quad \log u_B = \frac{\partial \log w''_{a,b}}{\partial b}, \tag{6.121}$$

and are both intensive functions (homogeneous in a and b with degree 0). We may now express the distribution in Eq. (6.119) in the equivalent form

$$\tilde{p}_{a|k} = \binom{k}{a} (\gamma_A \phi_A)^a (\gamma_B \phi_B)^{k-a}, \tag{6.122}$$

where γ_A and γ_B are defined as

$$\gamma_A = \frac{c_A u_A}{\phi_A}, \quad \gamma_B = \frac{c_B u_B}{\phi_B}. \tag{6.123}$$

To appreciate the meaning of the factors γ_A and γ_B, we write Eq. (6.122) in the equivalent form

$$\tilde{p}_{a|k} = \tilde{p}^{\text{rnd}}_{a|k} \cdot \left(\gamma_A^a \gamma_B^{k-a}\right), \tag{6.124}$$

where $\tilde{p}^{\text{rnd}}_{a|k}$ is the binomial distribution for random mixing at mass fractions ϕ_A of component A. It is easy to show that for unbiased mixing, $w''_{a,k-a} = 1$, we obtain $\gamma_A = \gamma_B = 1$ and recover the binomial distribution. The factors γ_A and γ_B represent deviations from random mixing due to bias.

6.7.1 Entropy of Mixing

The partition function of the sieve cut ensemble is

$$\log \omega'' = S^{\mathrm{bi}}(\tilde{p}_{a|k}) + \log \tilde{w}''$$

where

$$S^{\mathrm{bi}}(\tilde{p}_{a|k}) = -k\,(\phi_A \log \gamma_A \phi_A + \phi_B \log \gamma_B \phi_B) - \log \tilde{w}'' \tag{6.125}$$

is the bicomponent entropy and $\log \tilde{w}''$ is the log of the sieve cut bias evaluated at the most probable distribution. Combining these results we obtain the partition function:

$$\frac{\log \omega''}{k} = -\phi_A \log \gamma_A \phi_A - \phi_B \log \gamma_B \phi_B \equiv \frac{\Delta S}{k}. \tag{6.126}$$

As in random mixing, the log of the partition function is the change in entropy upon forming a mixed population starting with the pure components.

Note 6.4 (Solution Thermodynamics)

The partition function in Eq. (6.126) can be written as the sum of a contribution from random mixing plus a term that represents deviations due to bias:

$$\log \omega'' = \log \omega''_{\mathrm{rnd}} + \Delta \log \omega'', \tag{6.127}$$

with

$$\frac{\log \omega''_{\mathrm{rnd}}}{k} = -\phi_A \log \phi_A - \phi_B \log \phi_B, \tag{6.128}$$

and

$$\frac{\Delta \log \omega''}{k} = -\phi_A \log \gamma_A - \phi_B \log \gamma_B. \tag{6.129}$$

In the language of solution thermodynamics, $\log \omega''$ is the ideal entropy of mixing, and $\Delta\omega''$ is the excess entropy. The factors γ_A and γ_B are the activity coefficients of components. These are defined in Eq. (6.123) and are intensive functions of composition via the compositional dependence of u_A and u_B (recall that $\log u_A$ and $\log u_B$ are derivatives of $\log w''$ with respect to a and b and they are functions of a and $b = k - a$). Also from Eq. (6.123) we have

$$c_A u_A = \gamma_A \phi_A, \quad c_B u_B = \gamma_A \phi_A.$$

In solution thermodynamics these products give the activity of component in the mixture. In random mixing (ideal solution) the activities converge to the mass fraction of the components.

With these results we have made full contact with solution thermodynamics. In this case the compositional bias is determined by molecular interactions between components. The bicomponent ensemble allows to view mixed populations in general as subject to the same mathematical treatment as molecular solutions, regardless of the physical mechanisms that cause the mixing of components.

★

6.7.2 Scaling

The homogeneity of the compositional bias in Eq. (6.111) allows us to sketch the behavior of large clusters by scaling arguments. We study this behavior in this section. The compositional MPD maximizes the microcanonical weight in sieve cut k, defined by the number of clusters N_k and the mass $M_{A,k} = \bar{a}_k k N_k$ in the sieve cut. The extensive partition function is

$$\log \Omega''_{M_k, N_k} = \beta'' M_{A,k} + \left(\log q''\right) N_k. \tag{6.130}$$

The sieve cut ensemble obeys homogeneity with respect to $M_{A,k}$ and N_k: if both increase at fixed ratio

$$\frac{M_{A,k}}{N_k} = \bar{a}_k = k\phi_A, \tag{6.131}$$

the compositional distribution $\tilde{p}_{a|k}$ remains unchanged. The intensive partition function $\log \omega'' = \log \Omega''_{M_{A,k}, N_k} / N_k$ is

$$\log \omega''(\bar{a}_k, k) = \beta'' \bar{a}_k + \log q''. \tag{6.132}$$

The partition function is a function of $\bar{a}_k$ as well as of k (the parameters β'' and q'' both depend on the size k of the sieve cut). By standard property of the partition function, the mean amount of A is given by the first derivative of $\log q''(\beta'')$,

$$-\frac{d \log q''}{d\beta''} = \bar{a}_k = k\phi_A \tag{6.133}$$

and its variance by the second derivative,

$$\frac{d^2 \log q''}{d\beta''^2} = \text{Var}\,(a_k)\,. \tag{6.134}$$

We now apply the homogeneity property of w'' in Eq. (6.111), which we have not used yet:

$$\log w''(a,b) = (a+b)\log w''(\xi_A, \xi_B), \qquad [6.111]$$

If both a and b increase by the same factor λ (cluster mass $k = a + b$ will also increase by the same factor), the compositional distribution expressed in terms of the mass fraction $\xi_A = a/k$ remains the unchanged. This further means that the partition function $\log \omega''/k$ is independent of k:

$$\frac{\log \omega''(k, k\phi_A)}{k} = \frac{\beta'' \bar{a}_k}{k} + \frac{\log q''}{k}, \tag{6.135}$$

and since $\bar{a}_k = k\phi_A$,

$$\frac{\log \omega''}{k} = \phi_A \beta'' + \frac{\log q''}{k}, \tag{6.136}$$

where both β'' and $\log q''/k$ are independent of k and are functions of ϕ_A only:

$$\beta'' = \beta''(\phi_A), \quad \log q'' = k \log q_0''(\phi_A). \tag{6.137}$$

With these expressions Eq. (6.133) for the mean amount of a_k becomes

$$\bar{a}_k = -k\frac{d \log q_0''}{d\beta''} = k\phi_A, \tag{6.138}$$

which reaffirms that the mean mass fraction of A in all size cuts is ϕ_A. Similarly, Eq. (6.134) for the variance becomes

$$\text{Var}(a_k) = -k\frac{d\phi_A}{d\beta''} = -\bar{a}_k\frac{d \log \phi_A}{d\beta''}, \tag{6.139}$$

The variance of $\xi_A = a_k/k$ is

$$\text{Var}(\xi_A) = \frac{\text{Var}\,(a_k)}{\bar{a}_k^2} \sim \frac{1}{\bar{a}_k} = \frac{1}{k\phi_A}. \tag{6.140}$$

For large k the variance goes to zero, which implies that the distribution of ξ_A becomes increasingly sharper around its mean value ϕ_A.

6.7.3 Phase Splitting

If the bias is such that mixed clusters are not favored, it is possible to obtain phase separation of components. Rather than presenting a formal stability analysis, we will demonstrate deviations from random mixing and phase separation by numerical simulation. The compositional bias for this example is given as

$$w''(a, k-b) = e^{-\chi a(k-a)/k}, \tag{6.141}$$

whose logarithm is homogeneous in a and k with degree 1. The parameter χ controls the strength of the bias. With $\chi = 0$ we obtain $w'' = 1$ (random mixing). For $\chi < 0$ the bias has a maximum at $a/k = 1/2$ and promotes the formation of mixed clusters. For $\chi > 0$ the bias has two maxima at $a = 0$ and $a = k$, and promotes the formation of segregated clusters. We perform Monte Carlo simulations in the sieve-cut ensemble using the binary reaction method. The simulation of random mixing was outlined in Note 6.3. Briefly, this is done as follows: we pick two clusters at random, combine their contents, randomize the order of components, and split the randomized cluster into two clusters of equal size. This process is schematically represented by the reaction

$$R_1 + R_2 \to P_1 + P_2,$$

where R_i are the reacting clusters and P_i are the products. To simulate bias we accept the outcome with probability

$$P_{\text{accept}} = \frac{w''(P_1)w''(P_2)}{w''(R_1)w''(R_2)},$$

where $w''(P_i)$ and $w''(R_i)$ is the bias of the products and reactants, respectively. With uniform bias ($w'' = 1$) this process converges to the binomial distribution.

Figure 6.3 shows the distribution obtained in the simulation for a cluster of size $k = 8$ at $\phi_A = 1/2$. For $\chi < 0$ the distribution of components is narrower than the binomial distribution: components mix more intimately than in random mixing. When χ is positive the distribution is wider than the binomial distribution, and for sufficiently large values there is complete separation of components. The results of the simulation are also compared with a direct calculation of the MPD. The calculation is done using Eq. (6.119) and the parameters c_A and c_B are evaluated numerically by applying the normalization condition and requiring the mean fraction of component A to be ϕ_A. This calculation, shown by the vertical sticks in Fig. 6.3, is in very good agreement in the region $\chi \leq 0$ and also for small

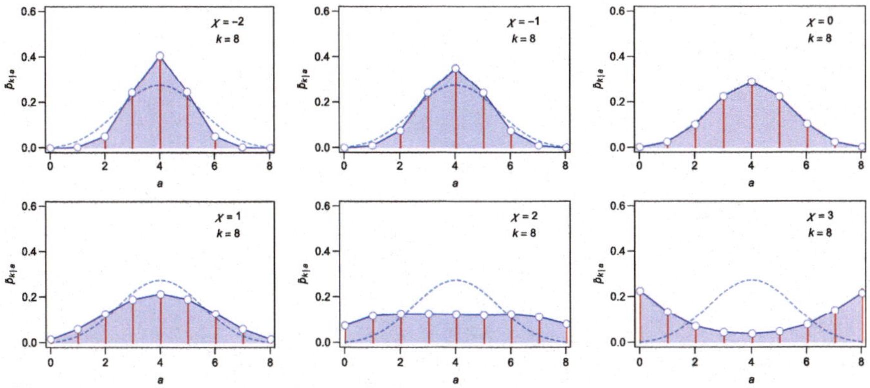

Fig. 6.3 Compositional distributions with bias $w''(a, k - b) = e^{-\chi a(k-a)/k}$ for a cluster with mass $= 8$ at $\phi_A = 1/2$. Open circles are the results of Monte Carlo simulations; vertical sticks represent the most probable distribution calculated numerically from Eq. (6.119); the dashed line is the binomial distribution with $\phi_A = 1/2$ (random mixing)

positive values of χ. It is possible to calculate the activity coefficients and construct the phase diagram of this system. We will not get into such details here. Our main goal in this chapter is to formulate the basis for the thermodynamic treatment of mixed populations as a natural extension of the cluster ensemble. In the rest of the book our focus will return to single component systems.

Chapter 7
Generalized Thermodynamics

The basis of all of our development up to this point has been the *cluster ensemble*, a discrete ensemble that generates every possible distribution of integers i with fixed zeroth and first order moments. Thermodynamics arises naturally in this ensemble when M and N become very large. In this chapter we will reformulate the theory on a mathematical basis that is more abstract and also more general. The key idea is as follows. If we obtain a sample from a given distribution h_0, the distribution of the sample may be, in principle, any distribution h that is defined in the same domain. This sampling process defines a phase space of distributions h generated by sampling distribution h_0. We will introduce a sampling bias via a selection functional W to define a probability measure on this space and obtain its most probable distribution. When the generating distribution h_0 is chosen to be exponential, the most probable distribution obeys thermodynamics. Along the way we will make contact with Information Theory, Bayesian Inference, and of course Statistical Mechanics.

7.1 Generating Distributions by Random Sampling

Suppose $h_0(x)$ is a probability distribution defined for $x \in (x_a, x_b)$, such that $h \geq 0$ inside this interval, $h = 0$ everywhere else, and h satisfies the normalization

$$\int_{x_a}^{x_b} h(x)dx = 1. \tag{7.1}$$

We construct a grid of points $x_i = x_a + (i-1)\Delta$, $i = 1, 2 \cdots K+1$ where $\Delta = (x_b - x_a)/K$. This grid discretizes the domain of h into K intervals of width Δ. For sufficiently small Δ the probability that a randomly sampled value from h falls in the ith interval is

$$p_i = h(x)\Delta. \tag{7.2}$$

T. Matsoukas, *Generalized Statistical Thermodynamics*,
Understanding Complex Systems, https://doi.org/10.1007/978-3-030-04149-6_7

We collect a random sample of N values from h and construct the frequency distribution $\mathbf{n} = (n_1, n_2 \cdots)$, where n_i is the number of sampled values in the ith interval. The probability to obtain distribution $\mathbf{n}$ is given by the multinomial distribution,

$$P(\mathbf{n}|h_0, N) = \mathbf{n}! \prod_i p_i^{n_i}. \tag{7.3}$$

We take the log of this probability using Stirling's formula for the factorials:

$$\log P(\mathbf{n}|h_0, N) = -N \sum_i \frac{n_i}{N} \log \frac{n_i}{N} + \sum n_i \log p_i + O(\log N). \tag{7.4}$$

We define $h(x_i) = n_i/(N\Delta)$ and use this definition along with Eq. (7.2) to write this probability as

$$\frac{\log P(\mathbf{n}|h, N)}{N} = -\Delta \sum_i h(x_i) \log \frac{h(x_i)}{h_0(x_i)} + \frac{O(\log N)}{N}. \tag{7.5}$$

In the limit $\Delta \to 0$, $N \to \infty$, $h(x_i)$ converges to a continuous distribution $h(x)$ and the above expression becomes[1]

$$\frac{\delta P(h|h_0, N)}{N} \to -\int h(x) \log \frac{h(x)}{h_0(x)} dx. \tag{7.6}$$

Here

$$\delta P(h|h_0, N) = \Phi(h|h_0, N)\delta h \tag{7.7}$$

is the probability that the distribution of a large random sample drawn from h_0 comes from region $(h, h+\delta h)$ of the continuous phase of distributions; accordingly, $\Phi(h|h_0, N)$ is the *probability distribution* of *probability distribution h*. The integral on the right-hand side of Eq. (7.7) defines the Kullback-Leibler divergence $D(h||h_0)$ of h relative to h_0, also known as relative entropy:

$$D(h||h_0) = \int h(x) \log \frac{h(x)}{h_0(x)} dx. \tag{7.8}$$

[1]From now on the integration limits will be assumed to be over the domain of h and will not be written explicitly.

Using $D(h||h_0)$, the probability of h is

$$\delta P(h|h_0, N) = e^{-ND(h||h_0)}. \tag{7.9}$$

The random sampling of h_0 produces a probability space of distributions. Any distribution $h(x)$ that is defined in the same domain or any sub-domain of h_0 may, in principle, be realized and its probability is $\delta P(h|h_0, N)$. We intuitively expect that most probable distribution in this space is h_0. Indeed this intuitive result can be derived rather easily. The probability in Eq. (7.9) satisfies the inequality $0 \le \delta P(h|h_0, N) \le 1$; then we must also have

$$D(h||h_0) \ge 0. \tag{7.10}$$

The distribution that maximizes $\delta P(h|h_0, N)$ minimizes $D(h||h_0)$ and from the definition of divergence we find $D(h_0||h_0) = 0$. Accordingly, h_0 minimizes the divergence, it is therefore the most probable distribution.[2] Let us compare the probability of h_0 relative to any other distribution. Using Eq. (7.9) and taking the limit $N \to \infty$ we have

$$\frac{e^{-ND(h_0||h_0)}}{e^{-ND(h||h_0)}} = e^{ND(h||h_0)} \to \infty. \tag{7.11}$$

The result states that h_0 is not only more probable, it is overwhelmingly more probable than any other distribution in the ensemble as N becomes large. Equation (7.9) expresses the sampling probability of h_0 in terms of relative entropy, a connection that has been pointed out in the literature of large deviations.[3] More important for our purposes is that the sampling of distribution h_0 establishes a sharply peaked probability space of distributions h defined over the same domain as h_0. The Gibbs inequality expresses the elementary fact that the most probable distribution in this space is h_0. We will generalize this probability space and the Gibbs inequality.

7.2 Biased Sampling

In random sampling of distribution h_0 the most likely outcome is h_0 itself. We now bias the sampling of h_0 with a bias functional $W[h]$ as follows: we obtain a random sample of size N from h_0 but accept it with probability proportional to $W[h]$, where

[2] Alternatively, we may formally maximize the functional in Eq. (7.9) with respect to h under fixed N and under the normalization constraint in Eq. (7.1).

[3] See for example Cover and Thomas (2006) and Touchette (2009).

h is the distribution in the random sample. We require the log of the bias to be homogeneous in h with degree 1:

$$\log W[\lambda h] = \lambda \log W[h]. \tag{7.12}$$

This condition ensures that the distribution of the sample will converge when $N \to \infty$. The probability of distribution h is now given as the product of the probability to obtain h in the random sample, given by Eq. (7.3) times the bias $W[h]$:

$$P(h|h_0, W, N) = \frac{W[h]}{r^N}\left(\mathbf{n}!\prod_i p_i^{n_i}\right). \tag{7.13}$$

where r^N is a normalization factor and r is constant.[4] Following the same steps that led to Eq. (7.6) we now obtain

$$\frac{\delta P(h|h_0, W, N)}{N} \to -\int h(x)\log\frac{h(x)}{w(x;h)h_0(x)}dx - \log r, \tag{7.14}$$

where $\log w(x; h)$ is the variational derivative

$$\log w(x;h) = \frac{\delta \log W[h]}{\delta h}. \tag{7.15}$$

We define the *probability functional* $\log \varrho$ by the right-hand side of the above result,

$$\log \varrho[h|h_0, W] = -\int h(x)\log\frac{h(x)}{w(x;h)h_0(x)}dx - \log r. \tag{7.16}$$

Using this functional the probability of h in this biased ensemble is

$$\delta P(h|h_0, W, N) = \varrho^N[h|h_0, W]. \tag{7.17}$$

Suppose that the most probable distribution is $f(x)$. We compare its probability to that of any other distribution:

$$\frac{\delta P(f|h_0, W, N)}{\delta P(h|h_0, W, N)} = \left(\frac{\varrho[f|h_0, W]}{\varrho[h|h_0, W]}\right)^N \to \infty. \tag{7.18}$$

Since $\varrho[h_0|h_0, W] > \varrho[h|h_0, W]$, this ratio goes to infinity as $N \to \infty$ for all h except f. As in random sampling, the most probable distribution is overwhelmingly

[4]In writing this probability we have anticipated the fact that the log of the normalization constant is homogeneous in N with degree 1.

more probable than any other. But then we must also have $\delta P(f|h_0, W) = 1$, or equivalently,

$$\varrho[f|h_0, W] = 1. \tag{7.19}$$

We obtain the most probable distribution by maximizing the probability functional under the normalization constraint in Eq. (7.1). The equivalent unconstrained maximization problem is

$$\max_h \left\{ -\int h(x) \log \frac{h(x)}{w(x; f) h_0(x)} dx - \log r - \lambda_0 \left(\int h(x) dx - 1 \right) \right\} \tag{7.20}$$

where λ_0 is the Lagrange multiplier for the normalization constraint. We set the functional derivative at $h = f$ under fixed r and λ_0 equal to zero:

$$0 = -\lambda_0 - 1 - \log h(x) + \log w(x; h) + \log h_0(x), \tag{7.21}$$

and solve for f:

$$f(x) = \frac{w(x; f) h_0(x)}{e^{1+\lambda_0}}. \tag{7.22}$$

We evaluate λ_0 by applying Eq. (7.19); we find

$$e^{1+\lambda_0} = r, \tag{7.23}$$

and with this result the most probable distribution becomes

$$f(x) = \frac{w(x; f) h_0(x)}{r}. \tag{7.24}$$

Ultimately, r is fixed by the normalization condition on f. If we choose $w(x; f) = h_0(x)/f(x)$, where f is some other probability distribution defined on the domain of h_0, then $f(x) = f(x)$.[5] Therefore we can construct a bias such that *any* distribution in the domain of h_0 is obtained as the overwhelmingly most probable distribution by biased sampling of h_0.

[5]With $w = h_0/f$ Eq. (7.24) gives $f = h_0/r$, and since both f and h_0 are normalized, we must have $r = 1$.

7.3 Canonical Phase Space

Any distribution h_0 may be used to generate any other distribution h in the same domain by biased sampling. We will now choose h_0 to be the normalized exponential distribution with parameter β:

$$h_0(x) = \beta e^{-\beta x}, \tag{7.25}$$

whose domain is $0 \le x \le \infty$. We call the probability space of distributions sampled from this exponential distribution *canonical*. The canonical space is a special case of the generic phase space of the previous section; all results will be obtained by substituting the exponential form in the results of Sect. 7.2.

We insert the exponential distribution in the probability functional in Eq. (7.16):

$$\begin{aligned} \varrho[h|W,\beta] &= -\int_0^\infty h(x)\log\frac{h(x)}{w(x;h)}dx + \int_0^\infty h(x)\log\beta e^{-\beta x}dx - \log r \\ &= -\int_0^\infty h(x)\log\frac{h(x)}{w(x;h)}dx - \beta\int_0^\infty xh(x)dx - \log(r/\beta). \end{aligned} \tag{7.26}$$

We define $q = r/\beta$ and write the above result in the more compact form

$$\log\varrho[h|W,\beta] = -\int_0^\infty h(x)\log\frac{h(x)}{w(x;h)}dx - \beta\bar{x} - \log q, \tag{7.27}$$

where $\bar{x}$ is the mean of $h(x)$. This defines the *canonical probability functional.* The most probable distribution is obtained from Eq. (7.24) with $h_0 = \beta e^{-\beta x}$ and $r = \beta q$:

$$f(x) = w(x;f)\frac{e^{-\beta x}}{q}. \tag{7.28}$$

This is the *canonical* representation of distribution f. From the normalization condition on f we have

$$q = \int_0^\infty w(x;f)e^{-\beta x}dx. \tag{7.29}$$

The derivative of $\log q$ with respect to β is

$$\frac{d\log q}{d\beta} = -\int_0^\infty x w(x;\, f)\frac{e^{-\beta x}}{q}dx + \int_0^\infty \frac{d\log w(x;\, f)}{d\beta} w(x;\, f)\frac{e^{-\beta x}}{q}dx. \tag{7.30}$$

The first integral on the right-hand side is the mean of $f(x)$; the second integral is zero by virtue of the Gibbs-Duhem equation (see section "Homogeneity" in the appendix). Therefore,

$$\frac{d\log q}{d\beta} = -\bar{x}. \tag{7.31}$$

This establishes the relationship between the canonical variables β, q, and the mean of the most probable distribution.

7.4 Microcanonical Phase Space

Distributions in the canonical space may have any mean between 0 and ∞. We define the *microcanonical* space as the subset of canonical distributions with fixed mean $\bar{x}$. The generating distribution is again exponential but now we write it as

$$h_0(x) = \frac{e^{-x/\bar{x}}}{\bar{x}}, \tag{7.32}$$

with $\bar{x}$ fixed. The probability of h is still given by Eq. (7.16) except that r must be replaced with r' in order to satisfy normalization in the microcanonical subspace. Using Eq. (7.32) this probability is

$$\begin{aligned}\frac{\delta P(h|h_0, W, N)}{N} &= -\int_0^\infty h(x)\log\frac{h(x)}{w(x;\, h)}dx + \int_0^\infty h(x)\log\left(\frac{e^{-x/\bar{x}}}{\bar{x}}\right)dx - \log r' \\ &= -\int_0^\infty h(x)\log\frac{h(x)}{w(x;\, h)}dx - \log\omega,\end{aligned} \tag{7.33}$$

where $\log\omega = -1 - \log\bar{x} - \log r'$. We define the log of the *microcanonical probability functional* by the right-hand side of the above result:

$$\log\varrho[h|W, \bar{x}] = -\int_0^\infty h(x)\log\frac{h(x)}{w(x;\, h)}dx - \log\omega. \tag{7.34}$$

Using this definition, the microcanonical probability of h is

$$\delta P(h|h_0, W, N) = \varrho^N[h|h_0, W, \bar{x}]. \tag{7.35}$$

For the same reasons that led to Eq. (7.19), the most probable distribution is overwhelmingly more probable than any other and satisfies

$$\varrho[f|W,\bar{x}] = 1. \tag{7.36}$$

To obtain the most probable distribution we maximize the microcanonical functional with respect to h under the conditions that h is normalized and its mean is fixed. The equivalent unconstrained maximization problem is

$$\max_h \left\{ -\int_0^\infty h(x)\log\frac{h(x)}{w(x;h)}dx - \log\omega \right.$$
$$\left. -\lambda_0\left(\int_0^\infty h(x)dx - 1\right) - \lambda_1\left(\int_0^\infty xh(x)dx - 1\right)\right\}. \tag{7.37}$$

We set the variational derivative with respect to h at $h = f$ equal to zero and solve for f:

$$f(x) = w(x;f)e^{-1-\lambda_0-x\lambda_1}. \tag{7.38}$$

With $\lambda_0 - \log q$, $\beta = \lambda_1$, this is identical to the canonical result:

$$f(x) = w(x;f)\frac{e^{-\beta x}}{q}. \tag{7.39}$$

It follows that the parameters q and β of most probable distribution in the microcanonical space satisfy Eq. (7.31). We insert the most probable distribution in the probability functional and apply the condition $\log\varrho[f|W,\bar{x}] = 0$. Noting that $f(x)/w(x;f) = e^{-\beta x}/q$, we have:

$$0 = \int_0^\infty h(x)\log\left(\frac{e^{-\beta x}}{q}\right)dx - \log\omega, \tag{7.40}$$

which leads to

$$\log\omega = \beta\bar{x} + \log q. \tag{7.41}$$

Substituting Eq. (7.31) into this result we have

$$\log \omega = \log q - \beta \frac{d \log q}{d\beta}. \tag{7.42}$$

This defines $\log \omega$ as the Legendre transformation of $\log q$ with respect to β. By Legendre properties we also have

$$\frac{d \log \omega}{d\bar{x}} = \beta. \tag{7.43}$$

In light of Eq. (7.41), the canonical functional (Eq. 7.27) and the microcanonical functional (Eq. 7.34) are seen to be the same. Indeed, the only difference is that in the microcanonical maximization the mean $\bar{x}$ is fixed while in the canonical case it is not. Both ensembles produce the same most probable distribution and their parameters q, β, and ω satisfy the same equations. Therefore the two ensembles are equivalent. We choose the microcanonical ensemble as the primary space because in this case all of the canonical parameters β, q, and ω are present and their interpretation is straightforward.

7.5 Entropy

Entropy has made no formal appearance in the theory so far. We introduce it now. We define the entropy of extensive distribution $H(x)$ as

$$S[H] = -\int H(x) \log \frac{H(x)}{J_0[H]} dx, \tag{7.44}$$

where J_0 is the zeroth moment,

$$J_0[H] = \int H(x) dx. \tag{7.45}$$

If $h(x)$ is normalized to unit area, Eq. (7.46) reverts to the familiar intensive functional,

$$S[h] = -\int h(x) \log h(x) dx. \tag{7.46}$$

We will establish the relationship of entropy to the multinomial coefficient by considering the sampling process introduced in Sect. 7.1. We begin with a continuous distribution $h(x)$ defined in the positive real axis. We discretize x in a grid of width Δ,

$$x_i = i\Delta, \quad i = 1, 2, \cdots . \tag{7.47}$$

We collect N random samples and form the distribution $\mathbf{n} = (n_1, n_2, \cdots)$ where n_i is the number of outcomes in the ith interval of the grid. The discrete approximation of $h(x)$ in the sample is

$$h(x_i) = \frac{n_i}{\Delta N}, \quad x_i = i\Delta, \tag{7.48}$$

and its limit is

$$\lim_{\substack{N\to\infty \\ \Delta\to 0}} \frac{n_i}{\Delta N} = h(x). \tag{7.49}$$

We also have

$$N = \sum_i \frac{n_i}{N} = \Delta \sum_i h(x_i) \to \int_0^\infty h(x)dx. \tag{7.50}$$

More generally, if $A(i)$ is any function of i, then

$$\sum_i A(i)\frac{n_i}{N} \to \int_0^\infty A(x/\Delta)\, h(x)dx. \tag{7.51}$$

With $A(i) = i$ we obtain

$$\bar{i}\,\Delta = \bar{x}, \tag{7.52}$$

where $\bar{i}$ is the mean value of i in $\mathbf{n}$. As the width of the grid decreases the value of $\bar{i}$ increases but the product $\bar{i}\Delta$ converges to $\bar{x}$. We now take the logarithm of the multinomial coefficient $\mathbf{n}!$:

$$\begin{aligned}\frac{\log \mathbf{n}!}{N} &= -\sum_i \frac{n_i}{N}\log\frac{n_i}{N} + \frac{O(\log N)}{N} \\ &= -\Delta \sum_i h(x_i)\left(\log h(x) - \log\Delta\right) + \frac{O(\log N)}{N} \\ &\to -\int_0^\infty h(x)\log h(x)dx - \log\Delta.\end{aligned} \tag{7.53}$$

Using $\Delta = \bar{x}/\bar{i}$, this becomes

$$\frac{\log \mathbf{n}!}{N} \to -\int h(x) \log h(x) dx - \log \frac{\bar{x}}{\bar{i}}. \tag{7.54}$$

We may write this result in the more symmetric form

$$\frac{\log \mathbf{n}!}{N} - \log \bar{i} = S[h] - \log \bar{x}. \tag{7.55}$$

This establishes the relationship between the multinomial coefficient of the sample collected from a distribution and the intensive entropy of that distribution. The distribution $\mathbf{n}$ on the left-hand side is that of the sample and is expressed in terms of the grid index i. This distribution depends on the discretization: as Δ decreases at fixed N, both $\mathbf{n}!$ and $\bar{i}$ increase,[6] but the difference

$$\frac{\log \mathbf{n}!}{N} - \log \bar{i}$$

converges to $S[h] - \log \bar{x}$. At fixed discretization, entropy is equal to $(\log \mathbf{n}!)/N$ to within an additive constant.

Suppose $H(x) = \lambda h(x)$ is an extensive homogeneous copy of $h(x)$; then,

$$H(x_i) = \lambda \frac{n_i}{\Delta N}, \tag{7.56}$$

and

$$J_0[H] = \int_0^\infty H(x) dx = \lambda. \tag{7.57}$$

Using $n_i = \Delta N H(x_i)/\lambda$, the log of the multinomial coefficient may be expressed in terms of H. Following the same steps as above, the equivalent form of (7.53) is

$$\frac{\log \mathbf{n}!}{N} \to -\frac{1}{\lambda} \int_0^\infty H(x) \log \frac{H(x)}{\lambda} dx - \log \Delta, \tag{7.58}$$

and since $\lambda = J_0[H]$,

$$\frac{\log \mathbf{n}!}{N} \to \frac{S[H]}{\lambda} - \log \Delta. \tag{7.59}$$

Equating this result with that in Eq. (7.53) we obtain

$$S[\lambda h] = \lambda S[h], \tag{7.60}$$

[6]Both n_i and N increase inversely proportional to Δ.

which restates the fact that the functional defined in Eq. (7.44) is homogeneous in h with degree 1.

Note 7.1 (Partition Function and Entropy)
The microcanonical functional may be expressed as

$$\log \varrho[h|W, \bar{x}] = S[h] + \log W[h] - \log \omega, \tag{7.61}$$

where $S[h]$ is the entropy functional,

$$S[h] = -\int h(x) \log h(x) dx. \tag{7.62}$$

Applying Eq. (7.61) to the most probable distribution we obtain a companion to Eq. (7.41):

$$\log \omega = S[f] + \log W[f]. \tag{7.63}$$

Equation (7.41) expresses log as a function of $\bar{x}$; Eq. (7.63) equates it to the value of the combined functional $S + \log W$ at the most probable distribution.

★

7.6 Curvature

The set of microcanonical relationships we obtained in the previous section follow from the maximization of the microcanonical functional with respect to h. This means that $\varrho[h|W, \bar{x}]$ is concave in h and further implies, as we will now show, that $\log \omega$ is a concave function of $\bar{x}$.

The microcanonical maximization problem can be defined via the microcanonical inequality

$$\varrho[h|W, \bar{x}] \le 1, \tag{7.64}$$

with the equal sign only for the most probable distribution f. This we can write in the equivalent form

$$S[h] + \log W[h] \le \log \omega(\bar{x}), \tag{7.65}$$

which applies to all distributions in $\mathbb{R}_+$ with fixed mean $\bar{x}$. Concavity of $\varrho[h|W, \bar{x}]$ with respect to h implies concavity of $S[h] + \log W[h]$. Since $S[h]$ is concave in h, it is sufficient that $\log W[h]$ be concave or linear in h. we now consider two microcanonical spaces with means $\bar{x}_1$ and $\bar{x}_2$, and most probable distributions f_1 and f_2, respectively. We form distribution h as

$$h = \alpha f_1 + (1-\alpha) f_2, \tag{7.66}$$

where $0 \le \alpha \le 1$. Its mean is

$$\bar{x} = \alpha \bar{x}_1 + (1-\alpha)\bar{x}_2. \tag{7.67}$$

We have the following series of results:

$$\log \omega(\bar{x}) = S[h] + \log W[h] \tag{7.68a}$$
$$\ge S[\alpha f_1] + \log W[\alpha f_1] + S[(1-\alpha)\bar{x}_2] + \log W[(1-\alpha)\bar{x}_2] \tag{7.68b}$$
$$\ge \alpha\,(S[f_1] + \log W[f_1]) + (1-\alpha)\,(S[f_2] + \log W[f_2]) \tag{7.68c}$$
$$= \alpha \log \omega(\bar{x}_1) + (1-\alpha) \log \omega(\bar{x}_2). \tag{7.68d}$$

Equation (7.68a) is the application of Eq. (7.63) to ensemble $(h; \bar{x})$; Eq. (7.68b) is application of the concave property of $S + \log W$; Eq. (7.68c) is application of the homogeneous property of S and $\lg W$; Eq. (7.68d) is the application of Eq. (7.63) to ensembles $(h_1, \bar{x}_1)$ and $(h_2, \bar{x}_2)$. The final result,

$$\log \omega(\bar{x}) \ge \alpha \log \omega(\bar{x}_1) + (1-\alpha) \log \omega(\bar{x}_2), \tag{7.69}$$

states the concave curvature of $\log \omega(\bar{x})$. We must have then,

$$\frac{d^2 \log \omega}{d\bar{x}^2} \le 0. \tag{7.70}$$

Using Eq. (7.43), an equivalent form of Eq. (7.70) is

$$\frac{d\beta}{d\bar{x}} \le 0. \tag{7.71}$$

The curvature of q is established as follows. Taking the derivative of Eq. (7.31) with respect to βwe have

$$\frac{d}{d\beta}\left(\frac{d \log q}{d\beta}\right) = -\frac{d\bar{x}}{d\beta} \ge 0, \tag{7.72}$$

with the last inequality obtained from Eq. (7.71). Therefore,

$$\frac{d^2 \log q}{d\beta^2} \ge 0, \tag{7.73}$$

and states that $q(\beta)$ is convex. This result could have been obtained directly from the properties of the Legendre transformation. In the sign convention used here, the transformation flips the curvature of the transformed function. Since $\omega(\bar{x})$ and $q(\beta)$ are Legendre pairs, they have opposite curvatures. Finally from Eq. (7.71) we obtain one more inequality. Writing this equation as $d\log q = -\bar{x}d\beta$ we have

$$\frac{d\log q}{d\bar{x}} = -\bar{x}\frac{d\beta}{d\bar{x}}, \tag{7.74}$$

then from Eq. (7.71) and $\bar{x} \geq 0$ we also have

$$\frac{d\log q}{d\bar{x}} \geq 0. \tag{7.75}$$

The parameters β and log therefore vary in opposite direction as a function of $\bar{x}$.

The concave curvature of the thermodynamic probability functional ensures that its extremum with respect to h is a maximum. If the concave condition is violated, the extremum is *not* the most probable distribution. In thermodynamic language the system is *unstable*. The example below demonstrates what happens in this case.

Note 7.2 (Stability of Linear Bias $w(x) = x^{-3}$)

As an example of stability analysis we consider the linear functional

$$\log W[h] = \int_1^0 h(x)\log x^{-3}dx, \tag{7.76}$$

with cluster function

$$w(x) = x^{-3}. \tag{7.77}$$

We solved this example in Chap. 5 (see Sect. 5.5) using the cluster ensemble. We will now obtain the solution using the variational theory of this chapter. The MPD is

$$f(x) = x^{-3}\frac{e^{-\beta x}}{q}, \tag{7.78}$$

with β and q given by the conditions

$$q = \int_1^\infty x^{-3}e^{-\beta x}dx = E_3(\beta), \tag{7.79}$$

$$\bar{x} = \frac{1}{q}\int_1^\infty x^{-2}e^{-\beta x}dx = \frac{E_2(\beta)}{E_3(\beta)}, \tag{7.80}$$

where $E_n(z)$ is the exponential integral function

$$E_n(z) = \int_1^\infty \frac{e^{-tz}}{t^n}dt.$$

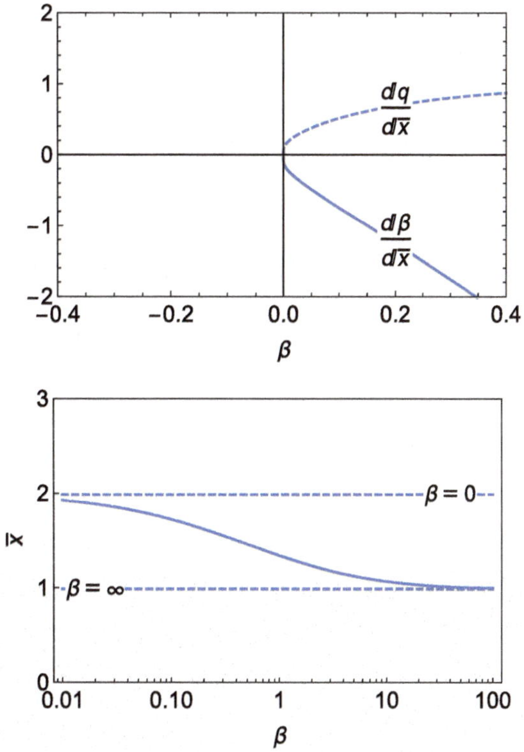

Fig. 7.1 The derivatives $dq/d\bar{x}$, $d\beta/d\bar{x}$ and the mean cluster size $\bar{x}$ as a function of β. Both derivatives satisfy the stability criteria in $0 < \beta < \infty$ and reach incipient instability at $\beta = 0$. The mean cluster size in the stable region ranges from 1 at $\beta = \infty$ to 2 at $\beta = 0$

The derivatives of β and q with respect to $\bar{x}$ are

$$\frac{d\beta}{d\bar{x}} = \frac{E_3(\beta)^2}{E_2(\beta)^2 - E_1(\beta)E_3(\beta)},$$

$$\frac{dq}{d\bar{x}} = \frac{E_2(\beta)E_3(\beta)^2}{E_2(\beta)^2 - E_1(\beta)E_3(\beta)}.$$

These are plotted in Fig. 7.1 as a function of β. Both derivatives satisfy the stability conditions in Eq. (7.71) and (7.75) in the region $0 < \beta < \infty$. They reach the boundary of stability at $\beta = 0$, where both derivatives become 0. Negative values of β are not possible because the integrals in Eqs. (7.79) and (7.80) do not converge. The value of $\bar{x}$ from Eq. (7.80) ranges from $\bar{x} = 1$ at $\beta = 0$ to $\bar{x} = 2$ at $\beta = \infty$. It is not possible to have a single sol phase with average size outside the range $1 \le \bar{x} \le 2$. If $\bar{x}$ falls in this range, the MPD is given by Eq. (7.78) with q and β from Eqs. (7.79) and (7.80), respectively. At the boundary of stability, $\beta = 0 \equiv \beta^*$, the MPD is

$$f^*(x) = \frac{2}{x^3}, \quad \bar{x}^* = 2, \quad \beta^* = 0, \quad q^* = 1/2, \tag{7.81}$$

with partition function

$$\log \omega^* = \bar{x}^* \beta^* + \log q^* = -\log 2. \tag{7.82}$$

We will now show that the MPD for $\bar{x} > 2$ is

$$f(x) = (1-\epsilon) f^*(x) + \epsilon f_g(x), \tag{7.83}$$

where $f^*(x)$ is the MPD of the sol at the boundary of stability, $f_g(x)$ is

$$f_g(x) = \delta(x - \bar{x}_g) \tag{7.84}$$

with

$$\bar{x}_g = \frac{\bar{x} - (1-\epsilon)\bar{x}^*}{\epsilon} \tag{7.85}$$

and $\epsilon \to 0$. The distribution is a linear combination of two MPDs, the sol MPD f^* in Eq. (7.81) with partition function $\log \omega^*$ in Eq. (7.82), and a Dirac delta with partition function

$$\log \omega_g = \log w(\bar{x}_g). \tag{7.86}$$

We confirm that f satisfies the conditions

$$\int_1^0 f(x)dx = (1-\epsilon) + \epsilon = 1,$$

$$\int_1^0 x f(x)dx = (1-\epsilon)\bar{x} + \epsilon \bar{x}_g = \bar{x},$$

therefore it is a member of the microcanonical space. The partition function that corresponds to Eq. (7.83) is

$$\log \omega = (1-\epsilon) \log \omega^* + \epsilon \log \omega_g = -(1-\epsilon) \log 2 + \epsilon \log w(\bar{x}_g), \tag{7.87}$$

with $w(\bar{x}_g) = \bar{x}_g^{-3}$. We will show that this partition function is maximized at $\bar{x} = \epsilon \bar{x}^* + (1-\epsilon)\bar{x}_g$. The maximization condition is

$$\frac{d \log \omega}{dx} = 0 = \epsilon \left(\frac{d \log \bar{x}_g^{-3}}{d\bar{x}_g} \right) \left(\frac{d\bar{x}_g}{d\bar{x}} \right) = -\frac{3}{\bar{x}_g^3}. \tag{7.88}$$

From Eq. (7.85) in the limit $\epsilon \to 0$ we find $\bar{x}_g \to \infty$. In this limit the maximization condition in Eq. (7.88) is indeed satisfied.

★

7.7 The Linearized Selection Functional

In the general case the selection functional is nonlinear and its variation derivative $\log w(x; h)$ depends on h. We now construct the linearized functional[7]

$$\log W_f[h] = \int h(x) \log w(x; f) dx, \tag{7.89}$$

which applies the variational derivative of $\log W[h]$ at $h = f$ to all distributions of the phase space. We use this linearized functional to construct a new microcanonical probability functional,

$$\log \varrho_f[h|W, \bar{x}] = -\int_0^\infty h(x) \log \frac{h(x)}{w(x; f)} dx - \log \omega. \tag{7.90}$$

We call this *linearized microcanonical probability functional* with the understanding that linearization refers to $\log W$.[8]

We will show that this functional satisfies the inequality

$$\varrho_f[h|W, \bar{x}] \le 1, \tag{7.91}$$

with the equal sign only for $h = f$. To find the distribution that maximizes the linearized functional we follow the same procedure as for $\varrho[h|W, \bar{x}]$ except that $w(x; f)$ is now constant. Its Euler equation is,

$$-\log f + \log w(x; f) - \beta x - \log q = 0,$$

and is maximized by the same distribution that maximizes $\varrho[h|W, \bar{x}]$:

$$f(x) = w(x; f) \frac{e^{-\beta x}}{q}, \qquad [7.39]$$

where β and q have the same meaning as in Eq. (7.39). We insert this result in Eq. (7.90):

[7]Functional $\log W_f[h]$ is linear with respect to all h with fixed $\bar{x}$; it is not a linear functional because $\log w(x; \bar{x})$ is not the same for *all* h (it depends on $\bar{x}$). For this reason we call $\log W_f$ *linearized* but not linear.

[8]While $\log W[h]$ is linear with respect to all h with fixed $\bar{x}$, functional $\log \varrho_f = S[h] + \log W_f[h] - \log \omega$ is not because $S[h]$ is not a linear functional.

$$\begin{aligned}\varrho_f[f|W,\bar{x}] &= -\int_0^\infty f(x)\log\frac{f(x)}{w(x;f)}dx - \log\omega \\ &= -\int_0^\infty f(x)\log\frac{e^{-\beta x}}{\bar{x}}dx - \log\omega \\ &= \beta\bar{x} + \log q - \log\omega = 0,\end{aligned} \tag{7.92}$$

from which we obtain,

$$\varrho_f[f|W,\bar{x}] = 1. \tag{7.93}$$

The linearized functional satisfies the same fundamental inequality as the microcanonical functional. The result shows that it is possible to construct more than one bias functional that delivers f as the most probable distribution. Indeed, there is an entire family of such functionals, as we show next.

7.7.1 The Inverse Problem

Once the bias is defined the most probable distribution is fixed and is expressed in terms of a set of canonical functions,

$$\{w,\ q,\ \beta,\ \omega\},$$

which are also fixed by W, that are interrelated via relationships we recognize as thermodynamics. In the inverse problem we begin with the distribution itself and reconstruct the set of canonical functions that reproduce the distribution. This reconstruction is possible but not unique, as we show next.

Given a distribution $f(x)$ defined in $\mathbb{R}_+$, we construct the linearized bias functional

$$\log W^*[h] = \int_0^\infty h(x)\log\frac{h(x)}{f(x)e^{a_0+a_1x}}dx, \tag{7.94}$$

whose variational derivative is

$$\log w(x;f) = f(x)e^{a_0+a_1x}, \tag{7.95}$$

where a_0 and a_0 are independent of x but otherwise unspecified. We construct the linearized functional

$$\log\varrho_f^*[h|W,\bar{x}] = -\int_0^\infty h(x)\log\frac{h(x)}{f(x)e^{a_0+a_1x}}dx - a_0 - a_1\bar{x}. \tag{7.96}$$

To determine the distribution that maximizes this functional we first write it in the equivalent form

$$\log \varrho_f^*[h|W,\bar{x}] = -\int_0^\infty h(x)\log\frac{h(x)}{f(x)}dx. \tag{7.97}$$

The right-hand side is the negative Kullback-Leibler divergence and it is maximized by $h = f$. This linearized functional therefore satisfies the inequality

$$\varrho_f^*[h|W,\bar{x}] \le 1, \tag{7.98}$$

with the equal sign only for $h = f$. Solving Eq. (7.95) for f we obtain

$$f(x) = e^{-a_0-a_1 x}. \tag{7.99}$$

If we set $\beta = x_1$, $q = e^{a_0}$, the above expression becomes the canonical representation of f. If moreover we require a_0 and a_1 to satisfy

$$\frac{da_0}{da_1} = -\bar{x}, \quad \frac{d^2a_0}{da_1^2} \le 0, \tag{7.100}$$

the set

$$\{w = w^*, q = e^{a_0}, \beta = a_1, \omega = e^{a_0+a_1\bar{x}}\}$$

is a canonical representation of f and satisfies all of the relationships obtained in Sects. 7.3 and 7.4. Even under the constraints of Eq. (7.100) there is an infinite number of pairs (a_0, a_1) that satisfy the above equations. The reconstruction of the canonical representation of a distribution is not unique.

7.7.2 Implications for the Inverse Problem

The inverse problem can be loosely posed as follows: If we know the MPD what can we say about the selection functional? It is impossible to identify the selection functional that produces the MPD if all we know is the MPD itself. This is because the MPD reveals partial information about the selection functional at the specific point $h = f$: namely, that the variational derivative of $S[h] + \log W[h]$ at $h = f$ is zero. At most we may recover the derivatives of $\log W$ (cluster function) at $h = f$. The linearized functional shows that this problem too is indeterminate and provides a recipe to calculate an entire family of linearized selection functionals. To understand this indeterminacy we revisit the inverse problem directly. If f is the MPD of unknown selection functional W, it maximizes $S[f] + \log W[f]$ under the constraints that fix the normalization and mean; it satisfies therefore the equation[9]

$$\frac{\delta}{\delta h}\Big(S[h] + \log W[h] - a_0 - a_1\bar{x}\Big)\Big|_{h=f} = 0, \tag{7.101}$$

with a_0 and a_1 undetermined Lagrange multipliers. In the forward problem W is known and the Lagrange multipliers are evaluated by back substitution into the constraints. In the reverse problem we do not have sufficient information to fix a_0 and a_1; they must remain undetermined. The additional conditions

$$\frac{da_0}{da_1} = -\bar{x}, \quad \frac{d^2 a_0}{da_1^2} \le 0,$$

ensure that the curvature of the partition function and all thermodynamic relationships are satisfied, though these relationships remain insufficient to fix a_0 and a_1. Equation (7.101) gives the selection functional in terms of entropy,

$$\log W[f] = -S[f] + a_0 + a_1\bar{x} \tag{7.102}$$

within an arbitrary constant $a_0 + a_1\bar{x}$.

The indeterminacy of the inverse problem, while at first sight inconvenient,[10] is a welcome result that establishes logical consistency. It would be strange indeed if knowledge of f would be sufficient to pinpoint W because it would then imply that all appearances of f in stochastic systems are associated with the *same* model W. The lack of uniqueness between f and W means that very different models could give rise to the same f. If we take the view that f is an observable quantity but W is not, then W cannot be known except partially. Any number of models in addition to the "true" model are consistent with the observed distribution.

---

[9]The complete functional to be maximized is

$$S[f] + \log W[f] + a_0\left(1 - \int f(x)dx\right) + a_1\left(\bar{x} - \int xf(x)dx\right)$$

but this is equivalent to

$$S[f] + \log W[f] - a_0 - a_1\int xf(x)dx,$$

which is the same as the functional in Eq. (7.101).

[10]It prevents the experimentalist with access to f from inferring W on the basis of f alone.

Example 7.1 (Canonical Representations of the Exponential Distribution)
In this example we demonstrate the construction of multiple linearized selection functionals all of which reproduce the exponential distribution,

$$f(x) = \frac{e^{-x/\bar{x}}}{\bar{x}}.$$

Canonical Set #1 We choose $a_0 = \log \bar{x}$, then use the first condition to obtain $a_1 = 1/\bar{x}$; this pair satisfies both conditions in (7.100). With $\beta^* = a_1, q^* = e^{a_0}$, we have the following canonical representation of the exponential distribution:

$$w^* = 1;\ \beta^* = 1/\bar{x};\ q^* = \bar{x};\ \log \omega^* = 1 + \log \bar{x}.$$

The corresponding functional W^* is obtained from the Euler relationship,

$$\log W^*[h] = \int h(x) \log w^*(x) dx = 0,$$

i.e., $W^*[h] = 1$ for all h (uniform functional).

Canonical Set #2 With $a_0 = \log \bar{x}^2$ we obtain the alternative set of functions

$$w^* = \bar{x} e^{x/\bar{x}};\ \beta^* = 2/\bar{x};\ q^* = \bar{x}^2;\ \log \omega^* = 2 + 2 \log \bar{x}.$$

This set corresponds to the entropic functional $\log W[h] = S[h]$ discussed in Sect. 4.4.
Additional sets can be obtained in the same manner. Table 7.1 summarizes just a few of them.

★

Table 7.1 Some possible canonical representations of the exponential distribution

$w^*(x, \bar{x})$	$\log \omega^*(\bar{x})$
1	$\log \bar{x} + 1$
$e^{x/\bar{x}} \bar{x}$	$2 \log \bar{x} + 2$
$e^{-\frac{x}{\bar{x}}} \left(\frac{\bar{x}-1}{\bar{x}}\right)^{1-x}$	$\log(\bar{x} - 1) - \bar{x} \log \frac{\bar{x}-1}{\bar{x}}$
$\frac{e^{x-2x/\bar{x}}}{\bar{x}} \left(\frac{\bar{x}-1}{\bar{x}}\right)^{1-x}$	$(1 - \bar{x}) \left(\log \frac{\bar{x}-1}{\bar{x}} - 1\right)$
$\frac{e^{\alpha x/\bar{x}}}{\alpha+1} \left(\frac{\bar{x}}{\alpha+1}\right)^{\alpha}$	$1 + \alpha + \log \left(\frac{\bar{x}}{\alpha+1}\right)^{\alpha+1}$

$$f = w^* \frac{e^{-\beta^* x}}{q^*} = \frac{e^{-x/\bar{x}}}{\bar{x}}$$

$$\beta^* = \frac{d \log \omega^*}{d\bar{x}};\ \log q^* = \log \omega^* - \frac{d \log \omega^*}{d\bar{x}}$$

7.8 What *Is* Thermodynamics?

We have formulated a variational theory that provides a thermodynamic description for any probability distribution whose domain is the positive real axis. The theory is self-contained and makes no reference to the discrete cluster ensemble. It is based on the principle that any probability distribution maximizes the probability functionals $\varrho[h|W]$ under an appropriate functional W. This variational principle is expressed by the inequality

$$\log \omega \le S[h] + \log W[h], \tag{7.103}$$

which becomes an exact equality for $h = f$. The maximization of this functional defines f and identifies it as the most probable distribution of the space. The maximization problem transforms f into the set a functions,

$$f(x,\bar{x}) \longleftrightarrow \left\{ w(x,\bar{x}),\ \omega(\bar{x}) \right\} \longleftrightarrow \left\{ w(x,\bar{x}),\ \beta(\bar{x}),\ q(\bar{x}) \right\}, \tag{7.104}$$

that are interconnected through a network of mathematical relationships summarized in Table 7.2. Distribution f can be reconstructed from its canonical parameters as

$$f = w\frac{e^{-\beta x}}{q} = \frac{w}{\omega} e^{-(x-\bar{x}) d \log \omega / d\bar{x}}. \tag{7.105}$$

To reconstruct f we need the cluster function $w(x,\bar{x})$ and the partition function $\omega(\bar{x})$; or the cluster function and the canonical parameters β and q. In the special case that the ensemble is linear the cluster functions can be obtained by inverse-transform of the Laplace relationship

$$q(\beta) = \int_0^\infty w(x) e^{-\beta x} dx. \tag{7.106}$$

In this case—and this one only—knowledge of the partition function as a function of $\bar{x}$ is sufficient to reproduce f. In all other cases the cluster function must be known, in addition to the partition function, in order to reconstruct f. Entropy is missing from these equations. The fundamental thermodynamic functional is

$$S[h] + \log W[h], \tag{7.107}$$

Table 7.2 Summary of results in the canonical and microcanonical phase spaces

	Canonical space	Microcanonical space
Phase space	$\int h(x)dx = 1$	$\int h(x)dx = 1$, $\int xh(x)dx = \bar{x}$
Generating function	$\beta e^{-\beta x}$	$e^{-x/\bar{x}}/\bar{x}$
Probability functional	$\log \varrho[h\|W, \beta]$ $S[h] + \log W[h] - \beta\bar{x} - \log q$	$\log \varrho[h\|W, \bar{x}]$ $S[h] + \log W[h] - \log \omega$
Gibbs inequality (second law)	$\varrho[h\|W, \beta] \le 1$	$\varrho[h\|W, \bar{x}] \le 1$
Most probable distribution (statistical thermodynamics)	$f(x) = w(x; f)\dfrac{e^{-\beta x}}{q}$ $\varrho[f\|W, \beta] = \varrho[h\|W, \bar{x}] = 1$ $S[f] + \log W[f] = \log \omega(\bar{x})$	
Most probable distribution (classical thermodynamics)	$\log \omega = \beta\bar{x} + \log q$ $\dfrac{d \log \omega}{d\bar{x}} = \beta; \quad \dfrac{d \log q}{d\beta} = -\bar{x}.$ $\dfrac{d^2 \log \omega}{d\bar{x}^2} \le 0.$	

$S[h]$ is the Gibbs-Shannon entropy; $W[h]$ is the sampling bias functional (selection functional); $\log w(x; h)$ is the variational derivative of $\log W[h]$ with respect to h; and
$S[h] + \log W[h] = -\int_0^\infty h(x) \log \frac{h(x)}{w(x; h)} dx$

which combines entropy and the selection functional. It is the maximization of this functional that produces the network of canonical relationships. At $h = f$ we have

$$\omega(\bar{x}) = S[f] + \log W[f], \tag{7.108}$$

and in the special case $W = 1$ (unbiased ensemble) we obtain $S = \log \omega$. In general, however, entropy and partition function are distinctly different from each other.

What we call *thermodynamics* in the common usage of the term is the network of canonical relationships associated with probability distribution f. Through historical accident, first contact with thermodynamics was made in the problem of the equilibrium state of matter. In the convoluted path of science, typical of all scientific discovery, the relationships between the variables we call ω, β, and q were discovered first, through laborious but ingenious arguments involving thermometers, pressure gauges, and steam engines, by the likes of Carnot, Helmholtz, Thomson, Clausius, and Maxwell.[11] The association of these variables with a probability distribution was made later, first tentatively by Boltzmann and then decisively by

[11] A nice historical account of the development of thermodynamics is given by Müller (2007).

Gibbs. The parallel development of ideas in thermodynamics and in theories of matter tied the two fields inextricably to the point that it is commonly accepted that thermodynamics is a theory of matter. We can now say that the mathematical shell of thermodynamics is a universal calculus of probability distributions and is not unique to matter. For lack of a better name we continue to use the dated term *thermodynamics* with the adjective generalized to describe this calculus independently of the specific problem to which it may be applied. Generalized thermodynamics accompanies any distribution, it is therefore applicable to stochastic processes in general. The equilibrium state of matter is only one such problem. Two specific problems to be discussed in the rest of the book are binary merging and binary fragmentation, both examples of irreversible processes whose advance in time is unidirectional.

7.9 Contact with Statistical Mechanics

The obvious way to make contact with statistical mechanics is to interpret f as the canonical probability of microstate. Then, $w = 1$, x is the energy E_i of microstate i, β is $1/kT$, q is the canonical partition function, ω is the partition function, and all equations developed here map to well-known thermodynamic relationships. The condition $w = 1$ is the mathematical statement of the postulate of equal a priori probabilities that assigns equal probability to all microstates with the same energy, volume, and number of particles. Since in this case $\varrho = e^{S[h]}/\omega$, the canonical probability f maximizes entropy. We have, therefore, complete correspondence with statistical thermodynamics.

This bottom-up derivation is not the only one possible. Generalized thermodynamics applies to any distribution and we may choose f to be the probability to find a *macroscopic* system of fixed (T, V, N) at energy E. We can do this even though the energy of a macroscopic system is asymptotically a deterministic variable because all this means is that its distribution is asymptotically a Dirac delta. We begin by writing the energy distribution in the form of Eq. (7.28) with w, β, and q to be determined. From Eqs. (7.31) and (7.41) with $\bar{x} = \bar{E}$ we have

$$\bar{E} = -\left(\frac{\partial \log q}{\partial \beta}\right)_{VN}; \quad \log \omega = \beta \bar{E} + \log q,$$

and comparison with established relationships of classical thermodynamics leads to the identifications $\beta \to 1/kT$, $\log q \to -F/kT$ (free energy), $\log \omega \to$ *thermodynamic entropy*. To identify w we require input from physics, and this comes via the observation that the probability density of macroscopic energy E is asymptotically a Dirac delta function. Then $S[f] = 0$ (this is the entropy of the *energy* distribution f, not to be confused with thermodynamic entropy, which refers to the probability of *microstate*).[12] With $(E') = \delta(E' - E)$ we then have

$$\omega(E) = \log W[f] = \int f(E') \log w(E', f) dx = \log w(E, f) \tag{7.109}$$

and conclude that $\log w$ is the thermodynamic entropy. This establishes full correspondence between generalized and macroscopic (classical) thermodynamics. If we make the assumption that $w(E)$ is the number of microstates under fixed volume and number of particles, we establish the microscopic connection. This is a *physical* assumption and must come outside thermodynamics. Since in this case $f(E)$ is proportional to the number of microstates with energy E, we may as well ascribe equal probability to each microstate. Thus we recover the postulate of equal a priori probabilities (statistical thermodynamics). Finally, by adopting a physical model of microstate, classical, quantum, or other, we obtain classical statistical mechanics, quantum statistical mechanics, or yet-to-be-discovered statistical mechanics, depending on the choice of the model. In all cases the thermodynamic calculus is the same, only the enumeration of microstates (that is, W) depends on the physical model.

7.10 Contact with the Maximum Entropy Principle

Generalized canonical thermodynamics shares a common origin with Jaynes's maximum entropy principle (MEP), also known as maximum entropy method (MEM) or simply MaxEnt. MEP is a method of statistical inference that seeks the unknown probability distribution $f(x)$ on the basis of incomplete information that prevents the unique determination of f. The central thesis of MEP is that the indeterminacy may be resolved by maximizing the entropy functional

$$S[h] = -\int h(x) \log h(x) dx, \tag{7.110}$$

with respect to all h in the same domain as the unknown f under the constraint that h is normalized,

$$\int h(x) dx = 1, \tag{7.111}$$

[12] While the entropy *functional* may be applied to *any* distribution, what we call *thermodynamic entropy* (i.e., the quantity measured as reversible heat over temperature) refers specifically to the application of the entropy functional to the probability distribution of microstates. The distinction is not always made clear in the literature, as Jaynes had to point out (Jaynes 1965).

plus any additional constraints that represent our knowledge about f. The classical example is that of a distribution defined in $0 \le x \le \infty$ whose mean $\bar{x}$ is known. The additional constraint that expresses our knowledge of $\bar{x}$ is

$$\int xh(x)dx = \bar{x}. \tag{7.112}$$

The maximization of entropy under these two integral constraints produces the exponential distribution,

$$f(x) = \frac{e^{-\beta x}}{q}, \tag{7.113}$$

just as generalized thermodynamics does. To obtain other distributions by MEP, one has to use additional constrains. For example, if we add the constraint

$$\int x^2 h(x)dx = \mu_2, \tag{7.114}$$

which fixes the second moment, maximization of entropy under Eqs. (7.111), (7.112), and (7.114) produces a Gaussian distribution with mean $\bar{x}$ and variance $\mu_2 - \bar{x}^2$, as one can easily confirm using Lagrange multipliers. By careful construction of constraints it is possible to obtain virtually any distribution as a maximum entropy distribution.[13]

A weakness of MEP is that it offers no organized methodology to convert what we know about the unknown distribution into an integral constraint. One way to address this problem is by implementing the extended form of entropy,

$$S[h] = -\int h(x) \log \frac{f(x)}{m(x)} dx. \tag{7.115}$$

The function $m(x)$ inside the logarithm was introduced by Jaynes to address the lack of invariance of the standard entropy functional under change of variables (Jaynes 1968). In practice $m(x)$ is interpreted as a prior probability. For example, if it is known that distribution $f(x; k)$ represents mixing of two components where x is the mass of component 1 and k is the total mass, one might set

$$m(x; k) = \binom{k}{x},$$

and in this case the maximum entropy method recovers the binomial distribution. This however merely shifts the problem, as no general method exists to determine the appropriate prior.

[13] Kapur (1989) gives several examples.

7.10.1 Priors and the Selection Functional

MEM has been very successful with problems for which our state of knowledge can be translated into a prior, proper, or improper, plus a set of constraints. Our results in this chapter can explain why. The linearized probability functional,

$$\log \rho_f[h|f] = -\int h(x) \log \frac{h(x)}{w(x;\, f)} dx - \log \omega \tag{7.116}$$

is of the form in Eq. (7.115) (as a constant, the partition function in Eq. (7.116) is of no consequence). In this form we identify $m(x)$ in Eq. (7.115) as the cluster function, essentially as the bias of x. The unbiased case $w(x;\, f) = 1$ corresponds to $m(x) = 1$, which in MEM represents the *uninformed prior*, an improper prior distribution that does not satisfy normalization in $x \in [0, \infty)$. In our view, $w = 1$ is "uninformed *bias*," a function that represents our state of being "uninformed" over the entire domain of the unknown distribution. It is not a prior because $w(x;\, f)$ is not a distribution. Using $w(x|f) = qe^{\beta x}/f(x)$ in Eq. (7.116) the result is

$$\log \rho_f[h|f] = -\int h(x) \log \frac{h(x)}{f(x)} dx. \tag{7.117}$$

This is the same probability functional as in (7.116) but now the quantity that appears in the denominator inside the logarithm is $f(x)$, a *proper prior*. Apart from the constant $\log \omega$, Eqs. (7.116) and (7.117) have the same mathematical structure: in one case we have the derivative of the bias and in the other the prior itself, both occupying the same slot of the functional. Both functionals are maximized by the same function. Intuition lead Jaynes onto something real but without the theory to explain it, it is easy to see why a variational derivative that sits in the slot of a proper prior would be confused for an improper prior.

Note 7.5 (Invariant Measure)

Jaynes justified the presence of $m(x)$ as a mathematical requirement for passing from the discrete to the continuous domain. He argued that the continuous entropy functional,

$$S[h] = -\int h(x) \log h(x) dx,$$

is a mathematically improper extension of the discrete functional into continuous space and it is for this reason that it lacks invariance under transformation of variables. Starting with the discrete entropy functional,

$$S[h_i] = -\sum_i h_i \log h_i.$$

Jaynes defined $m(x)$ as the density of phase space when the discrete variable i is transformed into the continuous variable x. In the limit that the number of discrete points i in a finite interval of the continuous phase space x goes to infinity we obtain

$$S[h_i] \to -\int h(x) \log \frac{h(x)}{m(x)} dx \equiv S[h].$$

Jaynes took the integral on the right-hand side to be the proper analogue in continuous space of Shannon's discrete entropy. This derivation is detailed in Jaynes (1983, Brandeis Lectures) and the implications are further elaborated in Jaynes (1968). Jaynes argued that the presence of $m(x)$ was necessitated by invariance in continuous phase space, and called $m(x)$ invariant measure.

We do not agree with Jaynes on this. Invariance is satisfied provided that entropy is defined as

$$S[H] = -\int H(x) \log \frac{H(x)}{J_0[H]} dx,$$

with $J_0[H]$ the zeroth order moment of H. This integral is the limit of discrete entropy,

$$\log \mathbf{n}! = -\sum_i n_i \log \frac{n_i}{N}$$

when the first moment of $\mathbf{n}$ is much larger than the zeroth moment, and amounts to uniform density of discrete points per unit of continuous phase space. The apparent lack of invariance is an aftereffect of normalizing the distribution. In our view it is important to consider the normalized distribution as a special member of a homogeneous family of extensive distributions. Homogeneity is a crucial property that allows sampling to converge to the same distribution regardless of sample size, provided it is large enough.

Jaynes is in fact anticipating the linear cluster bias $w(x)$. If the density of phase space is uniform, the relevant functional is entropy and its maximization with no further constraints beyond normalization and fixed mean produces the exponential distribution. By allowing the density to be nonuniform it is possible to obtain different distributions. These may be viewed as priors and be constrained with additional conditions to infer the unknown distribution. It is wrong, however, to associate the biasing of phase space with continuous distributions exclusively. The microcanonical functional in the cluster ensemble is

$$-S[\mathbf{n}] + \sum_i n_i \log w_{i|\mathbf{n}} = -\sum_i n_i \log \frac{n_i}{w_{i|\mathbf{n}}},$$

which is the discrete form of Jaynes's continuous entropy with the invariant measure included, and is precisely the functional whose maximization produces the unknown distribution. Jaynes's intuition is strong and brings him very close to the answer. He is looking through the fog at three related functionals,[14]

$$-\int h(x)\log h(x)dx;$$

$$-\int h(x)\log\frac{h(x)}{w(x;f)}dx;$$

$$-\int h(x)\log\frac{h(x)}{f(x)}dx,$$

all relevant to the problem, but he does not have the formalism to properly distinguish between them.

★

7.10.2 *Inference Versus Deduction*

Even though the motivation for our work is based on ideas that originated with Jaynes, MEM and variational thermodynamics address two quite different problems. MEM deals with the problem of inferring an unknown probability distribution from a small number of observables and possibly other prior information about the system. With generalized thermodynamics we seek to establish the relationship between a model for a stochastic system, expressed by the functional W, and the resulting distribution. We are mostly interested in the forward problem, in which W is given and the unknown distribution must be evaluated. The problem shares a superficial similarity to inference in that in both cases the unknown is a distribution about which we have some information. But ours is a problem of *deduction*, not one of *inference*. Given W, f is evaluated unambiguously.

Despite such differences, MEM and variational thermodynamics share common machinery: in both cases the unknown distribution is obtained through the constrained maximization of a functional. How this is done differs in each case. In MEM the functional to maximize is entropy (with or without the invariant measure[15] $m(x)$) under a set of constraints. These constraints always include the normalization condition, plus any number of other integral constraints, as the problem requires. In generalized thermodynamics the functional that is maximized is

[14] A fourth functional that belongs with other three,

$$-\int h(x)\log\frac{h(x)}{w(x;h)}dx,$$

does not appear in Jaynes's treatment.

[15] The MEM literature is not very clear on exactly how to handle the invariant measure.

$$S[h] + \log W[h]$$

and the maximization is conducted under two constraints that are *always* the same: normalization, and the condition that fixes the mean. Assuming the invariant measure is set to 1, the maximization in MEM is equivalent to unbiased sampling of the phase space with the role of the bias relegated to constraints that limit the search in the region of phase space where the unknown function resides. In variational thermodynamics this is accomplished by the selection bias. Since it is possible to obtain the same distribution either through MEM or through the thermodynamic functional, it should be possible to trade the bias of our development for an appropriate constraint. We show one way to do this.

Consider the MEM problem

$$\max_h - \int h(x) \log h(x) dx \tag{7.118}$$

with two constraints, one that fixes the normalization,

$$\int h(x) dx = 1, \tag{7.119}$$

and one of the form

$$\int h(x) a(x) dx = \bar{a}, \tag{7.120}$$

where $a(x)$ is some function. The function that maximizes entropy under these constraints is

$$f(x) = e^{-\lambda_1 - \lambda_2 a(x)}. \tag{7.121}$$

We now write f in canonical form

$$f(x) = w(x, \bar{x}) \frac{e^{-\beta}}{q} = e^{-\log q - \beta(x - \log w(x, \bar{x})/\beta} \tag{7.122}$$

Comparing with (7.121) we see that if we set

$$\lambda_1 = -\log q, \quad \lambda_2 = \beta, \quad a(x) = \frac{\log w(x, \bar{x})}{\beta}, \tag{7.123}$$

then the constraint in Eq. (7.120) becomes

$$\int h(x) \frac{\log w(x, \bar{x})}{\beta} dx = \frac{\log W[f]}{\beta} = \text{fixed}. \tag{7.124}$$

Maximization of entropy under the constraints in Eqs. (7.119) and (7.120) produces f as its MPD. It is possible therefore to trade the bias (and the constraint on $\bar{x}$) for the new constraint in Eq. (7.124). The location of f in phase space can be pinpointed either through the selection functional or through an appropriate constraint under unbiased sampling.

7.11 What Is W?

Once the selection functional W is specified, the MPD is fixed. But what *is* W? The selection functional is where our knowledge about the problem resides. The type of problem we have in mind is a stochastic process with fixed rules of sufficient detail to prescribe the behavior of the system. We have seen one example already: statistical mechanics. The stochastic process is the microstate of a material system. The physical model is that all possible microstates with the same energy, volume, mass, and composition are equally probable. This translates mathematically to

$$W[h] = 1,$$

where h is a distribution of microstates in the molecular phase space defined by the position and momenta vectors of all particles. With the selection bias specified, all thermodynamics follows. The assignment $W[h] = 1$ does not come from thermodynamics. It is a *model* assumption[16] external to thermodynamics. The same approach may be applied to any stochastic process. We are not yet at the point where we can propose a general theory of thermodynamics for generic stochastic processes, but we may sketch the approach for the type of problems we will consider in the remainder of this book.

Binary clustering is the process by which two clusters of size i and j merge to form a cluster of size $i + j$. Binary fragmentation is the splitting of a cluster of size i into two clusters of size $i - j$ and j. Both processes can be viewed as a Markov chain that converts a parent distribution $\mathbf{n}'$ into an offspring distribution $\mathbf{n}$,

$$\mathbf{n}' \xrightarrow{(i),(j)} \mathbf{n}$$

with a transition probability that is specified by the model about the process. Both clustering and splitting are *irreversible* process in the sense that the Markov chain of distributions $\mathbf{n}$ is not stationary. If we implement binary clustering starting with a distribution of all monomers, $n_i = M\delta_{1,i}$—or any other distribution, for that matter—the process will continue until all mass is collected in a single cluster, $n'_i = \delta_{M,i}$. All states in this process are transient. The same is true for splitting. The

[16] We may call it a *physical* assumption, if we are not concerned about the philosophical distinction between physical *reality* and *models* about this reality.

Markov chain of parent-offspring relationships may be viewed as a sequence of cluster ensembles,

$$(M, N') \xrightarrow{(i),(j)} (M, N' + \delta N')$$

where M is the total mass (first moment) of $\mathbf{n}'$ and N' is the total number of clusters (zeroth moment). For merging $\delta N' = -1$, for splitting $\delta N' = +1$. This is an example of a quasistatic (M, N)–process on the Ω surface

$$\log \Omega = \beta M + (\log q) N$$

under fixed W. In this case, the model is in the transition probabilities from parent to offspring. The master equation for this transition is

$$P(\mathbf{n}) = \sum_{\mathbf{n}'} P(\mathbf{n}) P_{\mathbf{n}' \to \mathbf{n}}$$

where $P(\mathbf{n}|\mathbf{n}')$ is the transition probability, and the summation goes over all parents of $\mathbf{n}$. The canonical probabilities of parent and offspring are

$$P(\mathbf{n}') = \frac{\mathbf{n}'! W(\mathbf{n}')}{\Omega'}, \qquad P(\mathbf{n}) = \frac{\mathbf{n}! W(\mathbf{n})}{\Omega},$$

with

$$\Omega' = \Omega_{M,N-\delta N}; \qquad \Omega = \Omega_{M,N}.$$

Upon substitution of the canonical probabilities into the master equation we obtain

$$\frac{\Omega_{M,N-\delta N}}{\Omega_{M,N}} = \sum_{\mathbf{n}'} \frac{\mathbf{n}'!}{\mathbf{n}!} \frac{W(\mathbf{n}')}{W(\mathbf{n})} P_{\mathbf{n}' \to \mathbf{n}}.$$

We recognize the right-hand side is the discrete form of the canonical parameter q and so the above equation splits into two, one for the partition function

$$\frac{\Omega_{M,N-\delta N}}{\Omega_{M,N}} = q, \tag{7.125}$$

and one for the summation over parents,

$$\sum_{\mathbf{n}'} \frac{\mathbf{n}'!}{\mathbf{n}!} \frac{W(\mathbf{n}')}{W(\mathbf{n})} P_{\mathbf{n}' \to \mathbf{n}} = q. \tag{7.126}$$

Equation (7.125) is a recursion for the partition function and Eq. (7.126) a recursion for the selection functional. In both cases the current values at state (M, N) can be calculated from those at $(M, N - \delta N)$, provided that $q = q(M/N)$ is known. The splitting of the master equation into two separate recursions can be done within any multiplicative constant $a = a(M/N)$ that leaves $q(M/N)$ indeterminate. We will fix q by imposing *thermodynamic consistency* on the parent-offering transition: We require q to be such that when the solution is the *unbiased* ensemble, i.e., an ensemble in which all cluster configurations are equally probable, then we must have

$$W(\mathbf{n}) = 1 \tag{7.127}$$

for all $\mathbf{n}$, and all M, N. Returning to Eq. (7.126), we note that the summation is over all parents of fixed distribution $\mathbf{n}$, where $\mathbf{n}$ is any distribution in the (M, N) ensemble. We may express this equation in the form

$$W(\mathbf{n}) = \frac{1}{q} \sum_{\mathbf{n}'} \frac{\mathbf{n}'!}{\mathbf{n}!} W(\mathbf{n}') P_{\mathbf{n}' \to \mathbf{n}},$$

which we may also write as

$$W(\mathbf{n}) = \frac{1}{q} \sum_{\mathbf{n}'} W(\mathbf{n}') P_{\mathbf{n}' \to \mathbf{n}}\, e^{\Delta S_{\mathbf{n}' \to \mathbf{n}}} \tag{7.128}$$

with $\Delta S_{\mathbf{n}' \to \mathbf{n}} = \log \mathbf{n}! - \log \mathbf{n}'$. This now tells us exactly how to convert the model into a selection functional: via the transition probabilities $P_{\mathbf{n}' \to \mathbf{n}}$. This recursion—or differential equation in the continuous phase space—allows us to obtain the selection functional of distribution $\mathbf{n}$ in ensemble (M, N) from that of its parents $\mathbf{n}'$ in the parent ensemble $(M, N - \delta N)$. Equations (7.125) and (7.128) are the transport equations in phase space. They apply to any process in which M is fixed and N varies.

The procedure described in this section are meant as an outline of the method we will use in Chaps. 8, 9, and 10. There we consider specific examples and models that will make the details of this methodology more transparent.

Note 7.6 (On Thermodynamic Consistency)

We understand *thermodynamic consistency* in the following context. We assume that the model that assigns transition probabilities depends on parameters that, while fixed during the process, they may be varied to produce a new set of transition probabilities under the same general model. In the case of aggregation, this parameter is the aggregation kernel K_{ij}. It is on the manipulation of such parameters that we impose consistency: If for certain values of the parameters the cluster weights are a constant number a for all clusters, we require this constant to be $a = 1$ (in aggregation this is the case when $K_{i,j} = 1$). We revisit this topic in Chap. 8.

★

Appendix: Calculus of Variations

We give here a brief review of some tools from the calculus of variations that are useful in handling the functionals that appear in the continuous domain. The review is based on Gelfand and Fromin (2000), a recommended reference that provides a more detailed presentation at a level accessible to most readers with basic background in calculus. Variational calculus is the study of continuous functionals and the conditions that define their extrema. One of the most basic types of functionals is one that can be expressed in the form[17]

$$J[h] = \int F(x, h)dx, \tag{7.129}$$

where $h = h(x)$ is a function that we treat as a variable in J and $F(x, y)$ is some function (not functional) of its arguments. The functionals we encounter in the cluster ensemble are either of this form, or can be expressed in terms of such functionals. Here are some examples:

$F(x, y)$	$J[x]$	**common name**
$x^k y$	$\int x^k h(x)dx$	moment of order k
$-y \log y$	$-\int h(x) \log h(x)dx$	entropy
$y \log w(x, \bar{x})$	$\int h(x) \log w(x, \bar{x})dx$	$\log W$

[17] *Functional* is a very general term for any mapping between a function and scalar. Here are some examples that do not conform to Eq. (7.129):

$$J[h] = h(x_0); \quad J[h] = \max_x h(x); \quad J[h] = \exp\left(\int h(x)dx\right).$$

Variation of Functional

A functional $J[h]$ is linear in h if it satisfies the conditions,

$$J[\lambda h] = \lambda J[h], \quad J[h_1 + h_2] = J[h_1] + J[h_2],$$

for any scalar λ and any h, h_1, h_2 in the domain of admissible functions. An example of a linear functional is

$$J[y] = \int a(x)h(x)dx, \tag{7.130}$$

where $a(x)$ is some function of x. Other forms of linear functionals are possible.[18] However, if $J[h]$ is of the form in Eq. (7.129) *and* it is linear, then it must be of the form in Eq. (7.130).[19]

The variation δJ of functional J is the change in its value when function h changes by δh and is analogous to the differential of regular functions (Fig. 7.2). If we change h by dh to $h + \delta h$, the corresponding change in J is

$$\delta J[h] = J[h + \delta h] - J[\delta h]. \tag{7.131}$$

If the functional is linear, then

$$\delta J[h] = \int a(x)\delta h(x)dx, \tag{7.132}$$

---

[18]For example, the functional

$$J[h] = \int \left(a_0(x)h(x) + a_1(x)h'(x) + a_2(x)h''(x) + \cdots\right)dx,$$

where h' is the first derivative of h, h'' is the second derivative, and so on, is also linear in h. This form is not of any relevance to our work.

[19]Linearity requires $J[\lambda h] = \lambda J[h]$ for all h and if F is of the form in Eq. (7.129), then

$$\int F(x, \lambda h)dx = \lambda \int F(x, h)dx.$$

which implies that F is homogeneous in h with degree 1. The Gibbs-Duhem equation is

$$\frac{\partial F}{\partial h} = 0,$$

which requires F to be independent of h, or $F(x, y) = a(x)$.

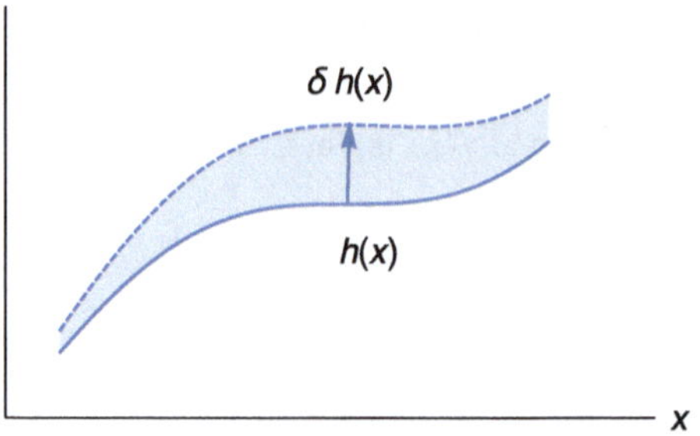

Fig. 7.2 Schematic representation of δh

We may interpret $a(x)$ as the derivative of linear functional with respect to h. We extend this to general functional. A functional J is differentiable if δJ in the limit $\delta h \to 0$ becomes a linear functional in δh. If we indicate this functional by $J'[h]$, in the vicinity of h we have

$$\delta J[h] = J'[x; h]\delta h. \tag{7.133}$$

We interpret $J'[h]$ as the derivative of the functional with respect to h. We may express this linear relationship in the form of Eq. (7.130),

$$J[y] = \int J'[x; h]y(x)dx, \tag{7.134}$$

where $y = h + \delta h$ is a function in the vicinity of h.[20] If we extend this functional to all functions y, we obtain a new functional,

$$\Phi[y; h] = \int J'[x; h]y(x)dx, \tag{7.135}$$

that is linear in y and has the same value *and* the same derivative at $y = h$ as the original functional $J[h]$. The functionals Φ and J are generally different from each other unless J is linear. Equation (7.135) represents a linear extrapolation of J from $y = h$.

Relevance to Generalized Thermodynamics

The fundamental functional in ensemble theory is log W. In general, log $W[h]$ is a nonlinear functional of distribution h, but since the cluster ensemble converges to the most probable distribution f, only distributions in the vicinity of f are relevant and for distributions in this narrow region we treat log W as linear.

[20]The notion of *vicinity* implies that we have some measure to determine the distance between two functions. There are various ways to define such measures, but we will not go into these details here. Interested readers are referred to Gelfand and Fromin (2000).

Functional Derivative

For functionals of the form in (7.129), the functional derivative is

$$\frac{\delta J[h]}{\delta h} = \left(\frac{\partial F(x,z)}{\partial z}\right)_{z=h(x)}. \tag{7.136}$$

This derivative is calculated as follows: treat the integrand of Eq. (7.129) as a regular function of h, and h as a regular variable. The derivative of the integrand with respect to h is the variational derivative. For example, the functional

$$J[h] = \int x^k h(x)dx,$$

is of the form in Eq. (7.129) with $F(x,z) = x^k z$. The functional derivative is

$$\frac{\delta J[h]}{\delta h} = \left(\frac{\partial (x^k z)}{\partial z}\right)_{z=h(x)} = x^k\Big|_{z=h(x)} = x^k.$$

In this case the derivative is independent of h because the functional is linear. As a second example we consider the intensive entropy functional

$$J[h] = -\int h(x)\log h(x)dx. \tag{7.137}$$

In this case $F(x,z) = -h\log h$ and the functional derivative is

$$\frac{\delta J}{\delta h} = -\log h - 1. \tag{7.138}$$

Relevance to Generalized Thermodynamics

The functional derivative is a function of x that depends on h. As we see in Eq. (7.138), the right-hand side is a function of x whose functional form depends on h. Our notation $w(x; h)$ expresses this connection to both x and h. If h is linear, its derivative is a pure function of x, the same function for all h, as we see in Eq. (7.137). We use the notation $w(x)$ to indicate linear functionals.

Homogeneity

Euler's theorem for homogeneous functions extends to homogeneous functionals. Let $J[h]$ be homogeneous in h with degree 1, i.e.,[21]

$$J[\lambda h] = \lambda J[h].$$

We discretize the x axis into a set of points $(x_1, x_2 \cdots)$ at which h receives the corresponding values $h_1, h_2 \cdots$. In this discretized space $J[h]$ becomes $J(h_1, h_2 \cdots)$ which may now be treated as a regular function of the h_i. Euler's theorem gives,[22]

$$J[h] = \sum h_i \frac{\partial J(h_1, h_2 \cdots)}{\partial h_i}, \tag{7.139}$$

where $\delta J[h]/\delta h$ is the derivative of the functional, namely, the change in J when h changes by δh. Passing from the discrete to the continuous limit,

$$J[h] = \int h(x) \frac{\delta J[h]}{\delta h} dx = \int h(x) a(x; h) dx, \tag{7.140}$$

where $a(x; h)$ is the functional derivative of J. This expresses Euler's theorem of homogeneous functional in h with degree 1.

Gibbs-Duhem Equation

Let us calculate the variation of J in Eq. (7.140) upon a small change δh:

$$\delta J[h] = \int a(x; h)\delta h(x) dx + \int h(x)\delta a(x; h) dx. \tag{7.141}$$

For small δh the variation δJ is given by the linear functional

$$\delta J[h] = \int a(x; h)\delta h(x) dx. \tag{7.142}$$

[21] The theorem extends to any degree of homogeneity.

[22] If $f(x_1, x_2 \cdots)$ is homogeneous in x_i with degree 1, then

$$f = x_1 f_1 + x_2 f_2 \cdots$$

where f_i is the derivative of f with respect to x_i.

Then we must have

$$\int h(x)\delta a(x; h)dx = 0. \tag{7.143}$$

This expresses the Gibbs-Duhem equation that is associated with the Euler Equation (7.140). Here is how to understand this result. The functional derivative is a function of x that depends on h. If h is changed by δh, a will also be changed. The total change integrated over h is not free to have any value, it must be zero. This relationship is imposed by the homogeneity condition.

The Gibbs-Duhem equation is satisfied for all variations in h. We may consider variations along some specific path by varying $h(x)$ via some parameter t. For example, we could take h to be the distribution $e^{-x/\bar{x}}/\bar{x}$ and use $\bar{x}$, or any function $t = t(\bar{x})$, as a parameter to vary h. Along this path a changes in response to changes in t. If we divide Eq. (7.143) by dt we obtain

$$\int h(x)\frac{\partial a(x; h)}{\partial t}dx = 0, \tag{7.144}$$

where we have interpreted $\delta a/dt$ as the derivative of a with respect to t, since the observed change in a is entirely due to dt. This can be expressed more simply as

$$\frac{\overline{\partial a(x; h)}}{\partial t} = 0, \tag{7.145}$$

where the bar indicates the mean operator over distribution h. This condition is a property of the homogeneous functional of which a is a derivative, not a property of h; it applies to any h along any path.

Relevance to Generalized Thermodynamics

The logarithm of the selection bias is homogeneous in h with degree 1. According to Eq. (7.140) we have,

$$\log W[f] = \int h(x)\frac{\delta \log W}{\delta h}dx = \int h(x)\log w(x; h)dx. \tag{7.146}$$

The derivative of $\log W$ with respect to h is the cluster function $w(x; h)$. If $\log W$ is linear, then $\log w$ is pure function of x, i.e., $w = w(x)$. If $\log W$ is not linear, $w(x; h)$ is a function of x and a functional of h.

Along the quasistatic path, $h = f$ and f is a parametric function of $\bar{x}$, i.e., $f = f(x, \bar{x})$. Applying Eq. (7.144) with $h = f$, $J = \log W$, and $t = \bar{x}$ we have

$$\int f(x, \bar{x})\frac{\partial \log w(x, \bar{x})}{\partial \bar{x}}d\bar{x} = 0. \qquad [7.143]$$

This result was used to obtain the relationship between $\log q$, β, and $\bar{x}$ in Eq. (7.31).

Functional Derivative of Extensive Entropy Functional

An important homogeneous functional in ensemble theory is the entropy functional, which we define as

$$S[h] = -\int h(x) \log \frac{h(x)}{J_0[h]} dx, \tag{7.147}$$

with

$$J_0[h] = \int h(x) dx. \tag{7.148}$$

This defines the entropy functional of extensive distribution h and involves a second functional, J_0, that represents the area under the distribution. We will calculate the functional derivative of this functional by allowing any variations δh without requiring the area under h to be constant. We refer to this as the unconstrained derivative of entropy to distinguish it from that when the normalization constraint is imposed. First we write the functional in the form

$$S[h] = -\int h(x) \log h(x) dx + J_0[h] \log J_0[h]. \tag{7.149}$$

We will calculate the derivative of each term separately. The first term is of the form in Eq. (7.129) with $F(x, z) = -z \log z$ and its derivative is

$$\frac{\delta}{\delta h}\left(-\int h(x) \log h(x) dx\right) = \left(\frac{\partial F}{\partial z}\right)_{z=h} = (-\log z - 1)_{z=h} = \\ -\log h(x) - 1. \tag{7.150}$$

For the second term we have

$$\frac{\delta (J_0 \log J_0)}{\delta h} = \left(\frac{\delta J_0}{\delta h}\right) \log J_0 + J_0 \left(\frac{\delta \log J_0}{\delta h}\right) = \\ \left(\frac{\delta J_0}{\delta h}\right) \log J_0 + \left(\frac{\delta J_0}{\delta h}\right) = \\ \left(\frac{\delta J_0}{\delta h}\right) (\log J_0 + 1). \tag{7.151}$$

J_0 is of the form in Eq. (7.129) with $F(x, z) = z$ and its derivative is

$$\frac{\delta J_0}{\delta h} = \left(\frac{\partial F}{\partial z}\right)_{z=h} = 1\Big|_{f=h} = 1. \tag{7.152}$$

Combining these results we obtain the functional derivative of entropy:

$$\frac{\delta S[h]}{\delta h} = -\log \frac{h(x)}{J_0[h]}. \tag{7.153}$$

Using this result the entropy functional can be expressed as

$$S[h] = \int h(x)\frac{\delta S[h]}{\delta h}dx, \tag{7.154}$$

which is a statement of Euler's theorem and demonstrates the applicability of the theorem to functionals.

The Gibbs-Duhem Equation for Entropy

We demonstrate the Gibbs-Duhem equation applied to entropy with an example. We take h to be the exponential distribution,

$$h(x) = \frac{e^{-\bar{x}}}{\bar{x}}, \tag{7.155}$$

and use $\bar{x}$ as a parameter, such that by varying $\bar{x}$ we allow h to trace a path in the phase space of distributions. The functional derivative of entropy for this choice of h is obtained by applying Eq. (7.153) to the exponential function (recall that in this case h is normalized to unit area)

$$a(x; h) = -\log h(x) = -\frac{x}{\bar{x}} - \log \bar{x},$$

and

$$\frac{\partial a}{\partial \bar{x}} = \frac{x}{\bar{x}^2} - \frac{1}{\bar{x}}.$$

We now calculate the integral

$$\int h(x)\frac{\partial a}{\partial \bar{x}}dx = \int \left(\frac{x}{\bar{x}^2} - \frac{1}{\bar{x}}\right)\frac{e^{-\beta x}}{\bar{x}}dx = \frac{\bar{x}}{\bar{x}^2} - \frac{1}{\bar{x}} = 0.$$

The result is zero in agreement with the Gibbs-Duhem equation given in Eq. (7.143). If we choose $t = t(\bar{x})$, where t any function of $\bar{x}$ we have

$$\int h(x)\frac{\partial a}{\partial t}dx = \frac{dt}{d\bar{x}}\int h(x)\frac{\partial a}{\partial \bar{x}}dx = 0,$$

which again is zero. We may try this with any other distribution: the Gibbs-Duhem equation is an identity by virtue of homogeneity, independently of the details of the distribution or the path.

Maximization

If $J[h]$ has an extremum (maximum or minimum) for some $h = h^*$, then its variation at that function is zero,

$$\delta J[h^*] = 0,$$

by analogy to the condition $dy = 0$ for regular functions. Whether this extremum is a maximum or a minimum is determined by the sign of the second variation; we will not get into the details of the second variation here and we will assume instead that we know that the extremum is a maximum. For the functional of the form in Eq. (7.129) this condition is equivalent to the Euler equation,

$$\left.\frac{\partial F(x,y)}{\partial y}\right|_{y=h^*} = 0.$$

This is easily extended to constrained maximization. Suppose we want the maximum of $J[h]$ with respect to h under the constraints,

$$\int h(x)dx = A; \qquad \int xh(x)dx = B. \tag{7.156}$$

Using Lagrange multipliers, the equivalent unconstrained problem is the maximization of the functional

$$\max_h \left\{ J[h] + \lambda_1\left(A - \int h(x)dx\right) + \lambda_2\left(B - \int xh(x)dx\right)\right\}, \tag{7.157}$$

where λ_1 and λ_2 are Lagrange multipliers. This functional has the same maximum with respect to h as the one below,

$$\mathcal{J}[h] = F(x,h) - \lambda_1 h - \lambda_2 xh.$$

This is of the form in Eq. (7.129) and its Euler equation is

$$\frac{\partial F(x,h)}{\partial h} - \lambda_1 - x\lambda_2 = 0. \tag{7.158}$$

The constrained maximization of a continuous functional then is not different from that in the discrete space, if the functional is of the form in Eq. (7.129).

Relevance to Generalized Thermodynamics

The MPD maximizes the functional

$$-\int f \log \frac{f}{w} dx,$$

which is of the form in Eq. (7.129) with

$$F(x, f) = -f \log f + f \log w,$$

and its derivative is

$$\frac{\partial F}{\partial h} = -\log f - 1 + \log w.$$

The Euler equation is obtained by combining this with Eq. (7.158),

$$-\log f - 1 + \log w - \lambda_1 - x\lambda_2 = 0,$$

and its solution is

$$f = w e^{\lambda_2 x - \lambda_1 - 1}.$$

With $\lambda_2 = \beta$, $e^{\lambda_1 + 1} = q$, we obtain the canonical form of the MPD.

Chapter 8
Irreversible Clustering

The merging of two clusters to produce a new cluster that conserves mass is one of the most basic mechanisms of size change in dispersed populations. The process is known by many different names: aggregation, agglomeration, coagulation, polymerization, flocculation and others, that refer to physical aspects of the particular problem at hand. We will use *aggregation*. In all cases the basic process is the stochastic irreversible merging of two clusters with masses i and j, respectively, to form a new cluster with mass $i + j$:

$$(i) + (j) \xrightarrow{K_{ij}} (i + j).$$

The reaction takes place irreversibly from left to right with a probability that depends on K_{ij} (*aggregation kernel*) and leads to ever increasing cluster sizes. In the model we will consider here the probability of the reaction between masses i and j within a population with distribution $\mathbf{n}$ is proportional to the number of i-j pairs in $\mathbf{n}$ with proportionality constant K_{ij} that generally depends on i and j. This very generic clustering process forms the basis of a mathematical model for many important physical processes. Its mathematical formulation was first developed by Smoluchowski in the context of aggregation of colloidal particles under Brownian control but was soon recognized as a generic model for merging processes that characterize disperse populations in a variety of physical contexts, such as polymerization, granular materials, social networks, of galactic matter. Our goal in this chapter is to obtain the canonical representation of the distribution of clusters undergoing irreversible binary clustering.

T. Matsoukas, *Generalized Statistical Thermodynamics*,
Understanding Complex Systems, https://doi.org/10.1007/978-3-030-04149-6_8

8.1 The Binary Clustering Graph

In aggregation we are dealing with a population in which particles of unit mass form clusters that are subsequently altered by the rules of aggregation. This cluster ensemble forms a natural formulation for this problem. When two clusters with masses i and j in distribution $\mathbf{n}$ merge, the result is a new distribution $\mathbf{n}$; schematically,

$$\mathbf{n}' \xrightarrow{i,j} \mathbf{n}.$$

The aggregation of a pair of clusters decreases the number of clusters by 1 and conserves the total mass in the distribution. If parent distribution $\mathbf{n}'$ is a member of the ensemble $(M, N+1)$, the offspring distribution $\mathbf{n}$ belongs to the ensemble (M, N). The aggregation process may then be viewed as transition between microcanonical ensembles,

$$(M, N+1) \longrightarrow (M, N).$$

Aggregation establishes a set of parent–offspring relationship between two ensembles with the same mass and whose number of clusters differs by 1. Every distribution in the parent ensemble has an offspring in the ensemble of the next generation. The only exception is the ensemble $(M, N=1)$, which represents the fully merged state and cannot produce offspring because it contains a single cluster. Conversely, every distribution in the offspring ensemble has a parent in the ensemble of the previous generation. The only exception is the ensemble $(M, N=M)$ (fully dispersed state) in which all clusters are of unit size and cannot be produced through the merging of smaller particles. Therefore, the complete set of offspring in the ensemble $(M, N+1)$ comprises the complete microcanonical ensemble (M, N); and vice versa: the complete set of parents of distributions in ensemble (M, N) comprises the complete ensemble $(M, N+1)$.

The parent–offspring relationships can be represented in the form of a directed graph. Figure 8.1 shows the aggregation graph for $M=6$ starting from a state of all monomers and ending with a fully merged state. Nodes represent distributions (they are shown as a collection of clusters in this figure) and edges are parent–offspring relationships. The graph is arranged so that all distributions at the same horizontal level have the same number of clusters and represent the ensemble at that N. The process begins at the top of the graph with $\mu C(M, M)$ (fully dispersed state) and ends at the bottom (fully merged). Ensembles are identified by the number of clusters they contain. It is convenient to identify ensembles by their generation g, which we define as

$$g = M - N. \tag{8.1}$$

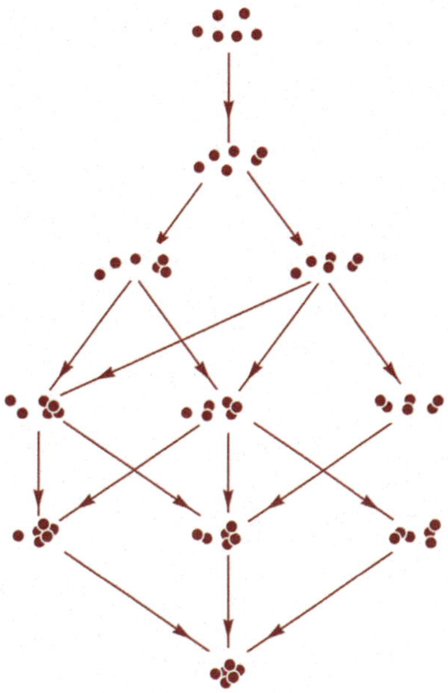

Fig. 8.1 The binary clustering graph for $M = 6$. Nodes are distributions, arrows indicate parent-to-offspring relationships. Distributions at the same horizontal location are members of the same ensemble. Distributions are represented by a configuration of clusters

Generation $g = 0$ is the fully dispersed state and has no parents; generation $g = M - 1$ is the fully merged state and has no offspring. All distributions within a generation satisfy

$$\sum n_i = M - g = N,$$

$$\sum i n_i = M,$$

which are the constraints that define the ensemble. Each generation, g contains the complete set of offspring of its previous generation $g - 1$ and forms the complete set of parents in generation $g + 1$. This is to say that every distribution is linked to at least one parent in the previous generation and at least one offspring in the next. The aggregation graph is connected and the unidirectional nature of the parent–offspring relationship means that any distribution can be reached from generation $g = 0$.

The binary clustering graph is the complete phase space of discrete finite clustering. A connected path from $g = 0$ to $g = M - 1$ is a possible trajectory and can be sampled via simulation. The transition probability from a parent to an offspring is solely determined by the state of the parent. Accordingly, aggregation is a nonstationary Markov chain of states along a connected trajectory in phase space. The goal is to obtain the probability distribution in this phase space under the rules that govern the stochastic merging of clusters.

8.2 Transition Probabilities

All physical information about the process is contained in the aggregation probability $\mathcal{P}_{i,j;\mathbf{n}}$ that cluster masses i and j in distribution $\mathbf{n}$ merge to form a cluster of mass $i + j$. By model assumption we take this probability to be proportional to the number of ways to choose cluster masses i and j out of all pairs that can be formed in distribution $\mathbf{n}$:

$$\mathcal{P}_{i,j;\mathbf{n}} = \begin{cases} C_{\mathbf{n}}\binom{n_i}{1}\binom{n_j}{1}K_{i,j} & \text{if } i \neq j; \\ C_{\mathbf{n}}\binom{n_i}{2}K_{i,i} & \text{if } i = j, \end{cases} \tag{8.2}$$

where $C_{\mathbf{n}}$ is a common factor for all pairs of clusters in distribution $\mathbf{n}$ and $K_{i,j}$ (*aggregation kernel*) is a function of the cluster masses that merge. The merging probability satisfies the normalization

$$\sum_{i=1}^{\infty}\sum_{j=1}^{i/2}\mathcal{P}_{i,j;\mathbf{n}} = 1, \tag{8.3}$$

which will be used to fix the value of $C_{\mathbf{n}}$. Equation (8.2) has a simple physical interpretation: the combinatorial terms express the probability to sample the particular pair of cluster masses, while the aggregation kernel $K_{i,j}$ biases the probability to merge the sampled pair into a new cluster. In physical terms, Eq. (8.2) represents a well-mixed system of clusters that merge via the elementary reaction

$$(i) + (j) \xrightarrow{K_{ij}} (i + j).$$

with rate constant $K_{i,j}$. The aggregation kernel $K_{i,j}$ is generally a function of the masses of the merging particles. It is a symmetric function of i, j, and is nonnegative. In addition, we normalize the kernel such that $K_{1,1} = 1$. The kernel $K_{i,j}$ embodies the physics of the process and in the literature we find numerous references to kernels derived for specific problems. We will leave the kernel general and unspecified, though certain special cases will be considered later in this chapter and in Chap. 9.

The aggregation probability in Eq. (8.2) can be expressed in the condensed form

$$\mathcal{P}_{i,j;\mathbf{n}} = C_{\mathbf{n}}\frac{n_i(n_j - \delta_{i,j})}{1 + \delta_{i,j}}K_{i,j}. \tag{8.4}$$

By substitution into Eq. (8.3) we obtain

$$\frac{C_{\mathbf{n}}}{2}\sum_{i=1}^{\infty}\sum_{j=1}^{\infty}n_i(n_j - \delta_{i,j})K_{i,j} = 1, \tag{8.5}$$

which fixes the factor $C_{\mathbf{n}}$. We define the mean aggregation kernel $\bar{K}(\mathbf{n})$ in distribution $\mathbf{n}$ as the mean value of $K_{i,j}$ over all pairs in $\mathbf{n}$:

$$\bar{K}(\mathbf{n}) = \frac{1}{N(N-1)} \sum_{i=1}^{\infty} \sum_{j=1}^{\infty} n_i (n_j - \delta_{ij}) K_{ij}. \tag{8.6}$$

Using this result in Eq. (8.5) we obtain the normalization constant in the form

$$C_{\mathbf{n}} = \frac{2}{N(N-1)\bar{K}(\mathbf{n})}. \tag{8.7}$$

The mean kernel $\bar{K}(\mathbf{n})$ generally varies among the distributions of the ensemble. Its average over all distributions of the ensemble is

$$\langle K \rangle_{M,N} = \sum_{\mathbf{n}} P(\mathbf{n}) \bar{K}(\mathbf{n}), \tag{8.8}$$

where $P(\mathbf{n})$ is the probability of distribution $\mathbf{n}$. Both $\bar{K}(\mathbf{n})$ and $\langle K \rangle_{M,N}$ will appear in subsequent results and it is important to distinguish clearly between the two:

- $\bar{K}(\mathbf{n})$ is the mean aggregation kernel within an individual distribution of the ensemble; it is a *functional* of distribution $\mathbf{n}$.
- $\langle K \rangle_{M,N}$ is the mean aggregation kernel in the entire ensemble; it is a *function* of M and N.

Note 8.1 (Counting Pairs)

The double summation in Eq. (8.3) goes over all $N(N-1)/2$ unordered pairs of cluster masses in distribution $\mathbf{n}$ (we understand $i/2$ to mean integer division). It is convenient to re-index the summation so that both i and j span the entire domain $1, \cdots \infty$:

$$\frac{1}{2} \sum_{i=1}^{\infty} \sum_{j=1}^{\infty} (1 + \delta_{i,j}) \mathcal{P}_{i,j;\mathbf{n}} = 1. \tag{8.9}$$

This summation treats $i-j$ pairs as ordered and applies the factor $(1+\delta_{i,j})/2$ as a correction for over counting. The result is a special case of an identity that applies to any symmetric function $A_{i,j} = A_{j,i}$:

$$\sum_{i=1}^{\infty} \sum_{j=1}^{i/2} A_{i,j} = \frac{1}{2} \sum_{i=1}^{\infty} \sum_{j=1}^{\infty} (1 + \delta_{i,j}) A_{i,j}. \tag{8.10}$$

We prefer to work with the summation on the right-hand side because it is symmetric in i and j and treats them as equivalent variables. As an example we set

$$A_{i,j} = \frac{n_i(n_j - \delta_{i,j})}{1 + \delta_{i,j}}.$$

We then obtain

$$\sum_{i=1}^{\infty}\sum_{j=1}^{i/2} \frac{n_i(n_j - \delta_{i,j})}{1 + \delta_{i,j}} = \frac{1}{2}\sum_{i=1}^{\infty}\sum_{j=1}^{\infty} n_i(n_j - \delta_{i,j}) = \frac{N(N-1)}{2},$$

which is the number of unordered pairs in distribution $\mathbf{n}$.

★

8.3 Parent–Offspring Relationships

A distribution $\mathbf{n}$ in the (M, N) ensemble has a finite number of parent distributions in the ensemble of the previous generation. We will adopt the convention that unprimed variables refer to the ensemble of interest, $\mu C(M, N)$, while primed variables refer to the parent ensemble $\mu C(M, N+1)$. The parent–offspring reaction that produces offspring $\mathbf{n}$ from parent $\mathbf{n}'$ via the aggregation of cluster masses $i - j$ and j is

$$\mathbf{n}' \xrightarrow{(i-j)+(j)} \mathbf{n}, \tag{8.11}$$

and represents the formation reaction of cluster size i in the offspring distribution from the merging of cluster masses $i - j$ and j in the parent distribution. We call this particular parent the "$(i - j, j)$-parent" of $\mathbf{n}$. We will now obtain the complete set of parent distributions of given distribution $\mathbf{n}$.

We start by noting that a cluster of mass $i > 1$ may be formed through the aggregation of masses $i - j$ and j, with $j = 1 \cdots i - 1$. We treat the reactants as an ordered pair and correct at the end for the fact that permutations in the order of the reactants point to the same offspring. The parent–offspring relationship can be flipped: a parent distribution of $\mathbf{n}$ is obtained by breaking a cluster with mass $i > 1$ into two integer nonzero pieces $(j, i - j)$ in that order. This process changes the number of clusters that contain i, j or $i - j$ monomers according to the following equations:

$$\begin{aligned} n'_i &= n_i - 1; \\ n'_{i-j} &= n_{i-j} + 1 + \delta_{i-j,j}; \\ n'_j &= n_j + 1 + \delta_{i-j,j}; \end{aligned} \tag{8.12}$$

All cluster masses other than i, j or $i-j$ are the same in the parent and the offspring. The complete set of the parents of $\mathbf{n}$ is obtained by letting i, j, span the range

$$i = 2, \cdots \infty; \quad j = 1, \cdots i/2. \tag{8.13}$$

The upper limit for j is $i/2$ because for $j > i/2$ the order of the fragments is reversed and we obtain the same parents as for $j < 1/2$.[1] These relationships produce the complete set of parents of distribution $\mathbf{n}$.

8.3.1 Systematic Calculation of Parents

Figure 8.2 demonstrates by example the relationships obtained here. This figure is the same as Fig. 8.1 but now the graph displays distributions rather than configurations, and the parent–offspring connections are labelled by the aggregation event that links them. Distributions are notated as vectors, $(n_1, n_2, n_3, n_4, n_5, n_6)$; the label (i, j) on a directed edge refers to the cluster masses whose aggregation transforms the parent into the corresponding offspring. As an example we obtain the parents of distribution

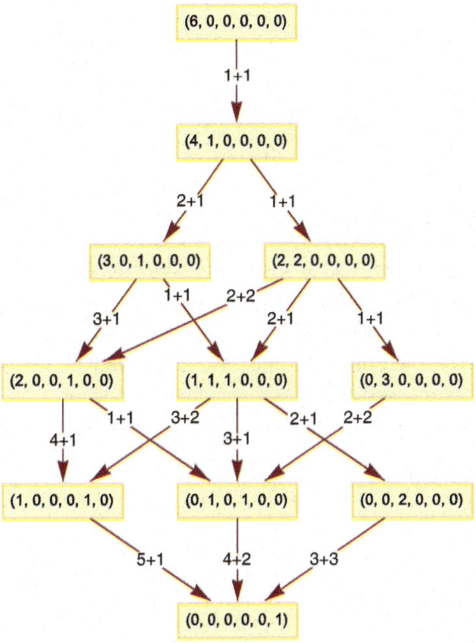

Fig. 8.2 Parent–offspring relationships on the aggregation graph for $M = 6$. Distributions are shown in vector format $(n_1, n_2, n_3, n_4, n_5, n_6)$ and the arrows indicate the cluster masses whose aggregation transforms the parent distribution into an offspring

[1] We read $i/2$ as an integer division.

$$\mathbf{n} = (1, 1, 1, 0, 0, 0)$$

for which $M = 6$, $N = 3$. The distribution contains one monomer, one dimer, and one trimer. Only the dimer and trimer can be broken up into smaller pieces. The possible fragments are

$$(1, 1), (2, 1)$$

and can be obtained systematically by breaking every cluster mass $i > 1$ with $n_i > 0$ into pairs $(i - j, j)$, with $j = 1 \cdots i/2$. Each pair points to a different parent (the pair (1,2) points to the same parent as (2, 1) and is not included). We construct the (1, 1) parent by decreasing the number of dimers by 1 and increasing the number of monomers by 2:

$$\mathbf{n}_{1,1} = (3, 0, 1, 0, 0, 0).$$

The $(1, 2)$ parent is obtained by decreasing the number of trimers by one and increasing the number of monomers and dimers by one each:

$$\mathbf{n}_{2,1} = (2, 2, 0, 0, 0, 0).$$

This calculation produces the edges

$$(3, 0, 1, 0, 0, 0) \xrightarrow{(1,1)} (1, 1, 1, 0, 0, 0)$$

and

$$(2, 2, 0, 0, 0, 0) \xrightarrow{(2,1)} (1, 1, 1, 0, 0, 0).$$

Continuing with every distribution in $\mu C(6, 3)$ we obtain the network of parent–offspring relationships with the parent ensemble $\mu C(6, 4)$; working recursively we construct the entire graph in a systematic way. This construction is numerically feasible only for small M. Nonetheless, the point of this example is to highlight the parent–offspring relationships in preparation for the analysis of the master equation in the next section.

8.3.2 *Phase Space in the ThL*

Aggregation graphs can be constructed by direct application of the parent–offspring relationships in Eq. (8.12). This can be easily programmed and is computed rather quickly for $M < 50$. Figure 8.3 shows the evolution of the aggregation graph with increasing total mass M, from $M = 6$ to $M = 14$, computed in *Mathematica*. To

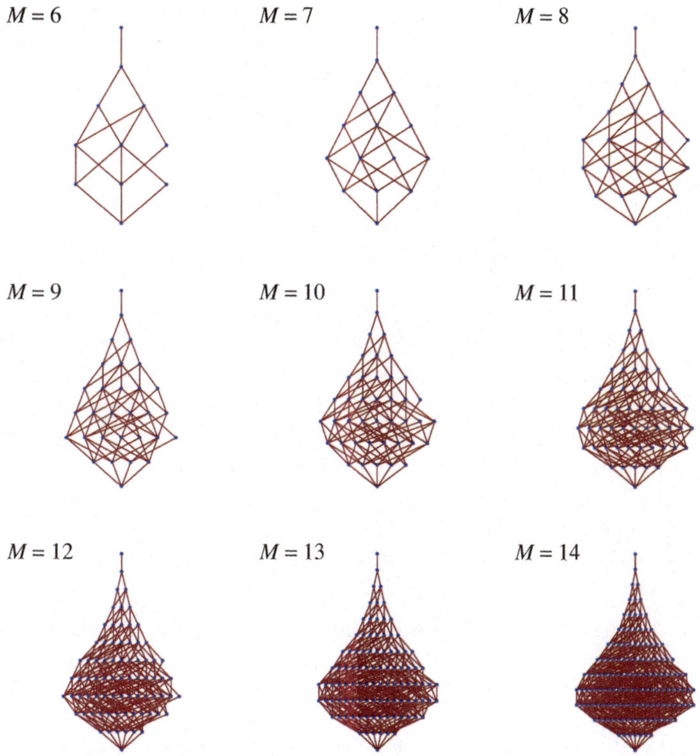

Fig. 8.3 Evolution of the aggregation graph with increasing M. Nodes represent distributions and edges represent parent–offspring relationships. These connections represent exact transitions and are calculated as explained in Sect. 8.3; the distributions themselves are not shown to avoid clutter

avoid clutter, the individual distributions are not identified in this graph but they are sorted from left to right such that they contain a larger number of smaller clusters. Generations within each M are vertically spaced by equal amounts. Similarly, distributions within a generation are spaced equally. Finally, the distance between the end states (top and bottom) and the width of the most populous generation are kept constant for all M (i.e., the density of points is the same in all generations). This amounts to scaling the vertical axis by M and the horizontal axis by the maximum value of $\mathcal{V}$. Several interesting observations can be made on this graph. The graph approaches a fixed shape and the density of states increases as M is increased, which fills up in a quasi-continuous manner as M is increased. A fixed vertical petition on any of these graphs represents a generation at the same mean cluster size. This is because the distance of a generation from the initial state is equal to N/M (recall that vertical distances are scaled by M). The width of generation N is proportional to the number of integer partitions of M into N numbers. Each partition represents a distribution but since the multiplicity of distribution is not shown, the width does not scale with the volume of the ensemble. A trajectory in phase space corresponds to a path that connects a point in one generation to a nearest neighbor in the next generation. The process begins at generation $g = 0$ at the top and ends at generation $g = M - 1$ at the bottom. There is a total of M generations, which motivates the introduction of a simple progress variables in the form of the normalized generation, $\theta \equiv g/M^2$:

$$\theta = \frac{g}{M} = \frac{M-N}{M} = 1 - \frac{1}{\bar{x}}. \tag{8.14}$$

Unlike the mean cluster mass, which diverges at the bottom of the graph when $M \to \infty$, θ collapses the entire process in the interval (0, 1). By fixing the height of all graphs to the same value, the vertical dimension of the graph, measured from the top, represents the progress variable θ. In this scaling, the approach to ThL is manifested by two features: (a) the density of states increases, which gives the graph its continuous form; and (b) the maximum average cluster size that can be reached increases indefinitely with increasing M.

As M increases, the tear-drop shape of the graph becomes stretched out towards the bottom. The width of the graph is the number of distributions in each generation normalized by the maximum width (in these figures the linear density of states is the same in all generations of a given graph). The shape of the graph reflects the number of restricted partitions of M into N parts.[3]

8.4 The Master Equation

We have established the phase space of distributions, the parent–offspring relationships from one generation to the next, and the transition probability from a parent to an offspring. We are ready to calculate the probability of distribution. We focus on a distribution $\mathbf{n}$ in the (M, N) ensemble with parents $\mathbf{n}'$ in the ensemble $(M, N+1)$. The propagation of probability $P(\mathbf{n})$ from parent to offspring is governed by the master equation

$$P(\mathbf{n}) = \sum_{\mathbf{n}' \to \mathbf{n}} P(\mathbf{n}|\mathbf{n}')P(\mathbf{n}'), \tag{8.15}$$

where $P(\mathbf{n}')$ is the probability of parent $\mathbf{n}'$ and $P(\mathbf{n}|\mathbf{n}')$ is the transition probability from parent $\mathbf{n}'$ to offspring $\mathbf{n}$ and the summation runs over all parents of $\mathbf{n}$. The transition probability is

[2] We have used θ already in Chap. 5. We now see that it has an intuitive interpretation.

[3] This number is also known as sequence A000009 in the online encyclopedia of integer sequences (see http://oeis.org/A000009).

$$P(\mathbf{n}|\mathbf{n}') = C_{\mathbf{n}'} \frac{n'_{i-j}(n'_j - \delta_{i-j,j})}{1 + \delta_{i-j,j}} K_{i-j,j}, \tag{8.16}$$

with

$$C_{\mathbf{n}'} = \frac{2}{N(N+1)\bar{K}(\mathbf{n}')}. \tag{8.17}$$

These are obtained from Eqs. (8.4) and (8.7) by replacing N with $N+1$. To continue we express the probabilities $P(\mathbf{n})$ and $P(\mathbf{n}')$ in the form:

$$P(\mathbf{n}) = \mathbf{n}! \frac{W(\mathbf{n})}{\Omega_{M,N}} \tag{8.18}$$

and

$$P(\mathbf{n}') = \mathbf{n}'! \frac{W(\mathbf{n}')}{\Omega_{M,N+1}} \tag{8.19}$$

where $\Omega_{M,N}$ and $\Omega_{M,N+1}$ are the partition functions in the offspring and parent ensembles, respectively, and W is a common functional that applies over the entire phase space represented by the binary aggregation graph. We insert Eq. (8.18) into the master equation and use the parent–offspring relationships to express the sum over $\mathbf{n}'$ as a double summation over the elements of the offspring distribution $\mathbf{n}$. The result is (the details of this derivation are given in appendix section "Equation (8.20)"):

$$\frac{\Omega_{M,N+1}}{\Omega_{M,N}} = \frac{1}{N} \sum_{i=2}^{\infty} \sum_{j=1}^{i-1} n_i \frac{K_{i-j,j}}{\bar{K}(\mathbf{n}')} \frac{W(\mathbf{n}')}{W(\mathbf{n})}. \tag{8.20}$$

In this form the summation over parents is conducted by stepping through the cluster masses of the offspring distribution $\mathbf{n}$ and systematically constructing its parents by splitting that mass into pairs. The inner summation in j goes over all binary aggregation events $(i-j, j)$ that form a cluster mass $i > 1$. Pairs in this summation are treated as ordered with corrections for over counting. Each term of the inner summation represents an ordered parent $(i-j, j)$ and contributes the factor

$$\frac{K_{i-j,j}}{\bar{K}(\mathbf{n}')} \frac{W(\mathbf{n}')}{W(\mathbf{n})}$$

where $\mathbf{n}'$ is the $(i-j, j)$-parent of distribution $\mathbf{n}$, $\bar{K}(\mathbf{n}')$ is the mean kernel of the parent, and $W(\mathbf{n}')$ is the selection bias of the parent.

8.5 Partition Function and Canonical Parameters

We express Eq. (8.20) in the equivalent form

$$\frac{\Omega_{M,N+1}}{\Omega_{M,N}} = \left(\frac{M-N}{N\,\langle K\rangle_{M,N+1}}\right) \times \sum_{i=2}^{\infty} \frac{n_i}{M-N} \left\{ \sum_{j=1}^{i-1} K_{i-j,j} \frac{\langle K\rangle_{M,N+1}}{\bar{K}(\mathbf{n}')} \frac{W(\mathbf{n}')}{W(\mathbf{n})} \right\}, \tag{8.21}$$

where $\langle K\rangle_{M,N+1}$ is the ensemble average kernel in the parent ensemble $\mu C(M, N+1)$. This equality is satisfied by all distributions $\mathbf{n}$ in $\mu C(M, N)$. Since the left-hand side is a function of M and N only, the same must be true for the double summation on the right-hand side (the factor outside the summation on the right is independent of $\mathbf{n}$ and a function of M and N only). Thus we must have[4]

$$\frac{\Omega_{M,N+1}}{\Omega_{M,N}} = \left(\frac{M-N}{N}\right) \frac{a_{M,N+1}}{\langle K\rangle_{M,N+1}}, \tag{8.22}$$

with

$$a_{M,N+1} = \sum_{i=2}^{\infty} \frac{n_i(i-1)}{M-N} \left\{ \frac{1}{i-1} \sum_{j=1}^{i-1} K_{i-j,j} \frac{\langle K\rangle_{M,N+1}}{\bar{K}(\mathbf{n}')} \frac{W(\mathbf{n}')}{W(\mathbf{n})} \right\}. \tag{8.23}$$

As we elaborate later in Sect. 8.6.3, the probability of distribution $P(\mathbf{n})$ is completely independent of the choice of $a_{M,N}$ and this implies that any choice for this function produces the correct statistics of the ensemble. To fix the value of $a_{M,N}$ we apply the condition of *thermodynamic consistency*, stated in Sect. 7.11, which requires the unbiased case to produce

$$W(\mathbf{n}) = 1$$

for all distributions $\mathbf{n}$, and all M and N. It will be shown that the function $a_{M,N}$ that satisfies thermodynamic consistency is

$$a_{M,N} = 1, \tag{8.24}$$

[4] For consistency with $\langle K\rangle_{M,N+1}$ we notate a as $a_{M,N+1}$ in Eqs. (8.22) and (8.23) to indicate its value in the parent ensemble.

but defer the proof until Sect. 8.6.3. With $a_{M,N}$ resolved, Eqs. (8.22) and (8.23) become two closed recursions, one for the partition function and one for the selection functional. To solve them we set the initial state to be the fully dispersed state $N = M$ with

$$\Omega_{M,M} = 1, \quad W(\mathbf{n}_0) = 1, \tag{8.25}$$

where $\mathbf{n}_0$ is a distribution of N unit masses (monodisperse at $i = 1$). The recursion for Ω in Eq. (8.22) is easily solved for Ω:

$$\Omega_{M,N} = \binom{M-1}{N-1} \prod_{g=0}^{M-N+1} \langle K \rangle_{M,M-g} \,. \tag{8.26}$$

The recursion for the selection functional is

$$W(\mathbf{n}) = \langle K \rangle_{M,N+1} \sum_{i=2}^{\infty} \frac{n_i}{M-N} \left\{ \sum_{j=1}^{i-1} \frac{K_{i-j,j}}{\bar{K}(\mathbf{n}')} W(\mathbf{n}') \right\}, \tag{8.27}$$

and gives the selection functional of distribution $\mathbf{n}$ in terms of the selection functional of its parents. It can be solved in certain special cases. In principle, the selection functional can be calculated by this equation working iteratively from the initial state.

Returning to Eq. (8.26), the binomial term on the right-hand side is the microcanonical volume. The second term is the product of the ensemble average kernel from the initial state (generation 0) up to the parent generation $g = M - N + 1$ of the current ensemble $\mu C(M, N)$. The right-hand side is a function of M and N (the ensemble average kernel K is a function of M and the number of clusters in the ensemble). We obtain the canonical variables β by differentiation with respect to M and N. It is easier to work with the finite difference versions of the derivatives. Thus we have

$$\beta = \log \frac{\Omega_{M+1,N}}{\Omega_{M,N}},$$

then using

$$\log \Omega_{M,N} = \log \binom{M-1}{N-1} + \sum_{g=0}^{M-N-1} \log \langle K \rangle_{M,N} \,,$$

we obtain

$$\beta = \log \frac{M}{M - N + 1} + \sum_{g=0}^{M-N-1} \log \frac{\langle K \rangle_{M+1,M-g}}{\langle K \rangle_{M,M-g}}. \tag{8.28}$$

Similarly, the discrete difference form for q is

$$\log q = \log \frac{\Omega_{M,N+1}}{\Omega_{M,N}},$$

and we find

$$q = \left(\frac{M - N}{N}\right) \frac{1}{\langle K \rangle_{M,N+1}}. \tag{8.29}$$

Equations (8.28) and (8.29) connect the canonical parameters β and q to the ensemble average kernel. The last element for the calculation of the MPD is the cluster function. This must be obtained from its definition,

$$\log \tilde{w}_i \frac{\partial \log W(\tilde{\mathbf{n}})}{\partial \tilde{n}_i}$$

with W from Eq. (8.27). The calculation of the cluster function $\tilde{w}_i$ will be possible only in special cases. We discuss them next.

8.6 Exact Results: Linear Ensembles

The coupling between $\langle K \rangle$, $W(\tilde{\mathbf{n}})$ and the MPD, makes the calculation of the MPD in closed form difficult for the general case. However we know that if the selection functional is linear, all important properties of the ensemble can be calculated in closed form. The question then is, is it possible to obtain linear selection functional under certain kernels? To answer this question we return to Eq. (8.27) and write as

$$\sum_{i=2}^{\infty} n_i \phi_i(\mathbf{n}) = M - N. \tag{8.30}$$

with

$$\phi_i(\mathbf{n}) = \sum_{j=1}^{i-1} K_{i-j,j} \frac{\langle K \rangle_{M,N+1}}{\bar{K}(\mathbf{n}')} \frac{W(\mathbf{n}')}{W(\mathbf{n})}. \tag{8.31}$$

We suppose $\log W$ to be a linear functional, i.e.,

$$\log W(\mathbf{n}) = \sum n_i w_i \tag{8.32}$$

for all $\mathbf{n}$. Considering that distribution $\mathbf{n}$ and its $(i-j, j)$-parent differ only in cluster masses $i-j$, j and i by $+1$ or -1, depending on whether the mass is on the product or reactant side of the aggregation reaction, we have[5]

$$\log W(\mathbf{n}) - \log W(\mathbf{n}') = \log w_i - \log w_{i-j} - \log w_j,$$

or

$$\frac{W(\mathbf{n})}{W(\mathbf{n}')} = \frac{w_{i-j} w_j}{w_i}. \tag{8.33}$$

Equation (8.31) now becomes

$$\phi_i(\mathbf{n}) = \sum_{j=1}^{i-1} K_{i-j,j} \frac{\langle K \rangle_{M,N+1}}{\bar{K}(\mathbf{n}')} \frac{w_{i-j} w_j}{w_i}. \tag{8.34}$$

$\phi_i(\mathbf{n})$ is a function of i and a functional of $\mathbf{n}$ through $\bar{K}(\mathbf{n}')$ on the right-hand side. Suppose that the kernel is such that

$$\bar{K}(\mathbf{n}') = \langle K \rangle_{M,N+1} \tag{8.35}$$

for all $\mathbf{n}'$ in the ensemble $(M, N+1)$. Then ϕ_i becomes a pure function of i:

$$\phi_i = \sum_{j=1}^{i-1} K_{i-j,j} \frac{w_{i-j} w_j}{w_i}. \tag{8.36}$$

This makes the right-hand side of Eq. (8.30) a linear functional. Expressing the right-hand side as

[5] This result can be confirmed for all $i > j \geq 1$ using the parent–offspring relationships.

$$M - N = \sum_i n_i(i-1), \tag{8.37}$$

we set

$$\phi_i = i - 1$$

and all equations are satisfied. From Eq. (8.36) then

$$w_i = \frac{1}{i-1}\sum_{j=1}^{i-1} K_{i-j,}, w_{i-j} w_j; \quad w_1 = 1. \tag{8.38}$$

which is a closed recursion for w_i, starting with $w_1 = 1$. The condition $w_1 = 1$ is fixed by the requirement $W(\mathbf{n}_0) = 1$ for the distribution in generation $g = 0$ and by the fact that this distribution is the product of N with a Kronecker delta at mass $i = 1$. Then

$$\log W(\mathbf{n}_0) = N \log w_1 = 0,$$

which gives $\log w_1 = 0$.

To summarize, if the kernel is of the form in Eq. (8.35) the resulting selection functional is linear and the cluster weights can be obtained from the closed recursion in Eq. (8.38).

8.6.1 Kernels That Produce Linear Ensemble

The condition in Eq. (8.35) that makes the linear functional an acceptable solution states that the mean kernel is the same in all distributions of the microcanonical ensemble and thus equal to the ensemble average kernel. This condition is satisfied by kernels of the form

$$K_{ij} = a_0 + a_1 \frac{i+j}{2} \tag{8.39}$$

where a_0 and a_1 are constants.[6] To calculate the mean kernel in distribution $\mathbf{n}$ we begin with the definition of $\bar{K}(\mathbf{n})$ in Eq. (8.6), which we write as

[6]Since we require the normalization $K_{ii} = 1$, these constants must satisfy $a_1 + a_2 = 1$.

$$\bar{K}(\mathbf{n}) = \frac{\displaystyle\sum_{i=1}^{\infty}\sum_{j=1}^{\infty} n_i n_j K_{ij} - \sum_{i=1}^{\infty} n_i K_{ii}}{N(N-1)} \tag{8.40}$$

Using K_{ij} from Eq. (8.39) the summations in the numerator give

$$\sum_{i=1}^{\infty}\sum_{j=1}^{\infty} n_i n_j K_{ij} = a_0 N^2 + a_1 N M, \tag{8.41}$$

and

$$\sum_{i=1}^{\infty} n_i K_{ii} = a_0 N + a_1 M.$$

Substituting these results in Eq. (8.41) we obtain

$$\bar{K}(\mathbf{n}) = \frac{a_0 N^2 - a_0 N + a_1 N M - a_1 M}{N(N-1)} = a_0 + a_1 \frac{M}{N},$$

which satisfies Eq. (8.35). Equation (8.39) gives the most general form of kernels that satisfy the uniform condition Eq. (8.35) for all distributions with fixed first moment M and zeroth moment N. Nonlinear combinations of i and j introduce additional moments in the summations of Eq. (8.40) that produce results that do not depend solely on M and N. Only the linear combination in Eq. (8.39) produces a mean kernel that is *uniformly distributed among all distributions of the ensemble.*

Equation (8.39) defines a family of kernels that encompasses two important cases: the constant kernel ($a_0 = 1, a_1 = 0$),

$$K_{ij} = 1; \tag{8.42}$$

and the sum kernel ($a_0 = 0$, $a_1 = 1$)

$$K_{ij} = \frac{i+j}{2} \tag{8.43}$$

We will obtain the solution for these two cases next.

8.6.2 Constant Kernel

The constant kernel, $K_{ij} = 1$, is the simplest kernel of the type in Eq. (8.39). It ascribes equal probability of merging to any two clusters regardless of size and satisfies

$$\bar{K}(\mathbf{n}) = \langle K \rangle_{M,N} = 1 \tag{8.44}$$

for all M and N. The recursion for the cluster function w_i in Eq. (8.38) gives

$$w_i = 1, \tag{8.45}$$

with

$$W(\mathbf{n}) = \prod_i w_i^{n_i} = 1, \quad \text{(for all } \mathbf{n}). \tag{8.46}$$

With $\langle K \rangle_{M,N} = 1$ the partition function in Eq. (8.26) is equal to the microcanonical volume,

$$\Omega_{M,N} = \binom{M-1}{N-1} \equiv \mathcal{V}_{M,N}. \tag{8.47}$$

The mean cluster distribution is calculated according to Eq. (2.54):

$$\frac{\langle n_k \rangle}{N} = \binom{M-k-1}{N-2} \Big/ \binom{M-1}{N-1}. \tag{8.48}$$

The parameters β and q are

$$\beta = \log \frac{M}{M-N+1} \to \log \frac{\bar{x}}{\bar{x}-1}, \tag{8.49}$$

$$q = \frac{M-N}{N} = \bar{x} - 1. \tag{8.50}$$

and are obtained from Eqs. (8.28) and (8.29) with $\langle K \rangle = 1$. The MPD is obtained by assembling w_i, β, and q into the canonical form,

$$\frac{\tilde{n}_i}{N} = \frac{1}{\bar{x}-1} \left(\frac{\bar{x}}{\bar{x}-1} \right)^{-i}. \tag{8.51}$$

It is an easy exercise to verify that this distribution is properly normalized and its mean is $\bar{x}$. The results can also be presented in terms of the progress variable θ:

$$\beta = -\log \theta; \tag{8.52}$$

$$q = \frac{\theta}{1-\theta}; \tag{8.53}$$

$$\frac{\tilde{n}_i}{N} = (1-\theta)\theta^{i-1}. \tag{8.54}$$

These are equivalent to Eqs. (8.49), (8.50), and (8.51). When $\bar{x} \gg 1$ the MPD simplifies to

$$f(x) = \frac{e^{-x/\bar{x}}}{\bar{x}}; \quad 0 \le x < \infty. \tag{8.55}$$

where $i \to x$ is the continuous cluster size. This distribution is normalized to unit area in $0 \le x < \infty$ and its mean is $\bar{x}$.

Note 8.2 (On the Significance of the Constant Kernel)

The constant kernel holds a special place in the mathematical and physical literature of aggregation. It is the simplest form of the aggregation kernel and represents the null hypothesis if nothing else about the particular system is known. It is often a reasonable approximation for the kernel that describes Brownian aggregation. Smoluchowski derived the following form for the aggregation kernel in a system of colloidal spheres of radius r and diffusion coefficient D_i:

$$K_{ij} = 4\pi \left(R_i + R_j\right)(D_i + D_j)$$

This rather simple result expresses the aggregation kernel as the product of the closest distance of approach between two spheres, $R_i + R_j$, times the mutual diffusion coefficient $D_i + D_j$. For spheres diffusing in a medium treated as a continuum fluid with viscosity η the diffusion coefficient is related to the radius by the Stokes-Einstein equation,

$$D = \frac{kT}{6\pi\eta},$$

and with this, the aggregation kernel can be expressed entirely in terms of the radii of the two spheres,

$$K_{ij} = \frac{2kT}{3\eta}\left(2 + \frac{R_i}{R_j} + \frac{R_j}{R_i}\right).$$

For equal size spheres this gives the constant value $8kT/3\eta$ regardless of the radius of the spheres. If one assumes that the aggregation rate is dominated by aggregation events between particles of the average size, the aggregation kernel may be approximately taken to assume this constant value. Smoluchowski used this approximation, formulated the governing equation for the cluster distribution, and obtained the solution. We discuss the Smoluchowski equation in Sect. 8.8, where we derive it in the context of the cluster ensemble.

★

8.6.3 Thermodynamic Consistency

With the constant kernel we obtain the unbiased ensemble with $W(\mathbf{n}) = 1$. Thermodynamic consistency is therefore satisfied, which justifies the selection $a_{M,N} = 1$ in Eq. (8.24). To prove that this is the only choice but also in order to understand the need for the requirement of consistency, we first examine how the choice of $a_{M,N}$ affects the partition function, the selection functional, and the microcanonical probability. Let Ω and W refer to the partition function and selection functional obtained with $a_{M,N} = 1$, and let Ω' and W' refer to those obtained using some arbitrary choice for $a_{M,N}$. In both cases the initial condition is given by Eq. (8.25). To maintain homogeneity $a_{M,N}$ must be an intensive function of M and N but will remain otherwise unspecified. The relationship between Ω and Ω' is established easily using Eq. (8.22):

$$\Omega'_{M,N} = \frac{\Omega_{M,N}}{A_{M,N}},$$

where

$$A(M,N) = \prod_{g=0}^{M-N+1} a_{M,M-g}. \tag{8.56}$$

First we show that $\log A_{M,N}$ in the ThL is homogeneous in M and N with degree 1. We take the logarithm of A and represent the sum with an integral,

$$\log A(M,N) = \sum_{g=0}^{M-N+1} \log a(M, M-g) \to \int_0^{M-N} \log a(M, M-g)dg.$$

Setting $M = \lambda M$ and $N = \lambda N$ in the above result we have

$$\log A(\lambda M, \lambda N) = \int_0^{\lambda(M-N)} \log a(\lambda M, \lambda M - g)dg.$$

With the substitution $g = z\lambda$ the above result becomes

$$\log A(\lambda M, \lambda N) = \int_0^{M-N} \log a(\lambda M, \lambda(M-z))d(\lambda z) = \lambda \int_0^{M-N} \log a(M, M-z)dz.$$

The integral on the far right is $\log A(M,N)$, therefore the final result is

$$\log A(\lambda M, \lambda N) = \lambda \log A(MN),$$

and proves that $\log A$ is homogeneous in M and N with degree 1. As a direct consequence of homogeneity we have the following results:

$$\log A = \beta_A M + (\log q_A)N, \tag{8.57}$$

$$\beta_A = \left(\frac{\partial \log A}{\partial M}\right)_N, \tag{8.58}$$

$$\log q_A = \left(\frac{\partial \log A}{\partial N}\right)_M, \tag{8.59}$$

$$\frac{d \log q_A}{d\beta_A} = -\frac{M}{N}. \tag{8.60}$$

Working with Eq. (8.23) recursively it is easy to establish that the relationship between W' and W is

$$W'(\mathbf{n}) = \frac{W(\mathbf{n})}{A_{M,N}} \tag{8.61}$$

where $A_{M,N}$ is the same as in Eq. (8.56). The microcanonical probability of distribution $\mathbf{n}$ is

$$P(\mathbf{n}) = \mathbf{n}!\frac{W(\mathbf{n})}{\Omega} = \mathbf{n}!\frac{W'(\mathbf{n})}{\Omega'}. \tag{8.62}$$

In both cases we obtain the *same* microcanonical probability. This is because the same factor $A(M, N)$ in the denominator of both W and Ω, leaving the ratio W/Ω unchanged. Therefore, the choice of $a_{M,N}$ has *no* effect on the statistics of the ensemble. It does affect parameters w_i', β', and q', as we show next.

We take the logarithm of both sides in Eq. (8.6.3) and evaluate the partial derivatives with respect to M and N. The result is

$$\beta' = \beta - \beta_A, \quad q' = q/q_A, \tag{8.63}$$

with β_A and q_A given by Eqs. (8.58) and (8.59). Differentiating $\log W'$ in Eq. (8.61) we obtain

$$\log w_i' = \log w_i - \left(\frac{\partial \log A}{\partial M}\right)\frac{dM}{dn_i} - \left(\frac{\partial \log A}{\partial N}\right)\frac{dN}{dn_i},$$

then using Eqs. (8.58) and (8.59) the final result is

$$\log w_i' = \log w_i - i\beta_A - \log q_A. \tag{8.64}$$

As Eqs. (8.63) and (8.64) show, the parameters β', q', and w'_i depend on $a_{M,N}$. The corresponding MPD is

$$\frac{\tilde{n}_i}{N} = w'_i \frac{e^{-\beta'_i}}{q'}, \tag{8.65}$$

but this, using Eqs. (8.63) and (8.64), is the same as

$$\frac{\tilde{n}_i}{N} = w_i \frac{e^{-\beta_i}}{q}. \tag{8.66}$$

While individually β', q', and w'_i all depend on $a_{M,N}$, the MPD they produce does not. This is not surprising since we established already that all probabilities are independent of $a_{M,N}$.

To summarize, the selection of $a_{M,N}$ has no effect on the microcanonical probabilities that are calculated from Eqs. (8.26) and (8.27) but affects the cluster function w_i and the canonical parameters β and q. Specifically, β, $\log q$ and $\log w_i$ are shifted linearly by amounts that depend on $a_{M,N}$. We encountered this situation in Chap. 7, where we showed that there is an infinite family of selection functionals that reproduce the same MPD. The condition of *thermodynamic consistency* removes the indeterminacy by fixing the selection functional relative to the unbiased functional expressed the normalized form $W(\mathbf{n}) = 1$. This sets $W = 1$ as the reference bias for all kernels.

In Sect. 7.11 we alluded to the presence of parameters in the model that allow us to reach the unbiased ensemble. In the aggregation model we are discussing here, this is achieved by setting $K_{i,j} = 1$. In certain problems we may be able to reach the condition $K_{i,j} = 1$ by manipulating some other parameter on which the kernel depends, for example temperature. Such "physical" manipulation is not necessary for the condition of *thermodynamic consistency*. To put it more concretely, the selection functional in the case of aggregation is fully determined by the kernel,

$$W = W(\mathbf{n}|K_{ij}).$$

If for some kernel K^* we have

$$W(\mathbf{n}|K^*_{ij}) = a,$$

where a is constant for all $\mathbf{n}$, then we must have $a = 1$.

8.6.4 Sum Kernel

The sum kernel, defined as

$$K_{ij} = \frac{i+j}{2}, \tag{[8.43]}$$

gives higher probability to merging events between large clusters. It is a prominent kernel in the mathematical literature of aggregation mainly because it is one of a handful of kernels for which the Smoluchowski equation can be solved analytically. The sum kernel satisfies the uniform condition with common value M/N:

$$\bar{K}(\mathbf{n}) = \langle K \rangle_{M,N} = \frac{M}{N}. \tag{8.67}$$

Using this expression for $\langle M, N \rangle$ in Eq. (8.26) we obtain the partition function of the sum kernel,

$$\Omega_{M,N} = N! \frac{M^{M-N}}{M!} \binom{M-1}{N-1}. \tag{8.68}$$

The cluster bias w_i is obtained from Eq. (8.38) with $K_{ij} = (i+j)/2$:

$$w_i = \frac{i^{i-1}}{i!}. \tag{8.69}$$

Equations (8.68) and (8.69) summarize the complete solution of discrete finite aggregation under the sum kernel. The cluster ensemble is linear and the probability of distribution $\mathbf{n}$ is

$$P(\mathbf{n}) = \frac{N!}{\Omega_{M,N}} \prod_i \frac{1}{n_i!} \left(\frac{i^{i-1}}{i!} \right)^{n_i}. \tag{8.70}$$

The mean distribution follows by application of Eq. (2.54):

$$\langle n_k \rangle_{M,N} = \frac{i^{i-1}}{i!} \cdot \frac{(M-k)^{M-N-k}}{(M-N-k+1)!} \cdot \frac{(N-1)(M-N)!}{M^{M-N-1}} \tag{8.71}$$

It is an exact result for any M and N.

Note 8.3 (Discrete Distributions for $M = 6$)

We demonstrate these results with a calculation in the cluster ensemble $M = 6$. We calculate the probability of the three distributions in $\mu C(6, 3)$,

$$\mathbf{n}_1 = (2, 0, 0, 1, 0, 0), \quad \mathbf{n}_2 = (1, 1, 1, 0, 0, 0), \quad \mathbf{n}_3 = (0, 3, 0, 0, 0, 0),$$

which represent all possible ways to distribute mass $M = 6$ into $N = 3$ clusters. The partition function from Eq. (8.68) is

$$\Omega_{6,3} = 3!\frac{6^{6-3}}{6!}\binom{5}{2} = 18$$

and the cluster functions are

$$w_i = 1, \ w_2 = 1, \ w_3 = 3/2, \ w_4 = 8/3, \ w_5 = 125/24, \ w_6 = 54/5.$$

Starting with (2, 0, 0, 1, 0, 0), its probability from Eq. (8.70) is

$$P(2, 0, 0, 1, 0, 0) = \frac{3!}{18}\left(\frac{1^2}{2!}\right)\left(\frac{8}{3}\right) = \frac{4}{9}.$$

By similar calculation we find

$$P(1, 1, 1, 0, 0, 0) = \frac{1}{2}, \quad P(0, 3, 0, 0, 0, 0) = \frac{1}{18},$$

and confirm that their sum is properly normalized to unity. We may calculate the mean distribution by ensemble averaging:

$$\langle \mathbf{n} \rangle = \mathbf{n}_1 P(\mathbf{n}_1) + \mathbf{n}_2 P(\mathbf{n}_3) + \mathbf{n}_3 P(\mathbf{n}_3).$$

For the monomer, for example, we have

$$\langle n_1 \rangle = \frac{4}{9}(2) + \frac{1}{2}(1) + \frac{1}{18}(0) = \frac{25}{18}.$$

By similar calculation we obtain the complete distribution:

$$(\langle n_1 \rangle, \langle n_2 \rangle, \langle n_3 \rangle, \langle n_4 \rangle) = \left(\frac{25}{18}, \frac{2}{3}, \frac{1}{2}, \frac{4}{9}\right)$$

i.e., the mean number of monomers is $25/18 = 1.389$, of dimers is $2/3 = 0.667$, of trimers is 0.5, of tetramers is $4/9 = 0.444$, and all other sizes are zero. We obtain the same result exactly from Eq. (8.71) with $M = 6$, $N = 2$.

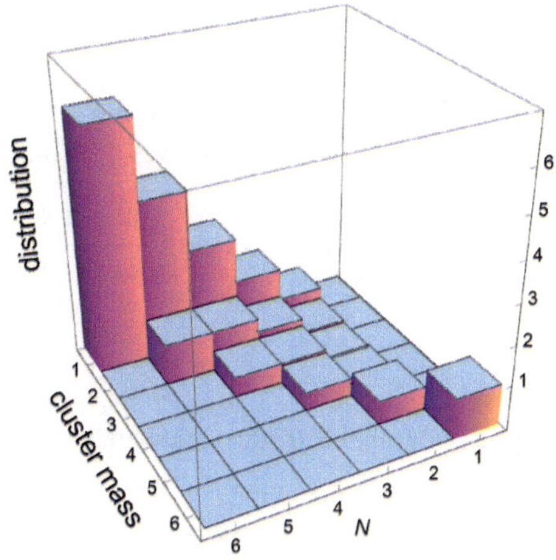

Fig. 8.4 Cluster distributions for the sum kernel in the discrete domain for $M = 6$ from Eq. (8.71). The initial state is $N = 6$ (all clusters are monomers) and the final state is $N = 1$ (a single cluster with 6 monomers)

Table 8.1 Probability of distribution for aggregation of $M = 6$ monomers under the sum kernel (the last column is the mean distribution in each generation)

Generation	N	$\Omega_{M,N}$	**n**	$P(\mathbf{n})$	$\bar{n}_1$	$\bar{n}_2$	$\bar{n}_3$	$\bar{n}_4$	$\bar{n}_5$	$\bar{n}_6$
0	6	1	(6,0,0,0,0,0)	1	6	0	0	0	0	0
1	5	5	(4,1,0,0,0,0)	1	4	1	0	0	0	0
2	4	12	(3,0,1,0,0,0)	1/2	$\frac{5}{2}$	1	$\frac{1}{2}$	0	0	0
2	4	12	(2,2,0,0,0,0)	1/2						
3	3	18	(2,0,0,1,0,0)	4/9						
3	3	18	(1,1,1,0,0,0)	1/2	$\frac{25}{18}$	$\frac{2}{3}$	$\frac{1}{2}$	$\frac{4}{9}$	0	0
3	3	18	(0,3,0,0,0,0)	1/18						
4	2	18	(1,0,0,0,1,0)	125/216						
4	2	18	(0,1,0,1,0,0)	8/27	$\frac{125}{216}$	$\frac{8}{27}$	$\frac{1}{4}$	$\frac{8}{27}$	$\frac{125}{216}$	0
4	2	18	(0,0,2,0,0,0)	1/8						
5	1	54/5	(0,0,0,0,0,1)	1	0	0	0	0	0	1

The mean distributions for all generations with $M = 6$ are shown in Fig. 8.4 and Table 8.1 summarizes the probabilities of all distributions in the aggregation graph for $M = 6$. The three-dimensional plot in Fig. 8.4 is a visual representation of the aggregation process. The horizontal plane (i, N) defined by the cluster mass i and the total number of clusters N contains defines the domain of the mean distribution. The aggregation process begins at the bottom left corner of this plane (all clusters are monomers) and ends at the upper right (a single cluster that contains all of the mass). It is interesting to notice that the distribution expands smoothly into the available range of cluster masses until in the last generation, where it abruptly collapses into the state of a single cluster. At $N = 2$, one generation before the last one, there is a hint of an imminent phase transition, with the mass of the distribution separated at the two ends (the mean distribution in Table 8.1 has two symmetric peaks at cluster masses 1 and 5). We will postpone the discussion of stability until the next chapter.

★

8.6.5 Scaling Limit

We obtain the parameters β and $\log q$ by differentiation of $\log\Omega$. Using the finite-difference form of the derivatives we obtain

$$\beta = (M-N)\log\left(\frac{M+1}{M}\right) + \log\left(\frac{M}{M-N+1}\right) \tag{8.72}$$

$$q = \frac{(N+1)(M-N)}{MN} \tag{8.73}$$

For large M and N these expressions simplify to[7]

$$\beta = \frac{M-N}{M} - \log\left(\frac{M-N}{M}\right) \tag{8.74}$$

$$q = \frac{M-N}{M}. \tag{8.75}$$

The MPD is obtained by constructing the canonical form using the above β and q along with w_i from Eq. (8.69). The result can be expressed more concisely in terms of the new progress variable

$$\theta = 1 - \frac{N}{M} = 1 - \frac{1}{\bar{x}}, \qquad [8.14]$$

defined earlier in this chapter. We then have

$$\beta = \theta - \log\theta, \tag{8.76}$$

$$q = \theta, \tag{8.77}$$

and the MPD is

$$\frac{\tilde{n}_i}{N} = \frac{i^{i-1}}{i!}\theta^{i-1}e^{-i\theta}. \tag{8.78}$$

[7]Use

$$\left(1+\frac{1}{M}\right)^M \to e; \quad \frac{M}{M-N+1} \to \frac{M}{M-N}; \quad \frac{N+1}{N} \to 1.$$

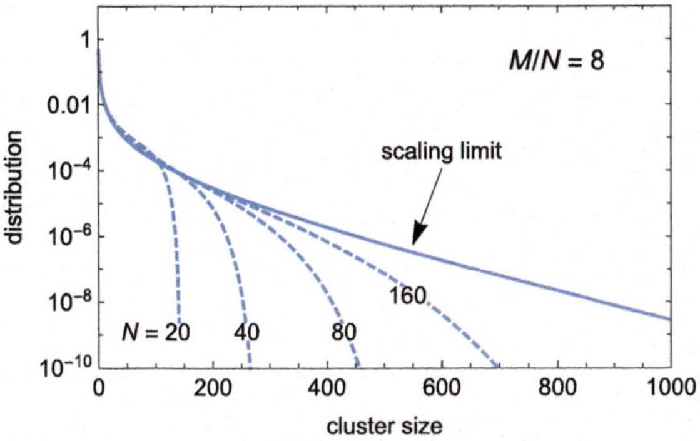

Fig. 8.5 The mean distribution for the sum kernel and its approach to the scaling limit. The dashed lines are from Eq. (8.71) at various N and fixed $M/N = 8$. The solid line is the MPD according to the scaling limit in Eq. (8.78). As N increases the mean distribution converges to the scaling form of the most probable distribution

Additional simplifications can be obtained for $\bar{x} \gg 1$. In this regime the cluster mass may be treated as a continuous function of cluster size. With the substitution $i \to x$, and the Stirling approximation for the factorial, the cluster function becomes

$$\frac{i^{i-1}}{i!} \to \frac{e^x}{x^{3/2}\sqrt{2\pi}},$$

and the most probable distribution takes the form

$$f(x) = \frac{\theta^{x-1}}{x^{3/2}} \cdot \frac{e^{-x\theta}}{\sqrt{2\pi}}. \tag{8.79}$$

Figure 8.5 illustrates the mean distribution at $M/N = 8$ ($\theta = 0.875$) and its approach to the scaling limit. The MPD drops sharply initially but then relaxes to a long exponential tail. The discrete solution from Eq. (8.71) is generally in good agreement with the scaling solution up until the cluster size is of the order of $M - N + 1$, which is the largest cluster size in the finite system. At $\bar{x} = 8$ Eq. (8.79) is indistinguishable from Eq. (8.78).

Note 8.4 (On the Scaling Form of the MPD)
The right-hand side of Eq. (8.78) satisfies the normalizations

$$\sum_{i=1}^{\infty} \frac{i^{i-1}}{i!}\theta^{i-i}e^{-i\theta} = 1,$$

$$\sum_{i=1}^{\infty} \frac{i^{i}}{i!}\theta^{i-i}e^{-i\theta} = \frac{1}{1-\theta} = \bar{x},$$

which confirm that the MPD is a properly normalized distribution with mean $\bar{x}$. The scaling expression in (8.79) treated as a continuous function of cluster size x

in the entire interval $(0, \infty)$ has somewhat problematic behavior. In particular, its first moment does not converge, due to the very rapid rise of the distribution in the vicinity of $x = 0$; its first moment, however, converges. This is an artifact of the passage into the continuous domain. In the discrete domain the distribution in Eq. (8.78) behaves properly with respect to both moments, as it should, since all distributions of the discrete cluster ensemble by definition have well-behaved zero- and first order moments.

★

8.7 Mean Distribution

Here we derive the governing equation for the evolution of the mean distribution from one generation to the next for any finite M and N.

The mean cluster distribution is

$$\langle \mathbf{n} \rangle = \sum_{\mathbf{n}} \mathbf{n} P(\mathbf{n}), \tag{8.80}$$

The mean number of clusters with mass k is the kth element of distribution $\langle \mathbf{n} \rangle$ in Eq. (8.80), i.e.,

$$\langle n_k \rangle = \sum n_k P(\mathbf{n}) \tag{8.81}$$

where $P(\mathbf{n})$ is the probability of the distribution to which n_k belongs. For convenience we define the transition matrix $T'_{i-j,j}$ as follows:

$$T'_{i-j,j} = \frac{n'_{i-j}(n'_j - \delta_{i-j,j})}{1 + \delta_{i-j,j}} \frac{K_{i-j,j}}{\bar{K}(\mathbf{n}')}, \tag{8.82}$$

so that the transition probability for the aggregation reaction of cluster sizes $i - j$ and j is

$$P(\mathbf{n}|\mathbf{n}') = \frac{T'_{i-j,j}}{N(N+1)}.$$

(We continue to follow the convention that primed quantities refer to the parent ensemble and unprimed quantities to the current (offspring) generation.) With this definition the master equation in Eq. (8.15) now becomes

$$P(\mathbf{n}) = \frac{2}{N(N+1)} \sum_{i=2}^{\infty} \sum_{j=1}^{i/2} P(\mathbf{n}') \, T'_{i-j,j}, \tag{8.83}$$

with the double summation on the right-hand side going over all aggregation events that form distribution **n** from its parents. Consider now how n_k changes when an aggregation event takes place between cluster masses $i - j$ and j. From the parent–offspring relationships in Eq. (8.12) we have

$$n_k = n'_k + \delta_{k,i} - \delta_{k,i-j} - \delta_{k,j}. \tag{8.84}$$

This result summarizes the cases in Eq. (8.12) into a single equation and gives the offspring distribution in terms of the parent distribution following the reaction

$$(i - j) + (j) \to (i).$$

We multiply this with the probability of distribution **n** from Eq. (8.83) and sum both sides over all **n**:

$$\sum_{\mathbf{n}} n_k P(\mathbf{n}) = \frac{2}{N(N+1)} \sum_{\mathbf{n}} \sum_{i=2}^{\infty} \sum_{j=1}^{i/2} P(\mathbf{n}') \left(n'_k + \delta_{k,i} - \delta_{k,i-j} - \delta_{k,j}\right) T'_{i-j,j}.$$

On the left-hand side this summation yields the mean value of n_k over the offspring ensemble.[8] On the right-hand side we obtain the sum of *all* aggregation events from *all* distributions $\mathbf{n}'$ of the *parent* ensemble $(M, N + 1)$. This is because each distribution **n** in the offspring ensemble corresponds to a *single* aggregation event in the parent ensemble. Therefore, the summation on the right-hand side over **n** amounts to a summation over $\mathbf{n}'$. And since the summation in the right-hand side is weighted by $P(\mathbf{n}')$, it represents the ensemble average of the quantity in parenthesis, evaluated over the *parent* ensemble:

$$\langle n_k \rangle_{M,N} = \frac{2}{N(N+1)} \left\langle \sum_{i=2}^{\infty} \sum_{j=1}^{i/2} \left(n_k + \delta_{k,i} - \delta_{k,i-j} - \delta_{k,j}\right) T_{i-j,j} \right\rangle_{M,N+1}, \tag{8.85}$$

where $T_{i-j,j}$ without the prime indicates that the transition matrix is applied to the offspring distribution:

$$T_{i-j,j} = \frac{n_{i-j}(n_j - \delta_{i-j,j})}{1 + \delta_{i-j,j}} \frac{K_{i-j,j}}{\bar{K}(\mathbf{n})}. \tag{8.86}$$

Now that the corresponding ensembles are indicated explicitly by the subscript on the ensemble average, the primes are no longer necessary and we treat **n** as a summation variable within the indicated ensemble. The quantity on the right-hand side is understood to be calculated first within *each* distribution of the parent

[8] The set of all parents of all offspring is the complete parent ensemble. Therefore, a summation over all parents of all offspring is a summation over the ensemble of parents.

ensemble, then averaged over all distributions of that ensemble with a weight equal to the probability of the distribution. Upon expanding the summations, the final result is (see appendix section "Smoluchowski Equation (8.87)" for further details)

$$\langle n_k\rangle_{M,N+1} - \langle n_k\rangle_{M,N} = \frac{2}{N(N+1)}\times \left\langle \frac{1}{2}\sum_{j=1}^{k-1} n_{k-j}(n_j - \delta_{i-j,j})\frac{K_{k-j,j}}{\bar{K}(\mathbf{n})} - \sum_{j=1}^{\infty} n_k(n_j - \delta_{k,j})\frac{K_{k,j}}{\bar{K}(\mathbf{n})}\right\rangle_{M,N+1} \tag{8.87}$$

This equation governs the change in the mean number of clusters of size k from the parent ensemble $(M, N+1)$ to the offspring ensemble (M, N). It is exact and applies to any M and N.

In the ThL the ensemble converges to a single distribution, the MPD. The mean distribution as well as all ensemble averages over $\mathbf{n}$ converges to the corresponding expression for the MPD. We may drop the Kronecker deltas because they represent an infinitesimal correction on n_i when $n_i \gg 1$, and similarly we replace $N + 1$ by N. Incorporating these simplifications into Eq. (8.87) we obtain

$$\frac{d\tilde{n}_k}{dN} = -\frac{2}{\bar{K}N^2}\left(\sum_{j=1}^{k-1}\tilde{n}_{k-j}\tilde{n}_j K_{k,j} - \frac{1}{2}\sum_{j=1}^{\infty}\tilde{n}_k\tilde{n}_j K_{k,j}\right) \tag{8.88}$$

where we identified the ratio of finite differences,

$$\frac{\tilde{n}_{k;M,N} - \tilde{n}_{k;M,N+1}}{\Delta N}, \tag{8.89}$$

where $\Delta N = -1$, as the derivative $d\tilde{n}_i/dN$.

8.8 Connection to the Smoluchowski Theory

The aggregation problem has a rich mathematical history that started with Smoluchowski's work in the early 1900s. Smoluchowski derived the rate equation for the distribution of clusters which forms the basis of the standard mathematical description of aggregation and is now known as the *Smoluchowski coagulation equation*.[9] In this section we derive the Smoluchowski equation in the cluster ensemble and establish the connection between our method and the standard approach in the literature that is based on rate models in the time domain.

The Smoluchowski equation is a rate equation for the evolution of the cluster size distribution, expressed as concentration c_i (number of clusters of size i per unit volume). The particle concentration satisfies the conditions,

$$\sum_i c_i = C_N, \quad \sum_i i c_i = C_M, \tag{8.90}$$

where C_N is the number concentration and C_M is the mass concentration of the particles in volume V. The rate equation is a statement of mass conservation for the reaction

$$(i) + (j) \xrightarrow{K_{i,j}} (i + j)$$

that produces cluster mass $i + j$-mer through the combination of masses i and j with rate $K_{i,j} c_i c_j$ for all $i, j \geq 1$. The Smoluchowski equation is

$$\frac{dc_i}{dt} = \frac{1}{2} \sum_{j=1}^{i-1} K_{i-j,j} c_{i-j} c_j - \sum_{j=1}^{\infty} K_{ij} c_i c_j, \tag{8.91}$$

and is derived as follows: The first term on the right-hand side is the rate of formation per unit volume of clusters with mass i through all possible integer combinations of smaller masses; the factor 1/2 corrects for the fact that the summation considers every pair of reacting masses twice.[10] The second term is the rate of disappearance per unit volume of clusters of mass i by reaction with all other clusters. No Kronecker delta's appear in these expressions because the number of clusters $N = VC_N$ is assumed to be infinite. The right-hand side in Eq. (8.91) is the net rate of formation per unit volume of clusters with mass i, represented by the derivative on the left-hand side. Equation (8.91) is a deterministic mean-field rate equation for the distribution of clusters. Along with an initial condition that described the distribution of clusters at time $t = 0$ describes the state of the population at any future moment in time. The mean-field character is reflected in the fact that a single distribution describes the state of the population, in other words, the parent of the distribution at time t is the same distribution at $t - dt$. In the continuous domain the Smoluchowski equation is an integro-differential equation in the cluster distribution $c(x, t)$.

[9] In the physical chemistry literature the clustering processes usually go by the name *coagulation*.

[10] In the case $i = 1$ there is no source term on the right-hand side because monomers cannot be produced by aggregation.

$$\frac{\partial c(x,t)}{\partial t} = \frac{1}{2}\int_0^x K(x-y,y)c(x-y,t)c(y)dy - \int_0^\infty K(x,y)c(x,t)c(y,t)dy. \tag{8.92}$$

All moments and their evolution can be obtained from the Smoluchowski equation. For the moment of order k we multiply both sides of the Smoluchowski equation by n_i^k and perform the summations or integrals over all cluster sizes. For C_N (zero order moment) and C_M (first order moment) these equations are

$$\frac{dC_N}{dt} = -C_N^2 \frac{\bar{K}(t)}{2}, \tag{8.93}$$

$$\frac{dC_M}{dt} = 0. \tag{8.94}$$

where $\bar{K}(t)$ is the mean aggregation kernel in the cluster population at time t:

$$\bar{K}(t) = \frac{1}{C_N^2}\sum_{i=1}^{\infty}\sum_{j=1}^{\infty} K_{ij}c_i c_j \to \frac{1}{C_N^2}\int_0^\infty\int_0^\infty K(x,y)c(x;t)c(y;t)dxdy. \tag{8.95}$$

The first moment is constant in time, since the aggregation reaction conserves mass. The number concentration decreases with rate proportional to the square of the concentration, because aggregation is a second order reaction in the number of clusters. Dividing the Smoluchowski equation by Eq. (8.93) for the number concentration we obtain the evolution of the size distribution in time-free form,

$$\frac{dc_i}{dC_N} = -\frac{2}{C_N^2 \bar{K}(C_N)}\left(\sum_{j=1}^{i-1} K_{i-j,j}c_{i-j}c_j - \sum_{j=1}^{\infty} K_{ij}c_i c_j\right). \tag{8.96}$$

In this form time is replaced by the number concentration C_N as an index of progress. Comparing this result with Eq. (8.88) we recognize the two equations to be the same, provided that we make the identifications

$$c_i \to \tilde{n}_i/V, \quad C_N \to N/V,$$

where V is the volume to which concentrations c_i and C_N refer.

Our derivation of the Smoluchowski equation in the discrete finite cluster ensemble allows us to interpret c_i more precisely as *the mean distribution in the limit that the cluster ensemble converges to a single distribution*. In this limit the Smoluchowski equation is the governing equation for the evolution of the MPD and contains all information about the population since fluctuations vanish. If, however, the population converges not to a single distribution but to a mixture of a sol phase and a gel cluster, the Smoluchowski equation ceases to describe the state. This situation arises under certain forms of the aggregation kernel and is discussed in Chap. 9.

8.8.1 Comparison to Known Solutions: Constant Kernel

The solution to the discrete Smoluchowski equation for monodisperse initial condition

$$c_i(0) = \delta_{i,1},$$

is[11]

$$c_i(t) = \frac{4}{(t+2)^2}\left(\frac{t}{t+2}\right)^{i-1}$$

subject to the normalizations

$$\sum_{i=1}^{\infty} c_i(t) = \frac{2}{t+2} \equiv C_N, \quad \sum_{i=1}^{\infty} t c_i(t) = 1 \equiv C_M,$$

and mean cluster size

$$\bar{x} = \frac{C_M}{C_N} = \frac{t+2}{2}.$$

We set

$$\theta = 1 - 1/\bar{x},$$

and express time as

$$t = \frac{2\theta}{1-\theta}.$$

Substituting this into the Smoluchowski solution we obtain

$$c(\bar{x}) = \frac{1}{\bar{x}^2}\left(\frac{\bar{x}}{\bar{x}-1}\right)^{1-i}.$$

Finally, we normalize to unit number concentration by multiplying both sides by $C_M/C_N = \bar{x}$,

$$\bar{x}c(\bar{x}) = \frac{1}{\bar{x}}\left(\frac{\bar{x}}{\bar{x}-1}\right)^{1-i} = \frac{1}{\bar{x}-1}\left(\frac{\bar{x}}{\bar{x}-1}\right)^{-i}.$$

The result is the same as Eq. (8.51) obtained previously in the ThL.

[11]See Eq. 4.10 in Leyvraz (2003).

8.8.2 Comparison to Known Solutions: Sum Kernel

The solution to the discrete Smoluchowski equation for the sum kernel with initial condition $c_i(0) = \delta_{i,1}$ is[12]

$$c_i(t) = \frac{i^{i-1}}{i!}\left(1 - e^{-t}\right)^{i-1} e^{-i(1-e^{-t})} e^{-t}, \tag{8.97}$$

subject to the normalizations

$$\sum_i ic_i = 1 \equiv C_M, \quad \sum_i c_i = e^t \equiv C_N; = 1/\bar{x}, \tag{8.98}$$

The mean cluster size is

$$\bar{x} = \frac{C_M}{C_N} = e^t.$$

We set $\theta = 1 - 1/\bar{x}$ and eliminate $\bar{x}$ in the above equation to express time in terms of the variable θ:

$$t = -\log(1-\theta).$$

We substitute into (8.98)

$$c_i(\theta) = \frac{i^{i-1}}{i!}(1-\theta)\theta^{i-1}e^{-i\theta} \tag{8.99}$$

and normalize the distribution to unit number concentration by multiplying both sides by $C_M/C_N = \bar{x} = 1/(1-\theta)$:

$$\bar{x}c_i(\theta) = \frac{i^{i-1}}{i!}\theta^{i-1}e^{-i\theta}.$$

The result is the same as Eq. (8.78).

⋆

8.9 Continuous Domain

In the ThL all intensive variables can be expressed as a function of a single intensive variable, which we may take to be the mean cluster size $\bar{x} = M/N$. To complete the discussion we obtain expressions for the canonical functions in terms of $\bar{x}$ in the continuous domain under the condition $\bar{x} \gg 1$. We use the following notation to indicate variables in the continuous domain:

[12]See Eq. 4.29 in Leyvraz (2003).

$$i \to x,$$

$$n_i/N \to f(x),$$

$$\Omega/N \to \omega,$$

Given the homogeneity condition

$$\log W[Nf] = N \log W[f], \tag{8.100}$$

we do not need a separate symbol for the intensive $\log W[Nf]/N$ since this is simply equal to $\log W[f]$. All intensive properties will be expressed as functions of the mean size, $\bar{x}$, which is

$$\bar{x} = \frac{M}{N} = \int_1^\infty x f(x) dx. \tag{8.101}$$

The mean size represents the progress variable and ranges from $\bar{x} = 1$ to ∞. The mean kernel is

$$\bar{K}(\bar{x}) \equiv \int\int K(x, y) f(x) f(y)\, dx\, dy, \tag{8.102}$$

with $\bar{K}(0) = 1$. As an intensive property, the mean kernel is a function of the mean cluster size $\bar{x}$.

We obtain the log of the partition function by writing $M = \bar{x}N$ and taking the limit $N \to \infty$ at constant $\bar{x}$. We find (see appendix section "Aggregation in the Continuous Domain" for the detailed derivation)

$$\log \omega = \bar{x} \log \bar{x} - (\bar{x} - 1) \log(\bar{x} - 1) + \bar{x} \int_1^{\bar{x}} \log \bar{K}(y) \frac{dy}{y^2}. \tag{8.103}$$

The corresponding β and $\log q$ may be obtained by taking the derivatives of $N \log \omega$ with respect to M and N. Alternatively, we may start with Eqs. (8.28) and (8.29) and take the limit $M, N \to \infty$ and $M/N = \bar{x}$. Both approaches will give the same answer (the details are given in the appendix section "Aggregation in the Continuous Domain"):

$$\beta = \log \frac{\bar{x}}{\bar{x} - 1} + \frac{\log \bar{K}(\bar{x})}{\bar{x}} + \int_1^{\bar{x}} \frac{\log \bar{K}(y)}{y^2} dy \tag{8.104}$$

$$q = \frac{\bar{x} - 1}{\bar{K}(\bar{x})}. \tag{8.105}$$

It is easy to confirm that these results satisfy the fundamental relationships

$$\log \omega = \bar{x}\beta + \log q, \tag{8.106}$$

$$\beta = \frac{d \log \omega}{d\bar{x}}, \tag{8.107}$$

$$\log q = \log \omega - \bar{x}\frac{d \log \omega}{d\bar{x}}. \tag{8.108}$$

As a further test, we calculate the derivatives of β and $\log q$ with respect to the mean size $\bar{x}$:

$$\frac{d\beta}{d\bar{x}} = \frac{1}{\bar{x}}\left(\frac{\bar{K}'}{\bar{K}} - \frac{1}{\bar{x}-1}\right),$$

$$\frac{d \log q}{d\bar{x}} = \frac{1}{\bar{x}-1} - \frac{\bar{K}'}{\bar{K}},$$

where $\bar{K}'$ indicates derivative with respect to $\bar{x}$. These satisfy the condition

$$\bar{x}\frac{d\beta}{d\bar{x}} + \frac{d \log q}{d\bar{x}} = 0, \tag{8.109}$$

which is the intensive form of the Gibbs-Duhem equation.

When $\bar{x} \gg 1$ we obtain additional simplifications. Using

$$\bar{x}\log\bar{x} - (\bar{x}-1)\log(\bar{x}-1) \to \frac{d\bar{x}\log\bar{x}}{d\bar{x}} = \log\bar{x} + 1,$$

Eqs. (8.103), (8.104), and (8.105) become

$$\log \omega = 1 + \log\bar{x} + \bar{x}\int_1^{\bar{x}} \log \bar{K}(y)\frac{dy}{y^2} \tag{8.110}$$

$$\beta = \frac{1 + \log \bar{K}(\bar{x})}{\bar{x}} + \int_1^{\bar{x}} \frac{\log \bar{K}(y)}{y^2}\, dy \tag{8.111}$$

$$q = \frac{\bar{x}}{\bar{K}(\bar{x})}. \tag{8.112}$$

These also satisfy the fundamental relationships in Eqs. (8.106)–(8.109).

To complete the description of the system we obtain expressions for the selection functional. The starting point is the recursion for W in Eq. (8.27). In the ThL we have the following special conditions:

$$\mathbf{n} \to \tilde{\mathbf{n}}, \quad \mathbf{n}' \to \tilde{\mathbf{n}}', \quad K(\tilde{\mathbf{n}}') \to \langle K \rangle_{M,N+1}, \quad \frac{W(\mathbf{n}')}{W(\mathbf{n})} = \frac{\tilde{w}_{i-j}\tilde{w}_j}{\tilde{w}_j}.$$

With these Eq. (8.27) becomes

$$\frac{1}{M-N} \sum_{i=2}^{\infty} \tilde{n}_i \sum_{j=1}^{i-1} K_{i-j,j} \frac{\tilde{w}_{i-j}\tilde{w}_j}{\tilde{w}_i} = 1 \tag{8.113}$$

and expresses a condition on the $\tilde{w}_i$ that is satisfied in every generation. In the continuous limit we make the substitutions $i \to x$, $\tilde{n}_i \to Nf(x)$, and express this result in integral form,

$$\frac{1}{\bar{x}-1} \int_1^{\infty} dx\, f(x) \int_1^{x-1} dy\, K(x-y, y) \frac{\tilde{w}(x-y)\tilde{w}(y)}{\tilde{w}(x)} = 1. \tag{8.114}$$

To obtain an expression for W, we return to Eq. (8.27) and write it as

$$W(\tilde{\mathbf{n}}) = \frac{1}{M-N} \sum_{i=2}^{\infty} \tilde{n}_i \sum_{j=1}^{i-1} K_{i-j,j} W(\tilde{\mathbf{n}}'). \tag{8.115}$$

On the left-hand side we have the MPD in the (M, N) ensemble and on the right-hand side we have a summation over all parents $\tilde{\mathbf{n}}'$ of $\tilde{\mathbf{n}}$. In the ThL the MPD has a single parent, the MPD in the previous generation $(M, N+1)$. Then we have

$$\begin{aligned} \frac{W(\tilde{\mathbf{n}})}{W(\tilde{\mathbf{n}}')} &= \frac{1}{M-N} \sum_{i=2}^{\infty} \tilde{n}_i \sum_{j=1}^{i-1} K_{i-j,j} \\ &\to \frac{1}{\bar{x}-1} \int_1^{\infty} dx\, f(x) \int_1^{x-1} dy\, K(x-y, y), \end{aligned} \tag{8.116}$$

where $\mathbf{n}'$ is understood to be the MPD of the previous generation. The result expresses the change in W from one generation to the next as a functional of the MPD that involves the aggregation kernel. Equations (8.104), (8.105), and (8.116) summarize the equations of change of the canonical MPD for a population undergoing binary aggregation.

8.10 Contact with the Literature

The mathematical literature on the Smoluchowski equation is extensive, as is the experimental literature on the same topic, given the ubiquitous presence of aggregation in physical and chemical systems. The mathematical literature focuses on the solution and its scaling under homogeneous kernels. An early review of the topic, motivated by applications in aerosol systems, is given by Drake (1972). A more recent review was given by Leyvraz (2003) and summarizes the state of the field as of 2003, including methodologies and known solutions for certain kernels. Aldous (1999) gives a review of the coagulation equation from a probabilistic perspective with focus on stochastic processes that are represented in the mean field by the Smoluchowski equation. All of these works begin with the Smoluchowski equation and do not intersect with our work except in comparing final solutions. Notable exceptions are the works of Marcus, Spouge, and Lushnikov.

Marcus (1968) was the first to explicitly consider aggregation in the discrete, finite space. He formulated the master equation in terms of time but conceded that the enumeration of states makes the problem intractable. Marcus gives a numerical example for the case $M = 6$ for the kernel

$$K_{ij} = \frac{\left(i^{1/2} + j^{1/2}\right)^2}{40}, \quad i = 1, 2$$

which permits reactions only among monomers and dimers. The phase space for this process is a truncated version of the graph in Fig. 8.1 that terminates with distributions that do not contain any monomers or dimers.[13]

Lushnikov started with Marcus's formulation and developed a mathematical procedure for the mean distribution of the phase space (Lushnikov 1978). This approach to coagulation is often called the *Marcus-Lushnikov* process. For the constant and sum kernels Lushnikov was able to obtain time-dependent solutions for discrete finite systems; for the product kernel, discussed in the next chapter, he was able to obtain the pre- and post-gel solution rigorously (Lushnikov 2004, 2005, 2006a,b, 2012, 2013). Apart from the treatment of the discrete phase space, Lushnikov made several contributions to the theory of the Smoluchowski equation (Lushnikov 1973, 1974, 1976, 2011); it is in fact in his mean field treatment of coagulation that the recursion for the cluster weight in linear ensembles, Eq. (8.38), makes its first appearance (Lushnikov 1973). Lushnikov's approach is highly mathematical and is complicated by the fact that time is treated explicitly. In our theory time is replaced by the mean cluster size as a progress coordinate, and this removes one layer of complexity. Time can be reintroduced, and the precise way to do this is given by Hendriks et al. (1985). And while Lushnikov's theory is not

[13]This graph is given on p139 in Marcus (1968).

limited by kernels, it is difficult to apply it in a meaningful way to kernels other than those in Eq. (8.117).

Spouge approached stochastic coagulation from a different perspective that is based on Stockmayer's treatment of polymerization.[14] Spouge obtained the solution to several aggregation kernels based on a combinatorial approach that enumerates the formation history (trajectory in phase space) of a cluster of known size (Spouge 1983a,b,c,d; Hendriks et al. 1985; Spouge 1985a,b). The kernels considered by Spouge are of the form

$$K_{ij} = A + B(i + j) + Cij, \tag{8.117}$$

where A, B, and C are constants. While his methodology does not invoke thermodynamics, as our theory does, it shares common elements that deserve a commentary. Spouge calculates the number of ways to form a cluster of size i and shows it to be

$$\frac{w_k}{k!},$$

where w_k is the cluster bias of our theory and derives the same recursion for w_k as Eq. (8.38).[15] Following Stockmayer (1943), Spouge writes the number of ways to form distribution $\mathbf{n}$ as

$$\frac{N!}{n_1!n_2!\cdots} w_1^{n_1} w_2^{n_2}\cdots, \tag{8.118}$$

and maximizes it with respect to n_i under the constraints,

$$\sum_i n_i = N, \quad \sum_i i n_i = M.$$

In a subsequent paper, Spouge and coworkers (Hendriks et al. 1985) proved that the combinatorial method is equivalent to the Smoluchowski equation in the limit that system becomes infinite. We recognize the kernels considered by Spouge as kernels that produce linear ensembles, provided we set $C = 0$[16] and the quantity in Eq. (8.118) as the microcanonical weight of distribution $\mathbf{n}$. The linearity of the ensemble is what allows Spouge to calculate w_k by a combinatorial calculation and obtain the most probable distribution. This method, however, is limited to kernels of the form in Eq. (8.117).

---

[14] We discuss Stockmayer's method in great detail in Chap. 9.

[15] Spouge in fact uses the symbol w_k for what we notate $w_k/k!$ To avoid confusion we translate Spouge's results in our own notation.

[16] With $C = 0$ we obtain Eq. (8.39), which guarantees that the ensemble is linear. The case $C > 0$ is discussed in Chap. 9, where we show that it produces a *linearized* ensemble that gives the correct distribution in the pre-gel state.

Our aim with this discussion is not to critique the relative merits of different approaches to discrete finite aggregation. Rather, it is to highlight the fact that our thermodynamic theory of distributions is indeed general and when applied to the problem of aggregation it reproduces the known solutions in the literature. Moreover, as we discuss in the next chapter, it allows us to treat gelation as a regular phase transition and to obtain the post gel solution using the familiar tools of constructing tie lines in vapor–liquid systems.

Appendix: Derivations

Equation (8.20)

Here we derive Eq. (8.20), which gives the evolution of the partition function during aggregation. We begin with

$$P(\mathbf{n}|\mathbf{n}') = C_{\mathbf{n}'} \frac{n'_{i-j}(n'_j - \delta_{i-j,j})}{1+\delta_{i-j,j}} K_{i-j,j}, \tag{8.16}$$

$$P(\mathbf{n}) = \mathbf{n}! \frac{W(\mathbf{n})}{\Omega_{M,N}} \tag{8.18}$$

We insert these into the master equation (8.15) and rearrange the result in the form,

$$\frac{W(\mathbf{n})}{\Omega_{M,N}} = \frac{2}{N(N+1)} \sum_{\mathbf{n}'} \frac{\mathbf{n}'!}{\mathbf{n}!} \cdot \frac{K_{i-j,j}}{\bar{K}(\mathbf{n}')} \cdot \frac{n'_{i-j}(n'_j - \delta_{i-j,j})}{1+\delta_{i-j,j}} \cdot \frac{W(\mathbf{n}')}{\Omega_{M,N+1}}$$

The ratio of the factorials involves the offspring distribution in the numerator, and the parent distribution in the denominator. These differ only in the number of cluster masses involved in the aggregation event:

$$\frac{\mathbf{n}'!}{\mathbf{n}!} = \frac{(N+1)!}{N!} \cdot \frac{n_i! n_j! n_{i-j}!}{n'_i! n'_j! n'_{i-j}!} = \frac{(N+1)n_i}{n'_{i-j}(n'_j - \delta_{i-j,j})}$$

with the last result obtained by application of the parent–offspring relationships in Eq. (8.12). Combining these results we obtain

$$\frac{W(\mathbf{n})}{\Omega_{M,N}} = \frac{2}{N} \sum_{i=2}^{\infty} \sum_{j=1}^{i/2} \frac{K_{i-j,j}}{\bar{K}(\mathbf{n}')} \cdot \frac{n_i}{1+\delta_{i-j,j}} \cdot \frac{W(\mathbf{n}')}{\Omega_{M,N+1}}.$$

Here, the summation over parents of **n** is organized as follows: for every cluster mass in **n** we first sum over the $i/2$ parents of that cluster mass (inner summation), then run the outer summation over all members of **n**. Cluster mass $i = 1$ cannot be formed by aggregation and thus the outer summation begins at $i = 2$. Given the symmetry of the kernel, $K_{ij} = K_{ji}$, the inner summation is extended over the range $j = 1$ to $j = i - 1$; this amounts to double-counting all terms of this summation and so the result is

$$\frac{W(\mathbf{n})}{\Omega_{M,N}} = \frac{1}{N}\sum_{i=2}^{\infty}\sum_{j=1}^{i-1} n_i \frac{K_{i-j,j}}{\bar{K}(\mathbf{n}')} \cdot \frac{W(\mathbf{n}')}{\Omega_{M,N+1}}.$$

This leads directly into Eq. (8.20) of the text. In this form, the summation over all $(i - j, j)$-parents of **n** is converted into a double summation in which the outer sum goes over all clusters $i > 1$ that can be formed by aggregation, while the inner sum goes over all possible binary events that can form the cluster mass i. The construction of this summation is illustrated with a numerical example below.

Note 8.5 (Construction of the Double Summation)

We consider the distribution

$$\mathbf{n} = (1, 1, 1, 0, 0, 0)$$

in the ensemble $M = 6$, $N = 3$ (see Fig. 8.6). The distribution consists of one monomer, one dimer, and one trimer. It has two parents, one that forms the trimer by the reaction $2 + 1$, and one that forms the dimer by the reaction $1 + 1$. The (2, 1)-parent is counted twice, therefore the number of ordered parent–offspring events is 3, also equal to $M - N$. The double summation in Eq. (8.20) contains the following terms:

$$n_2 \frac{K_{1,1}}{\bar{K}(\mathbf{n}_2)} \frac{W(\mathbf{n}'_2)}{W(\mathbf{n})} + n_3 \frac{K_{2,1}}{\bar{K}(\mathbf{n}_1)} \frac{W(\mathbf{n}'_1)}{W(\mathbf{n})} + n_3 \frac{K_{1,2}}{\bar{K}(\mathbf{n}_1)} \frac{W(\mathbf{n}'_1)}{W(\mathbf{n})}.$$

In this example, $n_3 = n_2 = 1$, and the summation finally expands to $M - N = 6 - 3 = 3$ terms. In general, if the multiplications by n_i are expanded to a sum of n_i identical terms, the total number of terms in the double summation will always be $M - N$. Each $(i - j, j)$-parent contributes the quantity

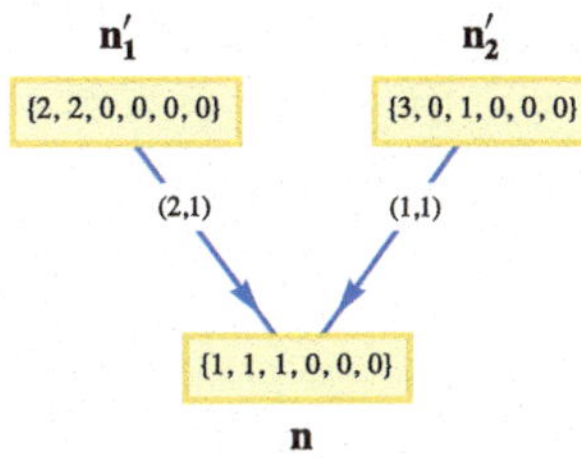

Fig. 8.6 Illustration of the summations in the recursion for the partition function in Eq. (8.20) for $M = 6$. Distribution **n** belongs in the ensemble $M = 6$, $N = 3$, and has two parents in the ensemble $M = 6$, $N = 2$

$$n_i \frac{K_{i-j,j}}{\bar{K}(\mathbf{n}')} \frac{W(\mathbf{n}')}{W(\mathbf{n})} (2 - \delta_{i-j,j})$$

to the summation. Then an alternative expression for the double sum in Eq. (8.20) is

$$\sum_{i=2}^{\infty} \sum_{j=1}^{i-1} n_i \frac{K_{i-j,j}}{\bar{K}(\mathbf{n}')} \frac{W(\mathbf{n}')}{W(\mathbf{n})} = \sum_{\mathbf{n}'} n_i \frac{K_{i-j,j}}{\bar{K}(\mathbf{n}')} \frac{W(\mathbf{n}')}{W(\mathbf{n})} (2 - \delta_{i-j,j}) \tag{8.119}$$

in which the summation is now over *all* $(i-j, j)$ *parents of distribution* $\mathbf{n}$.

Smoluchowski Eq. (8.87)

We begin with Eq. (8.85)

$$\langle n_k \rangle_{M,N} = \frac{2}{N(N+1)} \left\langle \sum_{i=2}^{\infty} \sum_{j=1}^{i/2} \left(n_k + \delta_{k,i} - \delta_{k,i-j} - \delta_{k,j} \right) T_{i-j,j} \right\rangle_{M,N+1}, \qquad [8.85]$$

and work the summation through the Kronecker deltas to express the result in the form

$$\langle n_k \rangle_{M,N} = \frac{2}{N(N+1)} \times$$
$$\left\langle n_k \sum_{i=2}^{\infty} \sum_{j=1}^{i/2} T_{i-j,j} + \sum_{j=1}^{k/2} T_{k-j,j} - \sum_{j=1}^{\infty} (1 + \delta_{kj}) T_{k,j} \right\rangle_{M,N+1} \tag{8.120}$$

All terms on the right-hand side refer to the parent ensemble, $\mu C(M, N+1)$, as indicated by the subscript on the ensemble average, therefore we drop the primes as unnecessary. The first term on the right-hand side reduces to the mean number of cluster mass k in the parent ensemble:

$$\frac{2}{N(N+1)} \left\langle n_k \sum_{i=2}^{\infty} \sum_{j=1}^{i/2} T_{i-j,j} \right\rangle_{M,N+1} =$$
$$\frac{2}{N(N+1)} \left\langle n_k \sum_{i=2}^{\infty} \sum_{j=1}^{i/2} \frac{n_{i-j}(n_j - \delta_{i-j,j})}{1 + \delta_{i-j,j}} \frac{K_{i-j,j}}{\bar{K}(\mathbf{n})} \right\rangle_{M,N+1} = \langle n_k \rangle_{M,N+1}$$

The second term is

$$\frac{2}{N(N+1)}\left\langle \sum_{j=1}^{k/2} T_{k-j,j} \right\rangle = \frac{2}{N(N+1)}\left\langle \sum_{j=1}^{k/2} \frac{n_{k-j}(n_j-\delta_{k-j,j})}{1+\delta_{k-j,j}}\frac{K_{k-j,j}}{\bar{K}(\mathbf{n})} \right\rangle$$

$$= \frac{1}{N(N+1)}\left\langle \sum_{j=1}^{k-1} n_{k-j}(n_j-\delta_{k-j,j})\frac{K_{k-j,j}}{\bar{K}(\mathbf{n})} \right\rangle$$

The third term is

$$\frac{2}{N(N+1)}\left\langle \sum_{j=1}^{\infty} (1+\delta_{kj})T_{k,j} \right\rangle = \frac{2}{N(N+1)}\left\langle \sum_{j=1}^{\infty} n_k(n_j-\delta_{k,j})\frac{K_{k,j}}{\bar{K}(\mathbf{n})}, \right\rangle$$

Combining these results we obtain Eq. (8.87) of the main text.

Aggregation in the Continuous Domain

Derivation 1

The partition function is the product of the volume of the ensemble and a term that represents the effect of the kernel,

$$\Omega_{M,N} = \mathcal{V}_{M,N}\,\mathcal{K}_{M,N} \tag{8.121}$$

with

$$\mathcal{V}_{M,N} = \binom{M-1}{N-1} \tag{8.122}$$

and

$$\mathcal{K}_{M,N} = \prod_{N'=N+1}^{M} \langle K_{M,N'} \rangle . \tag{8.123}$$

In this form the partition function is expressed as a product of two terms, one that represents the unbiased ensemble ($\mathcal{V}$), and one that represents the effect of the aggregation kernel ($\mathcal{K}$). This factorization carries over to the parameters β and $\log q$, which may also be expressed as the product of the unbiased case times a factor that includes the effect of the kernel.

The log of the first term in the ThL is obtained using the Stirling approximation $\log x! \to x\log x - x$

$$\frac{\log \mathcal{V}_{M,N}}{N} = \bar{x} \log \bar{x} - (\bar{x} - 1) \log(\bar{x} - 1). \tag{8.124}$$

For large $\bar{x}$ this further reduces to the derivative of $\bar{x} \log \bar{x}$:

$$\frac{\log \mathcal{V}_{M,N}}{N} = 1 + \log \bar{x}. \tag{8.125}$$

For $\log \mathcal{K}$ we have

$$\log \mathcal{K}_{M,N} = \sum_{N'=N+1}^{M} \log \langle K_{M,N'} \rangle. \tag{8.126}$$

The ensemble average kernel in the ThL is an intensive function of $\bar{x} = M/N$:

$$\langle K \rangle_{M,N} = \int \int f(x) f(y) k(x, y)\, dx\, dy \equiv \bar{K}(\bar{x}). \tag{8.127}$$

The summation in (8.126) becomes an integral in dN'

$$\log \mathcal{K}_{M,N} = \int_{N+1, M=\text{const.}}^{M} \log \bar{K}(\bar{x})\, dN', \tag{8.128}$$

which is to be evaluated at constant M. Using $N' = M/\bar{x}$, we have

$$dN'\Big|_M = -M \frac{d\bar{x}}{\bar{x}^2}. \tag{8.129}$$

We insert into Eq. (8.128) and switch the integration variable to $y = \bar{x}$. Noting that the integration limits are now from $y = \bar{x}$ (current state) to $y = 1$ (initial state), we have

$$\log \mathcal{K}_{M,N} = -M \int_{\bar{x}}^{1} \log \bar{K}(y) \frac{dy}{y^2} \tag{8.130}$$

and finally

$$\frac{\log \mathcal{K}_{M,N}}{N} = \bar{x} \int_{1}^{\bar{x}} \log \bar{K}(y) \frac{dy}{y^2}. \tag{8.131}$$

Combining Eqs. (8.124) and (8.131) we obtain the full partition function:

$$\log \omega \equiv \frac{\log \Omega_{M,N}}{N} = \bar{x} \log \bar{x} - (\bar{x} - 1) \log(\bar{x} - 1) + \bar{x} \int_{1}^{\bar{x}} \log \bar{K}(y) \frac{dy}{y^2}. \tag{8.132}$$

We obtain β and $\log q$ from their relationship to the partition function:

$$\beta = \frac{d \log \omega}{d\bar{x}} = \log \frac{\bar{x}}{\bar{x}-1} + \int_1^{\bar{x}} \frac{d \log \bar{K}(y)}{y}, \tag{8.133}$$

$$\log q = \log \omega - \bar{x} \frac{d \log \omega}{d\bar{x}} = \log(\bar{x}-1) - \log \bar{K}(\bar{x}). \tag{8.134}$$

Identity The following identity was used:

$$\int_1^{\bar{x}} \frac{d \log \bar{K}(y)}{y} = \frac{\log \bar{K}(\bar{x})}{\bar{x}} + \int_1^{\bar{x}} \frac{\log \bar{K}(y)}{y^2} dy. \tag{8.135}$$

This is obtained by integration by parts noting that $\log \bar{K}(1) = 0$.

Alternative Derivation

As an exercise we try an alternative derivation. We start with the discrete quantities

$$\beta = \log \frac{\Omega_{M+1,N}}{\Omega_{M,N}}$$

$$\log q = \log \frac{\Omega_{M,N+1}}{\Omega_{M,N}}$$

to obtain their form in the thermodynamic limit. We may obtain each of these terms for each partition function, $\mathcal{V}_{M,N}$ and $\mathcal{K}_{M,N}$, separately.

By direct application of the multinomial term, the M-derivative is

$$\beta_V = \binom{M+1}{N} \Big/ \binom{M}{N} = \log \frac{M}{M-N+1} \to \log \frac{\bar{x}}{\bar{x}-1}. \tag{8.136}$$

Similarly, the derivative with respect to N is

$$\log q_V = \binom{M}{N+1} \Big/ \binom{M}{N} = \log \frac{M-N}{N} \to \log(\bar{x}-1). \tag{8.137}$$

The finite difference of $\log \mathcal{K}_{M,N}$ with respect to M is

$$\beta_{\mathcal{K}} = \log \frac{\mathcal{K}_{M+1,N}}{\mathcal{K}_{M,N}} = \log \langle K \rangle_{M+1,M+1} + \sum_{N'=N+1}^{M} \log \frac{\langle K \rangle_{M+1,N'}}{\langle K_{M,N'} \rangle},$$

which is obtained by application of Eq. (8.123) and careful arrangement of the terms in the summation. For the first term we have

$$\log \langle K \rangle_{M+1,M+1} = \log K_{11} = 0$$

because by normalization of the kernel, $K_{11} = 1$. The ratios in the summation turn into the partial derivative of $\log \langle K \rangle_{M,N}$ with respect to M:

$$\log \frac{\langle K \rangle_{M+1,N'}}{\langle K_{M,N'} \rangle} \to \left(\frac{\partial \log \langle K_{M,N'} \rangle}{\partial M} \right)_{N'}$$

and the summation becomes an integral in dN' at constant M[17]:

$$\sum_{N'=N+1}^{M} \log \frac{\langle K \rangle_{M+1,N'}}{\langle K_{M,N'} \rangle} \to \int_{N+1, M=\text{const.}}^{M} \left(\frac{\partial \log \langle K_{M,N'} \rangle}{\partial M} \right)_{N'} dN'.$$

Setting

$$y = M/N'$$

we obtain

$$\partial M = N' dy, \quad \partial N' = -M \frac{dy}{y^2}$$

and with these results, the integral in the ThL becomes

$$\int_{N+1, M=\text{const.}}^{M} \left(\frac{\partial \log \langle K_{M,N'} \rangle}{\partial M} \right)_{N'} dN' = \int_{\bar{x}}^{1} \frac{1}{N'} \frac{d \log \bar{K}(y)}{dy} \left(-M \frac{dy}{y^2} \right).$$

Finally, the result for $\beta_{\mathcal{K}}$ is

$$\beta_{\mathcal{K}} = \int_{1}^{\bar{x}} \frac{d \log \bar{K}(y)}{y} = \int_{1}^{\bar{x}} \frac{\log \bar{K}(y)}{y^2} dy + \frac{\log \bar{K}(\bar{x})}{\bar{x}}. \tag{8.138}$$

This can be expressed in the equivalent form by performing integration by parts:

$$\beta_{\mathcal{K}} = \frac{\log \bar{K}(\bar{x})}{x} \Big|_{1}^{x} - \int_{1}^{x} \log \bar{K}(\bar{x}) d\left(\frac{1}{y} \right),$$

[17] The similarity to the integral for the residual enthalpy is remarkable.

which gives

$$\beta_{\mathcal{K}} = \frac{\log \bar{K}(\bar{x})}{x} + \int_1^x \log \bar{K}(\bar{x}) \frac{dy}{y^2}. \tag{8.139}$$

The N-derivative of $\log \mathcal{K}_{M,N}$ is much easier and follows from Eq. (8.123):

$$\log q_{\mathcal{K}} = \log \frac{\mathcal{K}_{M,N+1}}{\mathcal{K}_N} = \log \frac{1}{\langle K_{M,N+1} \rangle} \to -\log \bar{K}(\bar{x}).$$

The final results for β and $\log q$ are obtained by combining the above results:

$$\beta = \beta_V + \beta_{\mathcal{K}} = \int_1^{\bar{x}} \frac{\log \bar{K}(y)}{y^2} dy + \frac{\log \bar{K}(\bar{x})}{\bar{x}} + \log \frac{\bar{x}}{\bar{x} - 1},$$

$$\log q = \log q_V + \log q_{\mathcal{K}} = \log(\bar{x} - 1) - \log \bar{K}(\bar{x}).$$

These results agree with those in Eqs. (8.132), (8.133), and (8.134).

Chapter 9
Kinetic Gelation

Binary aggregation usually leads to stable populations whose mean size increases indefinitely as aggregation advances, but under special conditions it is possible to observe a phase transition that produces a giant cluster of macroscopic size. In the polymer literature this is known as *gelation* and is observed experimentally in some polymer systems. Polymerization is perhaps a standard example of aggregation: starting with monomers, polymer molecules of linear or branched structure are formed by joining smaller units together, one pair at a time. The aggregation kernel in this case is determined by the functionality and reactivity of the chemical sites where polymers can join by forming bonds. A kernel that is known to lead to gelation is the *product kernel*,

$$K_{i,j} = ij, \tag{9.1}$$

The classical symptom of gelation is the breakdown of the Smoluchowski equation: within finite time the mean cluster size diverges and the Smoluchowski equation ceases to conserve mass. Gelation has the qualitative features of a phase transition and is often discussed qualitatively in the thermodynamic language of phase equilibrium.

In this chapter we treat gelation as a formal phase transition that takes place when the stability criteria, developed in Chap. 5, are violated. We will demonstrate that the equilibrium sol-gel state can be calculated as a tie-line using the same tools as in vapor–liquid equilibrium, and we will find out that under appropriate set of coordinates, the behavior sol-gel system is similar to the van der Waals loop in molecular thermodynamics.

T. Matsoukas, *Generalized Statistical Thermodynamics*,
Understanding Complex Systems, https://doi.org/10.1007/978-3-030-04149-6_9

9.1 Product Kernel

Unlike the constant and sum kernel, the product kernel does not satisfy the uniformity condition

$$\bar{K}(\mathbf{n}) = \langle K \rangle_{M,N}\,, \tag{[8.35]}$$

and does not produce a linear ensemble. It produces, a *quasilinear* ensemble that in certain though not all respects behaves as a linear ensemble. The mean product kernel in distribution $\mathbf{n}$ is calculated from Eq. (8.40) with $K_{ij} = ij$:

$$\bar{K}(\mathbf{n}) = \frac{\sum_{i=1}^{\infty}\sum_{j=1}^{\infty} n_i n_j ij - \sum_{i=1}^{\infty} n_i i^2}{N(N-1)} = \frac{M^2}{N(N-1)} - \frac{\sum_i i^2 n_i}{N(N-1)}$$
$$\sim \left(\frac{M}{N}\right)^2 - \frac{\langle i^2 \rangle}{N} \tag{9.2}$$

where $\langle i^2 \rangle$ is the normalized second moment,

$$\langle i^2 \rangle = \frac{\sum_i n_i i^2}{N}. \tag{9.3}$$

If the distribution decays properly at large i, the second moment scales as the square of the mean cluster size, $\langle i^2 \rangle \sim \bar{x}^2$, and[1]

$$\frac{\langle i^2 \rangle}{N} \sim \frac{\bar{x}}{N}.$$

In the thermodynamic limit ($N \to \infty$ at constant $\bar{x}$) the second moment in Eq. (9.2) vanishes and the mean kernel reduces to

$$\bar{K}(\mathbf{n}) \to \left(\frac{M}{N}\right)^2 \equiv \langle K \rangle_{M,N}\,. \tag{9.4}$$

In this limit the mean kernel is the same in all distributions of the (M, N) ensemble that do not contain a gel cluster and satisfies the uniformity condition in Eq. (8.35). Equation (9.4) unlocks all the results of the product kernel. The partition function is obtained from Eq. (8.26) with $\langle K \rangle_{M,N}$ from Eq. (9.4):

[1]This scaling excludes distributions that contain a gel cluster. A necessary though not sufficient condition for Eq. (9.4) is that the distribution must be a single sol phase.

$$\Omega_{M,N} = \binom{M-1}{N-1} \left(N! \frac{M^{M-N}}{M!} \right)^2 . \tag{9.5}$$

The parameters β and $\log q$ are obtained as the derivatives of $\log \Omega$. In the thermodynamic limit we find[2]

$$\beta = \left(\frac{\partial \log \Omega}{\partial M} \right)_N = 1 - \frac{N}{M} - 2 \log \left(1 - \frac{N}{M} \right), \tag{9.6}$$

and

$$\log q = \left(\frac{\partial \Omega}{\partial N} \right)_M = \frac{N}{M} \left(1 - \frac{N}{M} \right). \tag{9.7}$$

Since the product kernel satisfies asymptotically the condition $\bar{K}(\mathbf{n}) = \langle K \rangle_{M,N}$, we may calculate the linearized cluster biased using the recursion for linear ensembles, Eq. (8.38), with $K_{ij} = ij$:

$$w_i = \frac{1}{i-1} \sum_{j=1}^{i-1} (i-j) j \, w_{i-j} w_j . \tag{9.8}$$

The solution with $w_1 = 1$ is[3]

[2]Details are given in the appendix, section "Derivations—Power-Law Kernels."

[3]The recursion for the sum kernel is

$$w_i = \frac{i}{2(i-1)} \sum_{j=1}^{i-1} w_{i-j} w_j, \quad \text{(sum)}$$

Using $a_i = i w_i$ the recursion for the product kernel can be written as

$$a_i = \frac{i}{i-1} \sum_{j=1}^{i-1} a_i, \quad \text{(product)}$$

Comparing the two recursions we establish the relationship between the cluster functions of the two kernels,

$$w_i^{\text{prod}} = \frac{2^{i-1}}{i} w_i^{\text{sum}},$$

which allows us to obtain w_i of the product kernel from that of the sum kernel.

$$w_i = \frac{2(2i)^{i-2}}{i!}. \tag{9.9}$$

We assemble the MPD by combining w_i, β, and w_i into canonical form. The results can be expressed more compactly in terms of the variable $\theta = 1 - 1/\bar{x}$. The parameters β and q Eqs. (9.6) and (9.7) are

$$\beta = \theta - 2\log\theta; \tag{9.10}$$

$$q = \theta(1-\theta), \tag{9.11}$$

and with these results the MPD takes the form

$$\frac{\tilde{n}_i}{N} = \frac{2\theta}{1-\theta}\frac{(2\theta i)^{i-2}}{i!}e^{-2\theta i}. \tag{9.12}$$

Equations (9.5), (9.10), (9.11), and (9.12) summarize the solution to the product kernel under the condition that no gel is formed.

Note 9.1 (On the Quasilinear Nature of the Product Kernel)

To obtain a strictly linear cluster ensemble, Eq. (9.4) must be true for *all* distributions of the ensemble. In the case of the product kernel this equation is true for *almost* all distributions, excluding those distributions with a heavy tail towards large cluster sizes. One consequence of this is that it is not possible to obtain expressions for the mean distribution in the discrete finite domain. For linear ensembles the mean distribution is given by Eq. (2.54),

$$\frac{\langle n_i \rangle}{N} = w_i \frac{\Omega_{M-i,N-1}}{\Omega_{M,N}}, \quad i = 1, \cdots M - N + 1. \tag*{[2.54]}$$

and is valid for all M, N. Applying this to the product kernel we obtain

$$\langle n_i \rangle_{M;N} \to \frac{2^{i-1} i^{i-2} M^{-2(M-N)} (M-i)^{2(-i+M-N+1)} M!(M-N)!}{i!(M-i)!(-i+M-N+1)!}, \tag{9.13}$$

with a condition: the result is asymptotically true (does not apply to finite M and N), provided that $M/N \le 2$. This condition alludes to the sol-gel instability discussed in the next section. Equation (9.13), while inexact in finite systems, asymptotically approaches the correct solution of the infinitely large system Fig. 9.1. For $M/N > 2$ Eq. (9.13) has interesting behavior: it develops a peak in the gel region, which suggests the formation of a gel phase. Quantitatively, however, Eq. (9.13) fails to

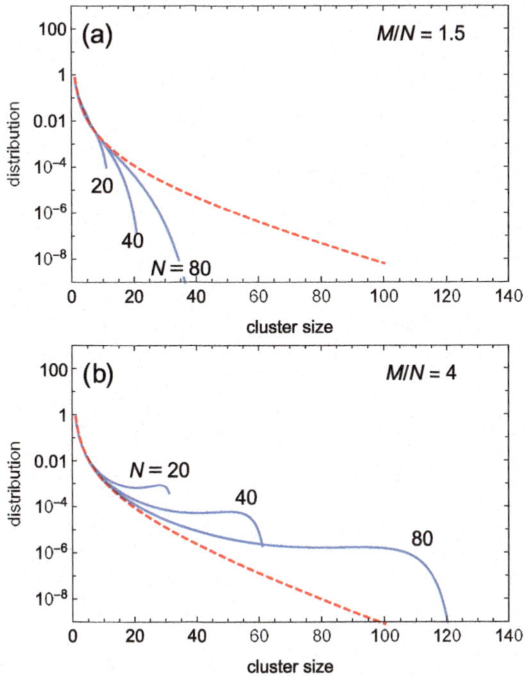

Fig. 9.1 Equation (9.13) is an approximation for the mean distribution that asymptotically approaches the true MPD (dashed line) as the size of the system increases (**a**). When $M/N > 2$ (**b**), Eq. (9.13) develops a peak in the gel region that indicates in a qualitative manner the coexistence of a gel and a sol phase. Quantitatively, Eq. (9.13) is a poor representation of the actual two-phase system and fails to reproduce N and M as moments of the distribution

reproduce N and M as the zeroth and first order moments and does not represent the correct distribution. As we will see, a gel phase indeed forms when $M/N > 2$, but its proper treatment must be based on stability analysis of the partition function.

★

9.2 Stability and Phase Diagram

To study the stability of the MPD we examine the behavior of β and q in Fig. 9.2. Stability requires

$$\frac{d\beta}{d\bar{x}} \leq 0; \quad \frac{d\log q}{d\bar{x}} \geq 0. \qquad [5.9]$$

or equivalently,

$$\frac{d\beta}{d\theta} \leq 0; \quad \frac{d\log q}{d\theta} \geq 0. \qquad (9.14)$$

Figure 9.2 shows that the stability conditions are met in the region $0 \leq \theta < 1/2$. Both β and q reach the point of incipient instability (zero slope) simultaneously at

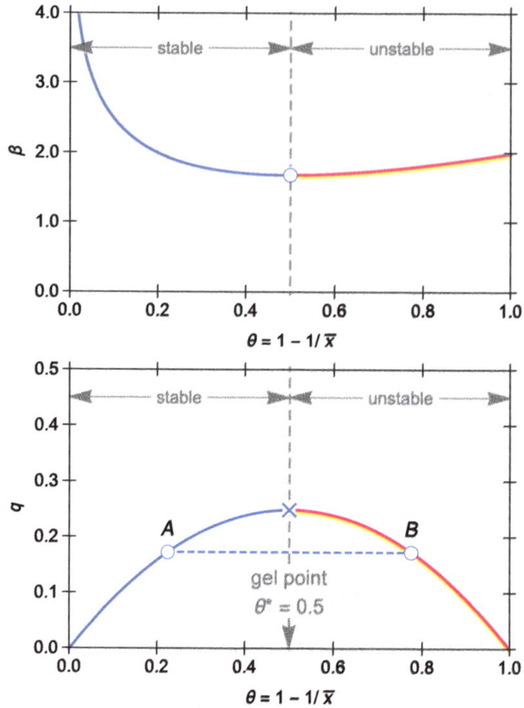

Fig. 9.2 Stability analysis of the product kernel. The stable region is $0 \le \theta \le 1/2$. At $\theta^* = 1/2$ the system reaches the limit of stability (gel point). In the region $1/2 < \theta \le 1$ the single-phase system is unstable and the system forms a sol-gel mixture

$\theta^* = 1/2$, which corresponds to mean cluster size $x^* = 2$, and both invert the sign of the slope in the region $1/2 < \theta < 1$, which is unstable. Aggregation always starts in the stable region at $\theta = 0$ and moves to the right. Up until point θ^* the system consists of a stable MPD, given by Eq. (9.12). Past θ^* the system forms a gel phase that exists in equilibrium with the sol. To determine the tie line, we recall the fundamental recursion for $\Omega_{M,N}$,[4]

$$\frac{\Omega_{M,N+1}}{\Omega_{M,N}} = \left(\frac{M-N}{N}\right)\frac{1}{\langle K\rangle_{M,N+1}} \equiv q. \tag{9.15}$$

This recursion governs the evolution of the partition function from one generation to the next and defines q in the thermodynamic limit. The sol phase must satisfy Eq. (9.15) as well as the stability condition

$$\frac{dq}{d\theta} \ge 0. \tag{9.16}$$

On the stable branch both Eqs. (9.15) and (9.16) are satisfied simultaneously. When the system passes into the unstable region (state B in Fig. 9.15) the stability

[4]See Eq. (8.22) on p. 252.

condition in Eq. (9.16) is violated but it can be restored if the sol has the properties of state A, which is defined by the intersection of the stable branch with the horizontal line that passes through B. State A has the same q as state B, and satisfies the stability condition. This graphical construction resolves the properties of the sol phase in the sol-gel region. Mathematically this is expressed by the equality

$$q(\theta_{sol}) = q(\theta), \tag{9.17}$$

where $\theta = 1 - 1/\bar{x} = 1 - M/N$ refers to the θ-value of the overall population and $\theta_{sol} = 1 - 1/\bar{x}_{sol} = 1 - M_{sol}/(N-1)$ refers to the θ-value in the sol phase. By virtue of the symmetry of q around θ, the solution to this equation in the unstable region ($\theta > 1/2$) is immediately identified to be

$$\theta_{sol} = 1 - \theta. \tag{9.18}$$

This condition defines the sol phase completely: the MPD of the sol is given by Eqs. (9.10), (9.11), and (9.12), with θ replaced by θ_{sol}:

$$\beta = \theta - 2\log\theta_{sol}; \tag{9.19}$$

$$q = \theta_{sol}(1-\theta_{sol}), \tag{9.20}$$

$$\frac{\tilde{n}_i}{N} = \frac{2\theta_{sol}}{1-\theta_{sol}}\frac{(2\theta_{sol}i)^{i-2}}{i!}e^{-2\theta_{sol}i}. \tag{9.21}$$

With $\theta_{sol} = 1 - 1/\bar{x}_{sol}$ in Eq. (9.18) we also obtain

$$\bar{x}_{sol} = 1/\theta, \tag{9.22}$$

which gives the mean size in the sol in terms of the overall θ-value.

The gel phase will be obtained by mass balance. In an extensive population with N clusters and total mass M, the gel phase consists of a single cluster ($N_{gel} = 1$) with mass m_{gel}, and the sol consists of $N_{sol} = N - 1$ clusters with mass $M_{sol} = M - m_{gel}$, which expresses the condition of mass conservation. We write mass conservation in terms of $\bar{x}_{sol} = M_{sol}/(N-1)$ and the gel fraction $\phi_{gel} = m_{gel}/M$:

$$\bar{x}N = \bar{x}_{sol}(N-1) + \phi_{gel}M = \bar{x}_{sol}(N-1) + \phi_{gel}\bar{x}N. \tag{9.23}$$

Letting $(N-1)/N \to 1$, we obtain a relationship between $\bar{x}$, $\bar{x}_{\rm sol}$, and $\phi_{\rm gel}$:

$$\frac{\bar{x}_{\rm sol}}{\bar{x}} + \phi_{\rm gel} = 1. \tag{9.24}$$

Using $\theta = 1 - 1/\bar{x}$ and $\theta_{\rm sol} = 1 - 1/\bar{x}$, this becomes

$$\phi_{\rm gel} = \frac{\theta - \theta_{\rm sol}}{1 - \theta_{\rm sol}}. \tag{9.25}$$

In combination with Eq. (9.18) we obtain the gel fraction as a function of θ in the sol-gel region ($\theta > 1/2$):

$$\phi_{\rm gel} = 2 - \frac{1}{\theta}. \tag{9.26}$$

If we combine this result with Eq. (9.18) for the mean cluster in the sol we obtain a remarkably simple relationship between the mean cluster size in the sol and the gel fraction:

$$\bar{x}_{\rm sol} + \phi_{\rm gel} = 2. \tag{9.27}$$

This remarkably simple and symmetric result is a statement of mass conservation in the sol-gel region: at the gel point $\phi_{\rm gel} = 0$, and $\bar{x}^*_{\rm sol} = 2$; at the end of the process $\phi_{\rm gel} = 1$ and $\bar{x}_{\rm sol} = 1$, which states that the sol phase ultimately retreats to its initial state. The pre-gel and post-gel results are summarized in Table 9.1.

Table 9.1 Complete ThL results for the product kernel

Pre Gel $(\theta \le 1/2)$		
Overall mean	$\bar{x} = 1/(1-\theta)$	
Distribution	$\frac{\tilde{n}_k}{N} = \frac{2\theta}{1-\theta}\frac{(2\theta k)^{k-2}}{k!}e^{-2\theta k}$	Eq. (9.12)
Post Gel $(\theta > 1/2)$		
Overall mean	$\bar{x} = 1/(1-\theta)$	
Mean in sol phase	$\bar{x}_{\rm sol} = 1/\theta$	Eq. (9.22)
Mass fraction of gel	$\phi_{\rm gel} = 2 - 1/\theta$	Eq. (9.26)
Sol distribution	$\frac{\tilde{n}_k}{N} = \frac{2\theta_{\rm sol}}{1-\theta_{\rm sol}}\frac{(2\theta_{\rm sol} k)^{k-2}}{k!}e^{-2\theta_{\rm sol} k}$	Eq. (9.21)
In all of cases	$\bar{x} = M/N,\ \theta = 1 - 1/\bar{x},\ \theta_{\rm sol} = 1 - \theta$	

9.3 Gel Branch and the Sol-Gel Tie Line

The gel phase is not represented in the phase diagram in Fig. 9.2, however it is possible to bring this branch into view. Let us start with the discrete form of $\log q$,

$$q = \frac{\Omega_{M,N+1}}{\Omega_{M,N}} = \frac{(M-N)(N+1)^2}{M^2 N} \tag{9.28}$$

and express it in terms of $\theta = 1 - M/N$ as

$$q = \theta\left(1 - \theta + \frac{1}{M}\right)\left(1 + \frac{1}{M(1-\theta)}\right). \tag{9.29}$$

For large M we obtain $q \to \theta(1-\theta)$ except near $\theta = 1$, where q diverges. This produces a curve that contains two stable branches with positive slope, one in the sol region ($\theta \to 0$), the other in the gel region ($\theta \to 1$), connected by an unstable branch. For finite M the stable branch on the gel side lies outside the physical range of θ, whose maximum possible value is when $N = 1$:

$$\theta_{\max} = 1 - 1/M.$$

As M increases, the gel branch becomes steeper and in the thermodynamic limit forms a vertical line at $\theta = 1$ that brings the gel branch within the admissible region $0 \le \theta \le 1$ (Fig. 9.3b). This construction restores stability in the entire (θ, q) plane. For $\theta < 1/2$ the system forms a stable single-phase sol whose q value is read off the stable branch (point A in Fig. 9.3b). A state in the unstable region $\theta > 1/2$ (point B) splits into two phases, a sol with the properties of state A, and a gel represented by point C. The line segment ABC is a tie line. As aggregation proceeds and state B advances to the right, the equilibrium sol recedes and retraces the pre-gel trajectory in the reverse direction, while losing mass to the gel phase.

Note 9.2 (On the Stability of the Gel Phase)

The unstable branch on the curve $q(\theta, M)$ is bounded by θ_1 and θ_2 such that

$$\left(\frac{\partial q}{\partial \theta_1}\right)_M = \left(\frac{\partial q}{\partial \theta_2}\right)_M = 0,$$

with $\theta_1 < \theta_2$. The stable branches are

$$\textit{sol:} \quad 0 \le \theta \le \theta_1$$

$$\textit{gel:} \quad \theta_2 \le \theta \le 1$$

We determine θ_1 and θ_2 by setting the derivative of Eq. (9.29) to zero:

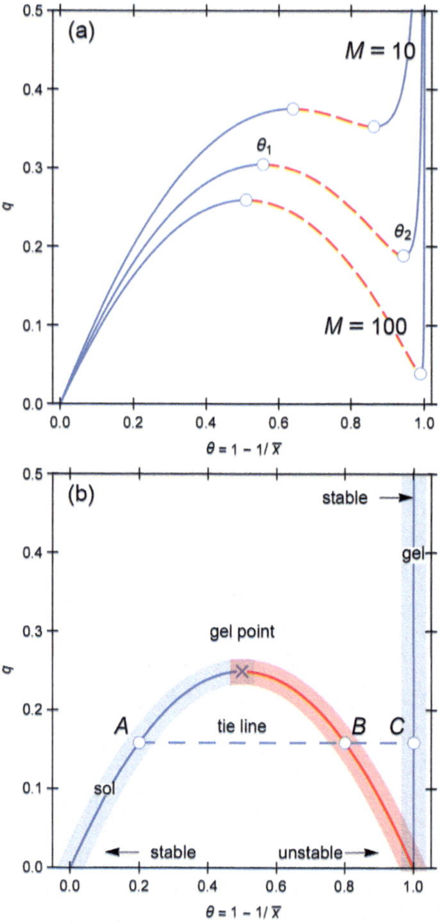

Fig. 9.3 Phase diagram and tie line on the (q, θ) plane. (**a**) The discrete-finite form of q in Eq. (9.29) reveals a structure that contains a stable sol region $(1 \le \theta\theta_1)$, unstable region to the right of the gel point $(\theta_1 < \theta < \theta_2)$, followed by yet another stable branch on the gel side $(\theta_2 \le \theta \le 1)$. (**b**) In the thermodynamic limit the gel branch collapses into a vertical line at $\theta = 1$. Point B in the unstable region splits into two stable phases, the sol A on the stable sol branch, and a gel at C. The line ABC is an equilibrium tie line

$$\theta_1 = \frac{1}{4}\left(3 - \sqrt{1 - \frac{8}{M}}\right), \quad \theta_2 = \frac{1}{4}\left(3 + \sqrt{1 - \frac{8}{M}}\right).$$

Letting $M \to \infty$ we obtain the asymptotic forms

$$\theta_1 \to \frac{1}{2} + \frac{1}{M}, \quad \theta_2 \to 1 - \frac{1}{M} \equiv \theta_{\max}(M).$$

The value of θ_1 converges to the gel point $\theta^* = 1/2$, while θ_2 approaches $\theta_{\max}$, the maximum possible value of θ in a system with finite M. For large M, therefore, $q(\theta)$ terminates at the minimum of q, leaving the stable gel branch in the unreachable region $\theta > 1/M$. When the system reaches the single cluster state $(\theta = 1 - 1/M)$, it is at a state of borderline stability. Asymptotically, the state $\theta = 1 - 1/M \to 1$ represents the stable gel.

★

9.4 Monte Carlo Simulation of Kinetic Gelation

We have obtained analytic expressions for the pre-gel and post gel solutions of the product kernel. We will compare these results against Monte Carlo simulation. In this simulation we produce a series of complete trajectories in phase space, starting from the fully dispersed state and ending at the fully gelled state using constant-volume Monte Carlo.[5] Unlike the method of the binary exchange reaction, which samples distributions within an ensemble of fixed M and N, constant-volume MC transitions to a distribution in the ensemble (M, N) from a distribution in $(M, N+1)$. The simulation keeps track of a configuration of cluster masses,

$$\mathbf{m} = (m_1, m_2 \cdots m_N)$$

The transition from current generation to the next is implemented as follows: Given configuration $\mathbf{m}$ in the current ensemble $\mu C(M, N)$, we chose a pair of clusters with mass m_i and $m_j j$ with probability

$$P[m_i, m_j] = \frac{2K(m_i, m_j)}{N(N-1)\langle K\rangle},$$

where $\langle K\rangle$ is the mean kernel in the configuration. Once the pair of clusters has been identified, cluster i is deleted and cluster j is replaced by $m_i + m_j$. The trajectory is initialized with a configuration of M monomers and ends when a single cluster forms. The process is repeated and the distributions in each generation are averaged to obtain the mean distribution in each generation over the entire aggregation graph at fixed M. To compare with theory, M must be large enough for the thermodynamic limit to apply. The values of M chosen here range from 40 to 200, as a compromise between the ThL requirement and computational time.

Given a configuration of N clusters with total mass M, the sol and gel regions are defined by the conditions

$$\text{sol:} \qquad 1 \le i < \frac{M-N+1}{2},$$

$$\text{gel:} \qquad \frac{M-N+1}{2} \le i \le M-N+1.$$

The mean cluster size of the sol, $\bar{x}_{\text{sol}}$, is the mean size in the sol region, and the gel fraction ϕ_{gel} is the mass fraction in the gel region. These are shown in Fig. 9.4 as a function of the overall progress variable

$$\theta = 1 - \frac{N}{M} = 1 - \frac{1}{\bar{x}}. \tag{9.30}$$

[5] See Smith and Matsoukas (1998).

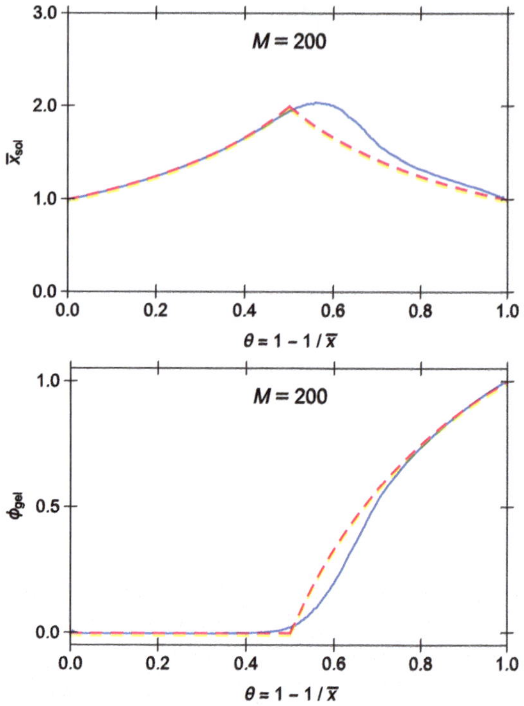

Fig. 9.4 Monte Carlo simulations (sold line) compared against theory (dashed line)

In the pre-gel region the mean sol size is equal to the overall mean, $\bar{x}_{\rm sol} = \bar{x} = 1/(1-\theta)$. In the post-gel region the mean size in the sol is $\bar{x}_{\rm sol} = 1/\theta$. These results can be summarized in a single equation,

$$\bar{x}_{\rm sol} = \frac{1}{\frac{1}{2} + \left|\frac{1}{2} - \theta\right|}, \tag{9.31}$$

which is symmetric about the gel point. The simulation follows this in the pre-gel region. The peak in $\bar{x}_{\rm sol}$ is somewhat delayed relative to Eq. (9.31) but the mean cluster size reverts to the theoretical line as $\theta \to 1$. The gel fraction in the simulation emerges more smoothly than theory predicts but once again theory and simulation converge as θ increases. The discrepancies near the gel point arise from the finite size of the system.

Figure 9.5 shows the evolution of the cluster distribution. These snapshots are taken at three points before the gel point ($\theta = 0.15$, 0.25 and 0.35) and at three symmetric points past the gel point ($\theta = 0.65$, 0.75 and 0.85). The theoretical distribution of the sol is calculated from Eq. (9.12) with

$$\theta_{\rm sol} = \frac{1}{2} - \left|\frac{1}{2} - \theta\right|,$$

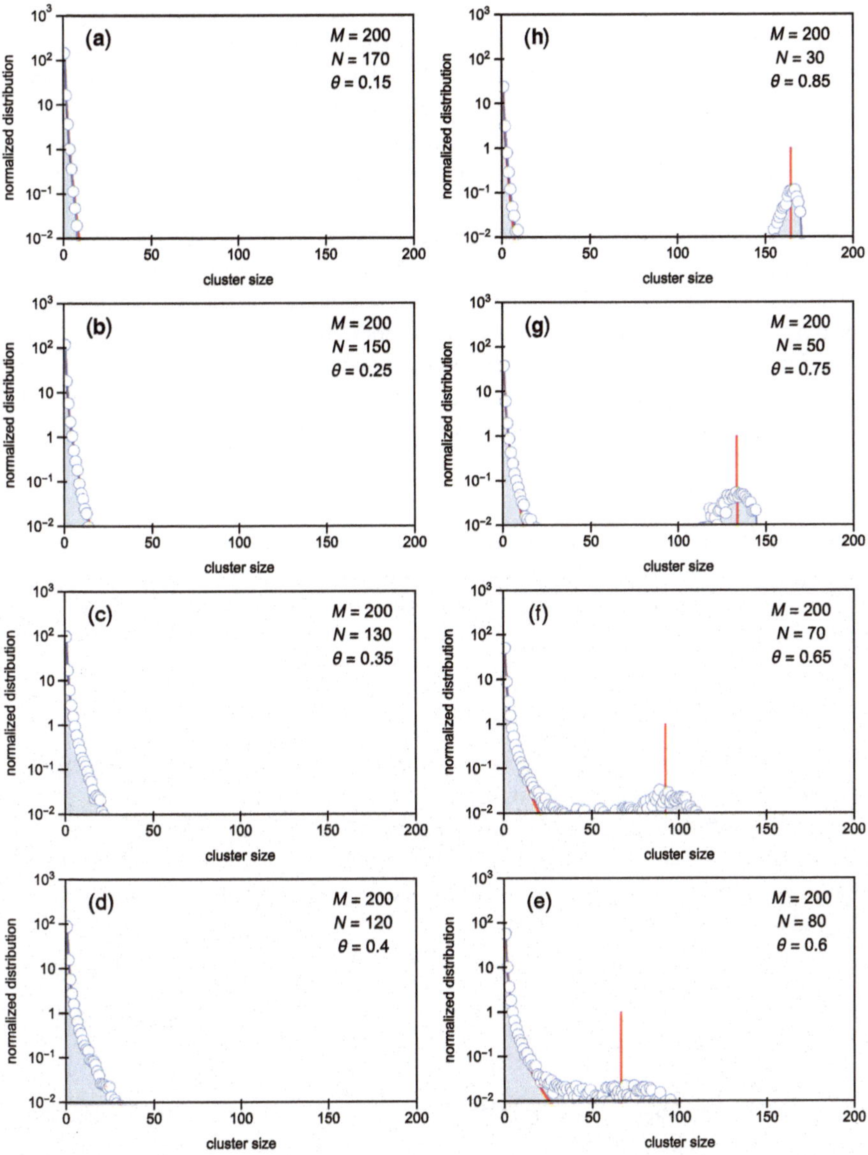

Fig. 9.5 Snapshots of the distribution at various symmetric points around the gel point $\theta^* = 1/2$. The sequence progresses from top to bottom in the first column, and from bottom to top in the second column. Snapshots on the same row are at symmetric states around $\theta^* = 1/2$ and except for the total number of clusters their theoretical sol distributions (solid lines) are identical

and the position of the gel cluster is calculated as $m_{gel} = M\phi_{gel}$ with ϕ_{gel} from Eq. (9.26). As theory predicts, past the gel point the distribution of the sol retraces its pre-gel state in the reverse direction. This behavior is quite remarkable: while aggregation proceeds irreversibly to ever increasing mean size, halfway through the process the sol fraction reverses direction and regresses to its earlier state. Indeed the only difference between the pre- and post-gel distributions of the sol is the total mass.

The position of the gel cluster agrees well with the peak position of the gel phase[6]—indeed thermodynamics makes statements about the *most probable* location of the gel mass. The gel peak in the post-gel region reflects fluctuations in the mass of the gel cluster between individual configurations of the cluster ensemble. The simulation samples the *mean* distribution and this in the case of the gel phase is not the same as the *most probable* distribution, which is a delta function. The difference between the two distributions is most pronounced around the gel point but vanishes as the size of the gel cluster increases. This is very similar to the behavior of the systems studied in Chap. 5.

Note 9.3 (Mean Aggregation Kernel)

It is an instructive exercise to calculate the mean aggregation kernel in the post-gel region. In the pre-gel region the mean kernel is equal to the square of the mean cluster mass,

$$\bar{K}(\tilde{\mathbf{n}}) = \bar{x}^2 \tag{9.4}$$

In the post-gel region the mean kernel must be calculated by considering aggregation events within the sol phase, and between the sol and gel phase. There are no aggregation events within the gel phase because this phase contains one cluster only. There are $N-1$ clusters in the sol phase that produce $(N-1)(N-2)$ ordered pairs that may undergo a clustering event; there are $2(N-1)$ ordered pairs between the sol and the gel. The mean kernel is

$$\bar{K} = \frac{1}{(N-1)(N-2)+2}\left\{\sum_{i,j\in\mathrm{sol}} K(i,j)\tilde{n}_i(\tilde{n}_j-\delta_{ij}) + 2\sum_{i\in\mathrm{sol}} K(m_{gel},i)\tilde{n}_i\right\},$$

where m_{gel} is the mass of the gel. With $K(i,j) = ij$ and dropping constants in relationship to N this becomes

$$\bar{K} = \frac{\sum_{\mathrm{sol}} ij\,\tilde{n}_i\tilde{n}_j}{N^2} + \frac{2m_{gel}M_{sol}}{N^2},$$

[6] On the other hand, the *average* mass of the gel cluster is somewhat lower than the position of the maximum, as indicated by the fact that in the vicinity of the gel point ϕ_{gel} from simulation lies below the theoretical value in Fig. 9.4.

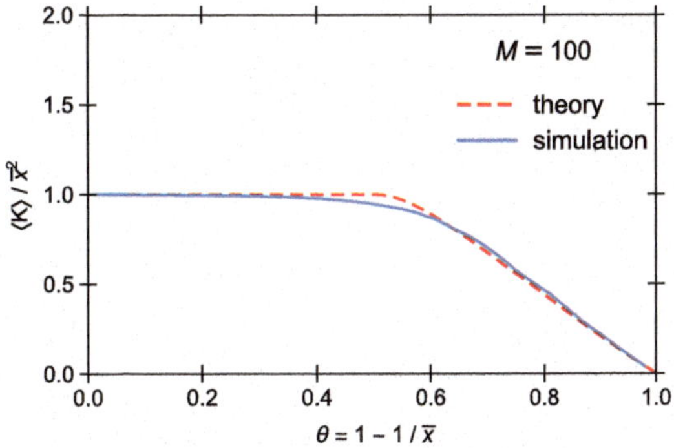

Fig. 9.6 Mean kernel in the pre-gel and post-gel regions (the simulation is done with $M = 100$ monomers)

where $M_{\text{sol}} = M - m_{\text{gel}}$ the mass of the sol. The first summation on the right-hand side is the mean kernel in the sol, $\bar{x}_{\text{sol}}^2$. Using $M_{\text{sol}}/N = \bar{x}$ and $m_{\text{gel}}/N = \bar{x}m_{\text{gel}}/M = \bar{x}\phi_{\text{gel}}$, this result becomes

$$\bar{K} = \bar{x}_{\text{sol}}^2 + 2\bar{x}\,\bar{x}_{\text{sol}}\phi_{\text{gel}}.$$

The result can be expressed as

$$\frac{\bar{K}}{\bar{x}^2} = \begin{cases} 1, & \theta \le 1/2; \\ 4/\theta - 1/\theta^2 - 3, & \theta > 1/2, \end{cases} \tag{9.32}$$

or equivalently in terms of cluster size,

$$\frac{\bar{K}}{\bar{x}^2} = \begin{cases} 1, & \bar{x} \le 2; \\ \dfrac{2\bar{x} - 3}{\bar{x}^2 - 1}, & \bar{x} \ge 2. \end{cases} \tag{9.33}$$

For $\bar{x} \gg 1$ we obtain the post-gels scaling

$$\langle K\rangle \sim 2\bar{x}.$$

The mean kernel in the post-gel region is *linear* function of the mean cluster, while in the pre-gel region it is quadratic. This behavior is checked in Fig. 9.6 against simulation and the agreement is indeed very good.

★

9.5 A Closely Related Linear Ensemble

In Chap. 5 we treated the formation of a gel phase in a generic ensemble and constructed the tie line by applying the isothermal condition

$$\beta^{\text{sol}} = \beta^{\text{gel}}. \tag{9.34}$$

In this chapter we constructed the tie line of the product kernel quite differently: we identified the equilibrium sol phase by equating the parameter q in the stable and unstable branches:

$$q^{\text{sol}} = q^{\text{gel}} = q. \qquad [9.17]$$

We must address this change of course.

Equation (9.34) is still valid but we do not have sufficient information to use it. In order to apply Eq. (9.34) we must know the cluster function of the gel. In a linear ensemble the cluster function of the gel is simply $w(m_{\text{gel}})$. The ensemble of the product kernel is *not* linear, however. We derived its cluster bias

$$w_i = 2\frac{(2i)^{i-2}}{i!}, \qquad [9.9]$$

under the condition

$$\bar{K}(\mathbf{n}) \to \left(\frac{M}{N}\right)^2 \equiv \langle K \rangle_{M,N}, \qquad [9.4]$$

which assumes that no gel cluster is present. Therefore the cluster bias of the gel phase *cannot* be obtained from Eq. (9.9). Let us suppose now that we do have a strictly linear ensemble with cluster functions from Eq. (9.9). We will construct the phase diagram of this model and compare it to that of the product kernel.

We set up the problem as follows: We are given a linear selection functional with the cluster functions in Eq. (9.9). We do not have expressions for the partition function or the parameters β and q, these will have to be obtained from the known cluster function. We write the MPD in canonical form,

$$\frac{\tilde{n}_i}{N} = w_i \frac{e^{-\beta i}}{q} = 2\frac{(2i)^{i-2}}{i!}\frac{e^{-\beta i}}{q}. \tag{9.35}$$

subject to the normalizations,

$$\sum_{i=1}^{\infty} \frac{\tilde{n}_i}{N} = 1; \quad \sum_{i=1}^{\infty} i\frac{\tilde{n}_i}{N} = \bar{x}. \tag{9.36}$$

Solving the normalization conditions for β and $\bar{x}$ we find

$$\beta = 1 - \frac{1}{\bar{x}} - 2\log\left(1 - \frac{1}{\bar{x}}\right) = \theta - 2\log\theta; \tag{9.37}$$

$$q = \frac{1}{\bar{x}}\left(1 - \frac{1}{\bar{x}}\right) = \theta(1-\theta). \tag{9.38}$$

These are the same as those for the product kernel. Therefore, the pre-gel distribution of the linear ensemble with cluster functions from Eq. (9.9) is identical to that of the product kernel. This should not come as a surprise since we have constructed the linear ensemble using the cluster bias of the MPD.

To determine the gel point we will apply the equilibrium condition in Eq. (9.34). First we use the Stirling formula for the factorial to express w_i as a continuous function of cluster size x:

$$\frac{2(2i)^{i-2}}{i!} \to \frac{1}{\sqrt{8\pi}}\frac{(2e)^x}{x^{5/2}} \equiv w(x), \tag{9.39}$$

where x stands for the cluster size in the continuous domain. Its logarithm is

$$\log w(x) = -\frac{5}{2}\log x + (1+\log 2)x - \frac{\log(8\pi)}{2}. \tag{9.40}$$

The temperature of the gel phase is the derivative of $\log w$ evaluated at m_{gel}:

$$\beta^{\text{gel}} = \left(\frac{d\log w}{dx}\right)_{m_{\text{gel}}} = -\frac{5}{2m_{\text{gel}}} + 1 + \log 2. \tag{9.41}$$

In the thermodynamic limit, $m_{\text{gel}} \to \infty$, and so

$$\beta^{\text{gel}} \to 1 + \log 2. \tag{9.42}$$

By the equilibrium condition, the temperature of the sol is

$$\beta^{\text{sol}} = \beta^{\text{gel}} = 1 + \log 2. \tag{9.43}$$

We obtain the gel point by equating this to the right-hand side of Eq. (9.37) that gives the temperature of the sol as a function of θ,

$$\theta^* + 2\theta^* = 1 + \log 2. \tag{9.44}$$

Its solution is

$$\theta^* = 1/2 \tag{9.45}$$

or equivalently, $\bar{x}^* = 2$. The gel point of the linear ensemble is the same as that of the product kernel. In the post gel region β is given by Eq. (9.43) and is constant at all times. It follows that the mean cluster size in the sol is constant, $\bar{x}_{\text{sol}} = 2$, at all times. The gel fraction is calculated from Eq. (9.25) with $\theta_{\text{sol}} = \theta^* = 1/2$:

$$\phi_{\text{gel}} = 2\theta - 1, \quad \theta \geq 1/2. \tag{9.46}$$

These results are summarized in Fig. 9.7 which shows the evolution of $\bar{x}_{\text{sol}}$ and ϕ_{gel}, and contrasts it with that for the product kernel. The sol-gel transition in the linear ensemble is isothermal, $\beta = \beta^*$ at all times past the gel point. By contrast, in the product kernel the state of the sol past the gel point retraces the pre-gel state in the reverse direction: β decreases up to the gel point, then increases. Taking $1/\beta$ to

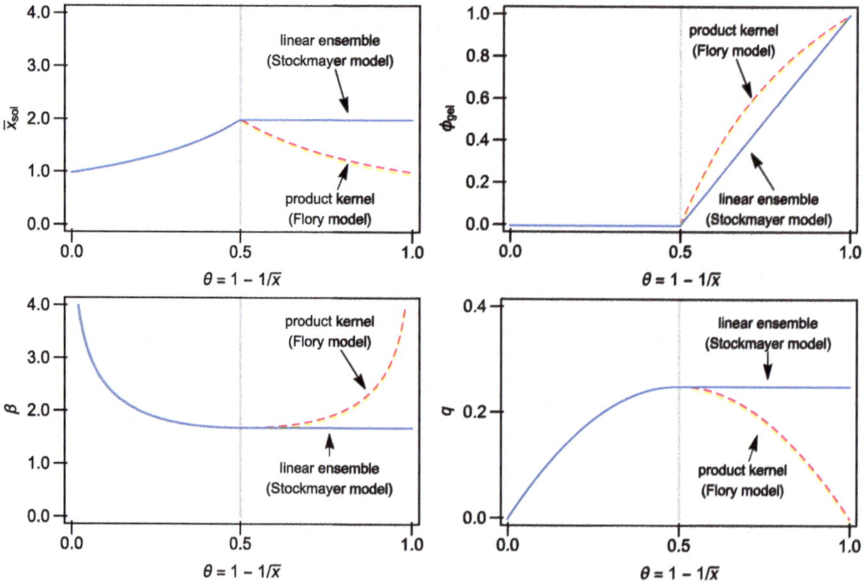

Fig. 9.7 Comparison between the product kernel (dashed line) and the linear ensemble with the cluster functions of the product kernel (solid line). The pre-gel behavior of the two models is identical and both have a gel point at $\theta^* = 1/2$. In the post-gel region the linear ensemble maintains constant $\bar{x}_{\text{sol}} = 2$, and the gel fraction increases linearly. The two ensembles are related to gelation models by Flory (product kernel) and Stockmayer (linear ensemble)

represent temperature, the system cools down as more sol is converted into gel. This is akin to an adiabatic process.

The above solution was first obtained by Stockmayer (1943) through a different derivation. Our derivation highlights the nonlinear nature of the ensemble that is produced by the product kernel versus the strictly linear ensemble in Stockmayer's treatment. The two models have distinctly different behavior of the strictly linear ensemble in the post-gel region. And yet, the complete agreement in the pre-gel region suggests that the ensemble of the product kernel is only mildly nonlinear. It is *quasilinear*, i.e., linear enough to produce cluster functions that are constant from one generation to the next to be calculated from the recursion

$$w_i = \frac{1}{i-1}\sum_{j=1}^{i-1} K_{i,j}, w_{i-j}w_j; \quad w_1 = 1. \tag{9.47}$$

This recursion, as we recall from the previous chapter, is not a general result for all kernels. It is specific to kernels that satisfy the uniformity condition

$$\bar{K}(\mathbf{n}) = \text{function of } (M, N) \text{ for all } \mathbf{n} \text{ in } \mu C(M, N). \tag{9.48}$$

The product kernel satisfies the condition for almost all distributions, with the exception of only those that carry significant number of large clusters. This essentially covers the entire phase space of *sol* distributions and allows the cluster functions to be obtained by the recursion in Eq. (9.47). The condition in Eq. (9.48) is clearly not satisfied if the distribution contains a gel cluster. Therefore the cluster function of the gel phase is *not* given by Eq. (9.47). In the pre-gel region the distribution consists entirely of the sol phase and the product kernel produces results identical to the strictly linear ensemble. In the post gel region the linear and quasilinear systems diverge because they assign a different cluster function to the gel phase.

In the thermodynamic limit the system samples a narrow range in the vicinity of the MPD. All distributions within this sampled range are differential variations of the MPD and their mean kernel is the same as that of the MPD:

$$\bar{K}(\mathbf{n}) = \bar{K}(\tilde{\mathbf{n}}) = \text{function of } (M, N) \text{ for all } \mathbf{n} \text{ in } (\tilde{\mathbf{n}} \pm \delta\mathbf{n}). \tag{9.49}$$

This is true for all kernels, but is a much narrower condition than (9.48) because it is limited to the neighborhood of the MPD in the ThL and is not sufficient to produce the recursion in Eq. (9.47). This recursion requires the uniformity condition to be valid over the entire cluster ensemble or, as in the case of the product kernel, over almost the entire ensemble.

9.6 Flory and Stockmayer (But Mostly Stockmayer)

With the post gel solutions for the product kernel and the strictly linear ensemble with the same cluster functions we have made contact with two gelation models associated, respectively, with Flory and Stockmayer. Flory stipulated that the pre-gel solution applies in the post-gel region by applying an argument analogous to ours, though based on entirely different reasoning, to "reflect" the post-gel solution into the pre-gel region.[7] Stockmayer rejected this reflection and proposed what we now recognize as the solution of the linear ensemble.[8] In this solution the intensive properties of the sol phase remain fixed at the conditions of the gel point as mass is transferred from the sol to the gel. This behavior is not consistent with the kinetic model of clustering. It is Flory's model that gives the correct answer to kinetic gelation, even though Flory did not invoke kinetic arguments.

Even though Stockmayer's model conflicts with kinetics, it is quite fascinating to explore Stockmayer's thinking on this problem because it represents the first serious attempt, incomplete as it was, to describe *kinetic* gelation as a *thermodynamic phase transition*. Stockmayer considers a system of multifunctional monomers that bond to form polymer structures. Each monomer has f reactive sites that can form bonds and all unreacted sites are equally reactive regardless of the size of the polymer on which they are found. The size of the polymer is given by the number i of monomers that form it. The polymer may contain any number of branches but it is assumed that no cycles form. The number of ways to form a polymer of size i out of monomers with f functional sites is[9]

$$a_i = \frac{f^i (fi - i)!}{(fi - 2i + 2)!}.$$

and is calculated by treating monomers and functional sites as distinguishable. The number of ways to form a distribution $\mathbf{n} = (n_1, n_2 \cdots)$ from these polymers is obtained by a straightforward combinatorial counting:

$$\Omega(\mathbf{n}) = \mathbf{n}! \prod_i \left(\frac{a_i}{i!}\right)^{n_i}.$$

In our terminology Stockmayer's Ω is the microcanonical weight,[10]

$$\mathbf{n}! W(\mathbf{n}),$$

[7]Compare Flory's Fig. 2 in Flory (1941) to our Fig. 9.4.

[8]Compare Stockmayer's Fig 2 in Stockmayer (1943) to ϕ_{gel} in our Fig. 9.7.

[9]Stockmayer uses w_i for this number but here we use a_i to avoid confusion with the cluster function $w_i = a_i / i!$

[10]To economize in notation we adopt Stockmayer's Ω for the microcanonical weight, even though we have reserved it for the partition function. We do this knowing that in the thermodynamic limit their logarithms are equal.

and the product

$$\prod_i \left(\frac{a_i}{i!}\right)^{n_i} = W(\mathbf{n})$$

is the selection functional, a linear functional with cluster functions $w_i = a_i/i!$ Next Stockmayer obtains the most probable distribution by maximizing $\Omega(\mathbf{n})$ under fixed M and N and expresses it as

$$\tilde{n}_i = A w_i \xi^i. \tag{9.50}$$

In our notation we recognize $A = N/q$ and $\xi = e^{-\beta}$. This is the distribution one expects in a system in which polymers exchange mass reversibly according to the reaction

$$(i) + (j) \overset{K_{\text{eq}}}{\rightleftharpoons} (i') + (j')$$

where $i + j = i' + j'$, with equilibrium constant

$$K_{\text{eq}} = \frac{w_{i'} w_{j'}}{w_i w_j}.$$

Quoting Stockmayer:

> It will be observed that the above process [maximization of Ω] resembles one of maximizing an entropy at constant energy. It should therefore be valid for the polyesters, in which the size distribution is probably controlled by ester-interchange equilibria. The resulting distribution law, however, is by no means restricted to cases of equilibrium, but should be equally applicable to unidirectional processes, as long as assumption (b) [equal reactivity of all free sites] is fulfilled. (Stockmayer (1943), p. 47.)

In the kinetic problem, polymers of size $i + j$ are formed by the irreversible reaction of chains with sizes i and j,

$$(i) + (j) \xrightarrow{K_{ij}} (i + j).$$

Since all available bonding sites on a polymer are assumed equally reactive, the kernel is given by the product of the number of sites in each reacting chain. In the absence of cycles the number of sites in a chain with i monomers is[11]

$$(f - 2)i + 2,$$

[11] In the absence of cycles, it is easy to show that if we move a monomer to a different position on the chain the number of free sites (that is, sites that are bonded to other monomers) does not change. This means that the number of free sites is the same for all polymers with the same number of monomers, regardless of structure. We may then establish this number by looking at a linear chain: the $i - 2$ monomers in the interior of the chain have $f - 2$ free sites and the two monomers at the ends of the chain have $f - 1$ each. The total number is $(i - 2)(f - 2) + 2(f - 1)$.

regardless of the number of branches. The aggregation kernel therefore is

$$K_{i,j} = \big((f-2)i+2\big)\big((f-2)j+2\big). \tag{9.51}$$

Stockmayer showed that the MPD of the equilibrium model satisfies the Smoluchowski equation with the kernel given above: the equilibrium and kinetic approaches converge to the same distribution.

The summations that give the zeroth and first order moments converge only under the condition

$$\bar{x} \le \frac{2(f-1)}{f-2}, \tag{9.52}$$

where $\bar{x}$ is the mean cluster size. With $f = 2$, the convergence condition reads $\bar{x} < \infty$ and is of course satisfied for all finite $\bar{x}$. For $f \ge 3$ convergence is possible only up to a certain cluster size. With $f = 3$, for example, convergence is possible only up to $\bar{x} = 4$. Stockmayer associated the divergence of the power series with the emergence of the gel phase and defined the gel point as the maximum mean cluster size for which convergence is possible:

$$\bar{x}^* = \frac{2(f-1)}{f-2}. \tag{9.53}$$

In the size range $\bar{x} \le \bar{x}^*$ the system consist of a stable distribution given by the MPD in Eq. (9.51). In the region $\bar{x} > \bar{x}^*$ the system consists of a sol-gel mixture. Stockmayer obtains the post-gel solution by a mathematical argument. He points out that the divergence of the power series is almost entirely due to growth of the terms for large cluster sizes while lower terms appear to converge. He deduces from this that the distribution of the sol normalized by the total number of clusters remains fixed at all times and equal to its value at the gel point. This leads to linear increase of the gel fraction as a function of the progress variable $1 - 1/\bar{x}$.

9.6.1 *Critique of Stockmayer's Approach*

Stockmayer's approach is remarkable for establishing mathematical equivalence between a purely equilibrium and a pure kinetic treatment of clustering—at least in the pre-gel region. That the most probable distribution that emerges through a purely enumerative process of counting the number of ways to form a polymer is in full agreement with a kinetic model that produces ever larger clusters through irreversible aggregation is a powerful hint that the tools of equilibrium thermodynamics can be transferred to unidirectional processes in time. Stockmayer is aware of this relationship and his 1943 paper is replete with references to phase transitions and to the work of J. E. Mayer and M. G. Mayer[12] in statistical mechanics. He compares the sol-gel transition to vapor condensation, the parameter ξ to fugacity (stays constant during the entire sol-gel transition), and the parameter A in Eq. (9.50) to volume (decreases during the transition). Reading Stockmayer's work through the prism of the cluster ensemble, we recognize that he is solving two different problems—or perhaps one and a half. He provides the correct solution to both, but confuses them to be the same problem. The first, or rather *half* problem, is the pre-gel solution to the aggregation problem under the kernel in Eq. (9.51). He obtained the MPD in the form of Eq. (9.50), which may also be expressed more fully as

$$\frac{\tilde{n}_i}{N} = \frac{(1-\alpha^2)f}{\alpha\bar{x}} \frac{(fi-i)!(f\xi)^i}{(i-1)!(fi-2i+2)!} \tag{9.54}$$

with

$$\xi = \frac{\alpha(1-\alpha)^{f-2}}{f},$$

and

$$\alpha = \frac{2}{f}\left(1-\frac{1}{\bar{x}}\right).$$

He also obtained the complete pre- and post-gel solution to the *strictly linear ensemble* with cluster weights

$$w_i = \frac{f^i(fi-i)!}{i!(fi-2i+2)!}.$$

As we now know, the strictly linear ensemble converges to the kinetic clustering ensemble, but only in the absence of a gel phase—and only for special kernels.

Stockmayer came tantalizingly close to unlocking the thermodynamic connection to gelation. The door that led him into this direction was the enumeration of ways to form a distribution of polymers and the maximization that produces the most probable distribution. The door that led him astray was mathematics. By focusing on the radius of convergence of the lower moments, and by solving this mathematical problem successfully, he was able to infer the stability of the system without the need to appeal to thermodynamics. In Stockmayer's defense, his work was published in 1943, five years before Shannon brought entropy outside thermodynamics and fourteen years before Jaynes began to discuss statistical

[12]Mayer and Mayer (1940).

mechanics in a generalized context. Without the benefit of these groundbreaking ideas it would have been difficult to imagine thermodynamics in the absence of physical interactions such as intermolecular potentials or some form of coupling that establishes equilibrium. Others have followed Stockmayer's work and added solutions for different models of multifunctional monomers, but no consistent thermodynamic treatment of gelation has appeared in the years since, until now.

9.6.2 Relationship to the Product Kernel

Stockmayer's kernel in Eq. (9.51) can be written as

$$K_{ij} = (f-2)^2 ij + 2(f-2)(i+j) + 4 \tag{9.55}$$

and is a linear combination of the constant, sum, and product kernels. In the literature of aggregation this is known as the bilinear kernel. The constant and linear components satisfy the condition of kernel uniformity for all distributions, and the product term satisfies it for almost all distributions. This kernel, like the product kernel, produces an asymptotically linear ensemble with constant cluster functions that can be calculated by the recursion in Eq. (9.47). This was shown by Spouge (1985b) via a combinatorial derivation unrelated to our method in Chap. 8 that nonetheless produces the same recursion for w_i. For large clusters the kernel is dominated by the product term. If we treat f as a parameter and let it go to ∞, then the kernel is dominated by the product term for all cluster sizes. We may view the product kernel as the limiting form of the Stockmayer kernel, a kernel based on physical considerations, when the functionality of the monomer is infinite. This situation is unphysical but the resulting kernel is simpler and has become the standard model for gelation. In the continuous limit the cluster function of the Stockmayer model becomes

$$w_i \to \frac{4}{\sqrt{8\pi}} \frac{e^x f^{2x-2}}{x^{5/2}} \equiv w(x). \tag{9.56}$$

This is of the same form as Eq. (9.39) except for factors that arise from the different normalization of the kernel (the Stockmayer kernel is not normalized to $K_{1,1} = 1$ while the product kernel is) and which make identical contribution to the selection functional for all distributions with the same M N. The cluster functions in both Eqs. (9.39) and (9.56) are in fact functionally equivalent to

$$w'(x) = x^{-5/2}.$$

This is of the same power-law type as the example studied in Sect. 5.5.

9.7 Power-Law Kernels

The product kernel is *quasi uniform* in the sense that it satisfies the uniformity condition

$$\bar{K}(\mathbf{n}) \to \left(\frac{M}{N}\right)^2 \equiv \langle K \rangle_{M,N} \qquad [9.4]$$

almost everywhere in phase space. This behavior extends to the more general class of power-law kernels of the form

$$k_{ij} = (ij)^{\nu/2}, \qquad (9.57)$$

where ν is constant. This class defines a family of kernels that are homogeneous functions of i and j with degree ν. With $\nu = 0$ we obtain the constant kernel, which produces a strictly linear ensemble; with $\nu = 2$ we obtain the product kernel, which produces a quasilinear ensemble; and with $\nu = 1$ we obtain a homogeneous kernel with degree 1 that behaves remarkably similar to the sum kernel. Equation (9.57) provides an interpolation between these two cases that we may expect to behave quasilinear by analogy to the product kernel. The mean kernel in distribution $\mathbf{n}$ is

$$\bar{K}(\mathbf{n}) = \frac{\sum_i \sum_j (ij)^{\nu/2} n_i n_j - \sum i^{\nu/2} n_i}{N(N-1)} = \frac{M_{\nu/2}^2 - M_\nu}{N(N-1)} \qquad (9.58)$$

where M_k is the kth order moment of distribution $\mathbf{n}$. We may write

$$M_{\nu/2}^2 \sim \left(\frac{M}{N}\right)^\nu, \quad M_\nu \sim \left(\frac{M}{N}\right)^{\nu/2}, \qquad (9.59)$$

which are acceptable when the variance is small. For large mean cluster size, $M_\nu^2 \gg M_{\nu/2}^2$ and the kernel approximately scales as

$$\bar{K}(\mathbf{n}) \sim \left(\frac{M}{N}\right)^\nu \sim \langle K \rangle_{M,N} \qquad (9.60)$$

Using this in Eq. (8.26) the partition function is

$$\Omega_{M,N} = \left(N!\,\frac{M^{M-N}}{M!}\right)^\nu \binom{M-1}{N-1}, \qquad (9.61)$$

with corresponding canonical parameters

$$\beta = \nu\theta - \log\theta; \tag{9.62}$$

$$q = \theta(1-\theta)^{\nu-1}. \tag{9.63}$$

The cluster functions are given by the recursion

$$w_i = \frac{1}{i-1}\sum_{j=1}^{i-1} K_{i,j},\, w_{i-j}w_j; \qquad w_1 = 1, \qquad [9.47]$$

and the MPD is

$$\frac{\tilde{n}_i}{N} = w_i \frac{e^{-\beta i}}{q}. \tag{9.64}$$

These equations summarize the results for the power law-kernel. Equations (9.60), (9.61), (9.62), and (9.63) represent smooth interpolations between the results for the constant kernel ($\nu = 0$) and product ($\nu = 2$) kernel. They encompass the sum kernel as well ($\nu = 1$), even though the sum kernel is not of the power-law form. The power-law kernel with $\nu = 1$ shows "sum-like" behavior: both kernels are homogeneous with degree 1 and to a good approximation share the same Ω, β, and q, though their w_i are different.

To study the stability of the power-law kernel we examine the behavior of q in Fig. 9.8 as a function of $\theta = 1 - 1/\bar{x}$. For $\nu < 1$, q increases monotonically with θ and produces a stable sol at all θ. For $\nu > 1$, q has a maximum at

$$\theta^* = \frac{1}{\nu}. \tag{9.65}$$

This defines the gel point and corresponds to cluster size

$$\bar{x}^* = \frac{\nu}{\nu - 1}. \tag{9.66}$$

In the region $0 < \theta < \theta^*$ (or $1 \le \bar{x} \le \bar{x}^*$) the system exists as a single sol. Above θ^* (or $\bar{x} > \bar{x}^*$) it exists as a sol-gel system whose tie line is constructed by analogy to the product kernel: The mean cluster size in the sol, $\bar{x}_{\text{sol}}$, is obtained by solving

$$q(\bar{x}_{\text{sol}}) = q(\bar{x})$$

under the condition $\bar{x}_{\text{sol}} < \bar{x}$, which identifies $\bar{x}_{\text{sol}}$ as the mean cluster in the stable branch with the same value of q as the unstable state. The gel fraction is obtained by mass balance,

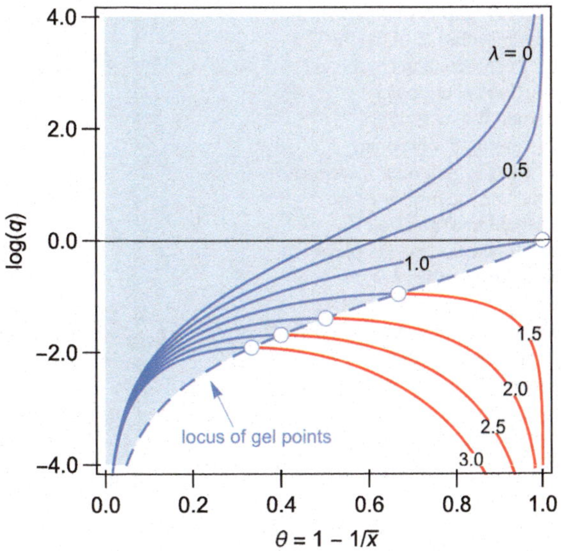

Fig. 9.8 The parameter q of power-law kernel as a function of the exponent ν. For $\nu < 1$ the system is stable at all θ. For $\nu > 1$ q has a maximum at $\theta^* = 1/\nu$ which marks the gel point. The locus of gel points divides the plane into a stable region, above the locus (shaded area), and an unstable region below (unshaded). In the stable region the state is a single sol phase. In the unstable region the state is a mixture of a stable sol, with the same q as the unstable state, plus a gel phase

$$\phi_{\text{gel}} = 1 - \frac{\bar{x}_{\text{sol}}}{\bar{x}} = \frac{\theta - \theta_{\text{sol}}}{1 - \theta_{\text{sol}}}. \tag{9.67}$$

As the exponent ν increases, the gel point moves to the left (it appears earlier). The locus of gel points passes through $\theta = 1$ at $\nu = 1$, and through $\theta = 0$ at $\nu \to \infty$. The corresponding behavior of $\bar{x}_{\text{sol}}$ and ϕ_{gel} is illustrated in Fig. 9.9. In the stable region we have a single sol phase ($\bar{x}_{\text{sol}} = \bar{x}$), and the mean cluster size is

$$\bar{x}_{\text{sol}} = \frac{1}{\theta - 1},$$

as shown by the highlighted line in Fig. 9.9a. Past the gel point $\bar{x}_{\text{sol}}$ decreases, as it traces the stable portion of q in the reverse direction. With increasing ν the gel point moves to earlier times, $\bar{x}_{\text{sol}}$ is closer to 1, and the gel fraction converges to the diagonal $\phi_{\text{gel}} = \theta$.

Note 9.4 (Comparison with Simulation: $\nu = 1.5$)

The results for the power-law kernel are semi exact and to test how well they hold we compare them against Monte Carlo simulation. In this example we use $\nu = 1.5$, which places the kernel in the unstable region, half-way between the border of stability ($\nu = 1$) and the product kernel ($\nu = 2$). We use $M = 200$ starting with $N = M$ (all clusters are monomers) and allow the system to aggregate until $N = 1$. According to theory the gel point is at

$$\theta^* = \frac{1}{\nu} = 0.666,$$

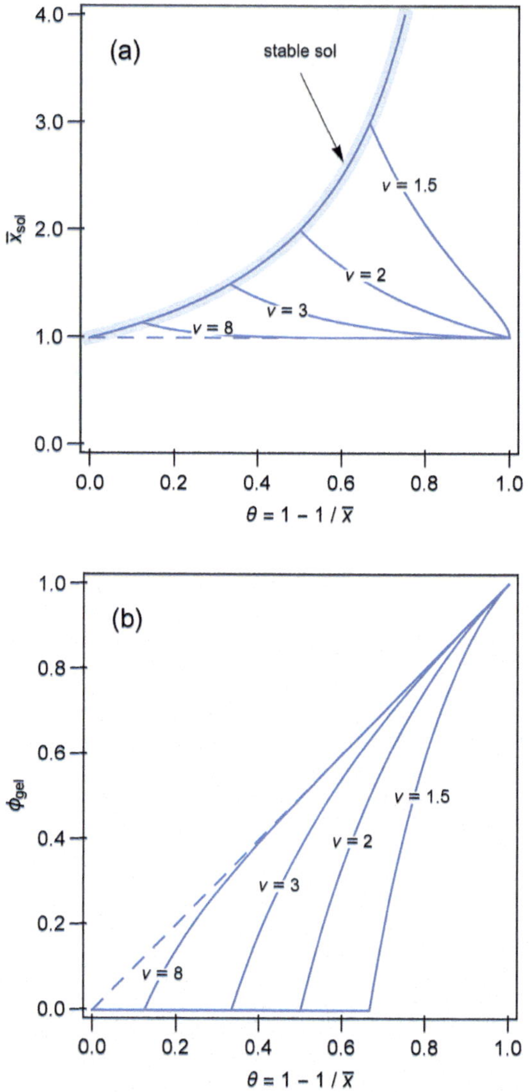

Fig. 9.9 Effect of exponent ν on the evolution of (**a**) the mean sol size and (**b**) the gel fraction under power-law kernel $K_{i,j} = (i\,j)^{\nu/2}$. Increasing the exponent ν moves the gel point to earlier times. In the limit $\nu \to \infty$ (dashed line) gelation is instantaneous: the gel phase is present at all times, the sol consists of monomers only, and aggregation amounts to continuous transfer of monomers to the gel phase

or $\bar{x} = 3$. The simulation (Fig. 9.10) gives a delayed gel point at $\theta^* = 0.74$, or $\theta_{\mathrm{sol}} = 3.25$. We have seen a similar delay of the gel point for the product kernel. Keep in mind that θ amplifies the early stages of growth and collapses the region $\bar{x} \gg 1$ to a narrow neighborhood near $\theta = 1$. Discrepancies are therefore exaggerated.

Once θ_{sol} is known, the mass fractions of the sol and the gel phase are obtained by mass balance

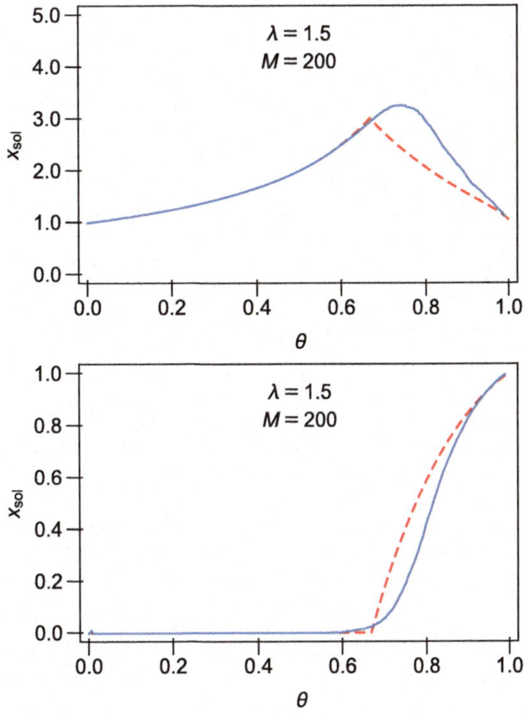

Fig. 9.10 Sol fraction for the power-law kernel with $\nu = 1.5$. Solid lines are Monte Carlo simulations, dashed lines are theory

$$\phi_{\text{sol}} = \frac{1-\theta}{1-\theta_{\text{sol}}}, \tag{9.68}$$

$$\phi_{\text{gel}} = \frac{\theta-\theta_{\text{sol}}}{1-\theta_{\text{sol}}}. \tag{9.69}$$

Figure 9.10 shows the sol fraction calculated by these equations. Also shown are results from a Monte Carlo simulation with $M = 160$. In the simulation, the mass fraction of the sol is calculated as the mass fraction of clusters in the sol range $1 \le i < k_{\max}/2$, with $k_{\max} = M - N + 1$. The agreement is good except that the sol fraction in the simulation lags behind the theoretical prediction, as we also saw in the case of the product kernel in Fig. 9.4.

Two sample distributions are shown in Fig. 9.11. At $\theta = 0.345$ ($\bar{x} = 1.527$) the system is below the gel point. Its sol distribution is calculated using θ from Eq. (9.62), q from (9.63), and w_i from (9.47) by numerical calculation. At $\theta = 0.895$ the system is above the gel point. We obtain the size of the equilibrium sol by solving

$$\theta_{\text{sol}}(1-\theta_{\text{sol}})^{\nu-1} = \theta(1-\theta)^{\nu-1} = 0.2900.$$

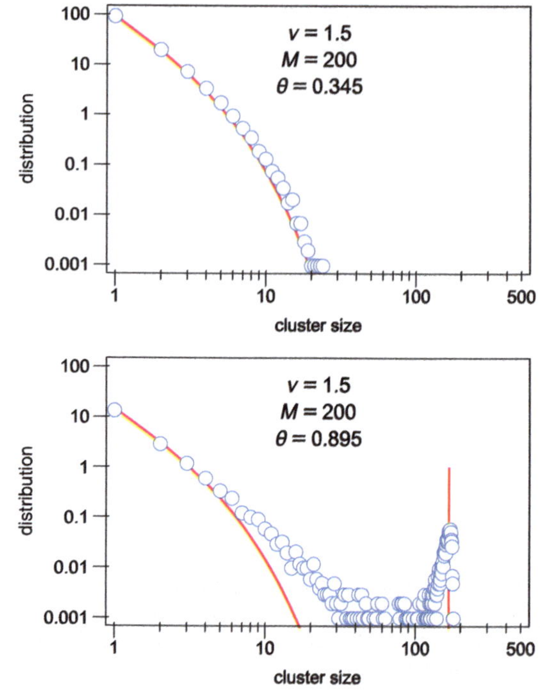

Fig. 9.11 Sample distributions in the pre-gel and post-gel region for the power-law kernel $K_{ij} = (i\,j)^{\nu/2}$ with $\nu = 1.5$. Lines are calculated from theory and symbols are from the simulation

We find $\theta_{\text{sol}} = 0.3635$, or $\bar{x}_{\text{sol}} = 1.571$. The gel fraction is

$$\phi_{\text{gel}} = \frac{\theta - \theta_{\text{sol}}}{1 - \theta_{\text{sol}}} = 0.835,$$

which places the gel cluster at

$$m_{\text{gel}} = M\phi_{\text{gel}} = 167.$$

These calculations, shown by the dashed lines in Fig. 9.11, are in fairly good agreement with simulation and imply that the scaling form of the mean kernel is essentially correct. This is not very surprising because the value of ν places the system between the constant and product kernel, both of which agree with the scaling. For ν larger than two we must be more careful because we are now in the extrapolation region and here the scaling may or may not hold. As ν increases, the MPD forms a longer tail into large clusters, especially as the state gets closer to the gel point, and the scaling in Eq. (9.59) is no longer correct. The scaling is always correct at $\theta \to 0$ and $\theta \to 1$, since in both cases the sol is nearly monodisperse, but may break down closer to the gel point if ν is sufficiently large. Therefore the phase diagram of the power law kernel in Figs. 9.8 and 9.9 should be viewed with caution if ν is outside the interval (0, 1).

★

9.8 Contact with the Literature

The connection between gelation and the Smoluchowski equation is a topic of continuing interest in the literature. The reviews by Aldous (1999) and Leyvraz (2003) discuss this topic extensively. The main questions in these studies are related to kernels that lead to gelation and the solution of the Smoluchowski equation in the pre-gel and post-gel regions. In the context of the Smoluchowski equation, gelation is manifested as a breakdown of mass conservation, a mathematical consequence of the formation of a single infinite cluster that is not captured by the size distribution but which nonetheless contains a finite amount of mass. Ziff and coworkers have shown that power-law kernels of the form in Eq. (9.57) lead to gelation for $\nu \geq 1$ (Ziff and Stell 1980; Ernst et al. 1984; Ziff et al. 1982). In the pre-gel region the Smoluchowski equation behaves properly (conserves mass) and produces the correct solution. To obtain solutions in the post-gel region one invokes assumptions about the interaction between the sol and the gel (Ziff et al. 1983). If the sol is allowed to interact with the gel we obtain the Flory solution, otherwise we obtain the Stockmayer solution (see discussion in Sect. 9.6). Spouge (1983d,c) was also able to obtain the post-gel solution by combinatorial counting (see discussion in Sect. 8.10) by invoking the assumption of the Flory model. The most complete solution, free of external assumptions regarding the interaction between sol and gel is due to Lushnikov (2004, 2005, 2006a,b, 2012, 2013).

Through our thermodynamic analysis of aggregation we have made contact with all results of this prior literature. We have obtained gelling states with the power-law kernel and $\nu \geq 1$ and confirmed that post-gel solution conforms with Flory's model, not Stockmayer's. The point we wish to stress is that we have obtained our results by applying the criteria of thermodynamic stability, a general theory that is not specific to gelation. We have shown that the construction of the tie line, using the familiar tools of phase equilibrium, leads to the same solutions obtained from the Smoluchowski equation, but in a manner that is self-consistent and does not require external assumptions about the nature of interactions between the sol and gel. In the cluster ensemble it is an elementary fact that the gel phase cannot react with itself because it consists of a single cluster, whereas aggregation is a reaction between two clusters.

Appendix: Derivations

Derivations: Power-Law Kernels

Here we derive the parameters β and q for the power-law kernel

$$K_{i,j} = (i\, j)^{\nu/2}. \tag{9.70}$$

This includes the product kernel as a special case for $\nu = 1$. For distributions without a gel phase, the mean kernel in distribution $\mathbf{n}$ is

$$\bar{K}(\mathbf{n}) = \left(\frac{M}{N}\right)^{\nu}. \qquad [9.58]$$

The partition function is

$$\Omega_{M,N} = \binom{M-1}{N-1} \prod_{g=0}^{M-N-1} \left(\frac{M}{M-g}\right)^{\nu} = \binom{M-1}{N-1} \left(\frac{M}{M}\frac{M}{M-1}\cdots\frac{M}{N+1}\right)^{\nu},$$

which condenses to

$$\Omega_{M,N} = \binom{M-1}{N-1} \left(\frac{M!}{N!} M^{M-N}\right)^{\nu}. \qquad (9.71)$$

We obtain β and q in finite-difference form. Starting with β we have

$$\beta = \log \frac{\Omega_{M+1,N}}{\Omega_{M,N}} = \nu(M-N)\log\left(\frac{M+1}{M}\right) - \log\left(\frac{M-N+1}{M}\right). \qquad (9.72)$$

Using

$$\log\left(\frac{M+1}{M}\right)^{M} \to 1,$$

we obtain the limiting form of β in the thermodynamic limit:

$$\beta = \nu\frac{M-N}{N} - \log\frac{M-N}{N},$$

or

$$\beta = \nu\theta - \log\theta. \qquad (9.73)$$

For q we find

$$q = \frac{\Omega(M, N+1)}{\Omega(M, N)} = \left(1 + \frac{1}{N}\right)\left(1 - \frac{N}{M}\right)\left(\frac{N+1}{M}\right)^{\nu-1}. \qquad (9.74)$$

In the thermodynamic limit this becomes

$$q = \left(1 - \frac{N}{M}\right)\left(\frac{N}{M}\right)^{\nu-1},$$

or

$$q = \theta(1-\theta)^{\nu}. \tag{9.75}$$

With $\nu = 0$, 1, or 2, Eqs. (9.71), (9.73), and (9.75) revert to those for the constant kernel, sum kernel, or product kernel, respectively.

Locus of Gel Points

The system is unstable if $q(\theta)$ contains a branch where $dq/d\theta < 0$. The onset of instability is defined by the condition,

$$\frac{dq(\theta)}{d\theta} = 0 = (1-\theta)^{\nu-2}(1-\theta\nu), \tag{9.76}$$

whose solution is

$$\theta^* = 1/\nu.$$

Since θ must be between 0 and 1, instability may occur only if $\nu > 1$. The condition that defines the locus of gel points is

$$\theta^* = 1/\nu, \quad \nu > 1. \tag{9.77}$$

The case $\nu = 1$ represents a singular case: the system forms a gel phase but the gel point is at $\theta = 1$, which corresponds to $\bar{x} = \infty$ and is not observed under finite $\bar{x}$.

Some Useful Power Series

Here are some useful results involving infinite series that are appear in the solution of the sum and the product kernel.

$$S_0 = \sum_{i=1}^{\infty} \frac{i^i}{i!} e^{-i(a-\log a)} = \frac{a}{1-a} \tag{9.78}$$

$$S_1 = \sum_{i=1}^{\infty} \frac{i^{i-1}}{i!} e^{-i(a-\log a)} = a \tag{9.79}$$

$$S_2 = \sum_{i=1}^{\infty} \frac{i^{i-2}}{i!} e^{-i(a-\log a)} = a(1-a/2) \tag{9.80}$$

These can be also expressed in equivalent form as

$$\sum_{i=1}^{\infty} \frac{(ia)^i}{i!} e^{-ia} = \frac{a}{1-a} \tag{9.81}$$

$$\sum_{i=1}^{\infty} \frac{(ia)^{i-1}}{i!} e^{-ia} = 1 \tag{9.82}$$

$$\sum_{i=1}^{\infty} \frac{(ia)^{i-2}}{i!} e^{-ia} = \frac{1-a/2}{a} \tag{9.83}$$

We outline the derivation of these results. We start with the power series[13]

$$\sum_{i=1}^{\infty} \frac{i^{i-1}}{i!} e^{-bi} = a, \tag{9.84}$$

where a is the solution of

$$e^{-b} = ae^{-a}. \tag{9.85}$$

Taking the log of this expression we have

$$b = a - \log a, \tag{9.86}$$

which proves Eq. (9.79). We define

$$C_k(i) = \frac{i^{i-k}}{i!}, \tag{9.87}$$

and express the summations in Eqs. (9.78)–(9.80) as

[13]Wolfram Research, Inc., *Mathematica*, Version 11.3, Champaign, IL (2018).

$$S_k(b) = \sum_{i=1\infty} C_k(i)e^{-bi}. \tag{9.88}$$

We now have

$$S_0(b) = -\frac{\partial S_1}{\partial b} \tag{9.89}$$

$$S_1(b) = a(b), \tag{9.90}$$

$$S_2(b) = -\int S_1 db \tag{9.91}$$

with the relationship $a = a(b)$ from Eq. (9.86). By differentiation and integration of $S_1(b)$ with respect to b we obtain Eqs. (9.78) and (9.80), respectively. Equations (9.81)–(9.83) follow by straightforward manipulation.

Chapter 10
Fragmentation and Shattering

Binary fragmentation is the reverse process of binary aggregation: a cluster with i monomers splits into two clusters with masses j and $i-j$ such that $1 \le j \le i-1$. This is represented schematically by the irreversible mass-conserving reaction

$$(i) \xrightarrow{B_{i-j,j}} (i-j) + (j). \tag{10.1}$$

The breakup kernel $B_{i-j,j} = B_{j,i-j}$ is a symmetric function of the cluster masses of the fragments and characterizes the probability of the particular outcome out of all possible fragmentation outcomes of mass i. In this chapter we formulate this problem in the context of the cluster ensemble, we obtain the partition function and show that fragmentation produces a phase transition, *shattering*, which is analogous to gelation.

10.1 The Process

The basic process is the formation of an ordered pair of cluster masses $(i-i, i)$ from a cluster with mass i (parent) within distribution $\mathbf{n}$. By model assumption, we take the probability for this transition to be proportional to the number of clusters with mass i with the proportionality constant given by the *fragmentation kernel* $B_{i-j,j}$:

$$\mathcal{P}_{i-j,j|\mathbf{n}} = C(\mathbf{n}) n_i B_{i-j,j}. \tag{10.2}$$

The factor $C(\mathbf{n})$ is common for all fragmentation events within the same distribution $\mathbf{n}$ and is fixed by the normalization condition

T. Matsoukas, *Generalized Statistical Thermodynamics*,
Understanding Complex Systems, https://doi.org/10.1007/978-3-030-04149-6_10

$$\sum_{i=2}^{\infty}\sum_{j=1}^{i-1} \mathcal{P}_{i-j,j|\mathbf{n}} = 1,$$

which leads to

$$\frac{1}{C(\mathbf{n})} = \sum_{i=2}^{\infty}\sum_{j=1}^{i-1} n_i B_{i-j,j}. \tag{10.3}$$

We define the mean fragmentation kernel $\bar{B}(\mathbf{n})$ as the weighted average

$$\bar{B}(\mathbf{n}) = \frac{\sum_{i=2}^{\infty}\sum_{j=1}^{i-1} n_i B_{i-j,j}}{\sum_{i=2}^{\infty}\sum_{j=1}^{i-1} n_i} = \frac{1}{M-N}\sum_{i=2}^{\infty}\sum_{j=1}^{i-1} n_i B_{i-j,j}. \tag{10.4}$$

To obtain the last result we used the identity

$$\sum_{i=2}^{\infty}\sum_{j=1}^{i-1} n_i = \sum_{i=2}^{\infty}(i-1)n_i = M-N,$$

which gives the total number of ordered fragmentation events in distribution $\mathbf{n}$.[1] Given this definition the factor $C(\mathbf{n})$ can be expressed more compactly as

$$C(\mathbf{n}) = \frac{1}{(M-N)\bar{B}(\mathbf{n})}, \tag{10.5}$$

and the probability to obtain ordered pair $(i-j,\, j)$ becomes

$$\mathcal{P}_{i-j,j|\mathbf{n}} = \frac{n_i B_{i-j,j}}{(M-N)\bar{B}(\mathbf{n})}. \tag{10.6}$$

[1] There are n_i clusters with mass i and each produces $i-1$ ordered fragments (fragmentation events). The total number of fragmentation events is

$$\sum_{i=1}^{\infty}(i-1)n_i = M-N.$$

The mean fragmentation kernel is a functional of distribution **n** and generally varies among distributions of the ensemble. The ensemble average fragmentation kernel is

$$\langle B\rangle_{M,N} = \sum_{\mathbf{n}} P(\mathbf{n})\bar{B}(\mathbf{n}), \tag{10.7}$$

where $P(\mathbf{n})$ is the probability of distribution n in the ensemble $\mu C(M, N)$. The ensemble average kernel is a function of M and N.

Note 10.1 (On the Limits of the Summations)
The double summation

$$\sum_{i=2}^{\infty}\sum_{j=1}^{i-1}(i-1)n_i$$

goes over all ordered pairs $(i-j, j)$ of fragments that can be produced in distribution **n**. The outer summation starts at $i = 2$, the smallest cluster mass that can produce fragments (monomers do not break).

⋆

10.1.1 Fragmentation Rate and Fragment Distribution

The fragmentation probability of cluster mass i is proportional to the sum of all fragmentation events in cluster weighted by the fragmentation kernel,

$$\sum_{j=1}^{i-1} B_{i-j,j} \equiv b_i. \tag{10.8}$$

This is the total fragmentation probability of mass i and we may view it as the rate constant for the fragmentation of cluster mass i. We will call b_i the *fragmentation rate* of cluster mass i, even though our treatment makes no explicit reference to time. The mean rate of fragmentation $\bar{b}(\mathbf{n})$ in distribution **n** is

$$\bar{b}(\mathbf{n}) = \frac{1}{N}\sum_{i=2}^{\infty} n_i b_i, \tag{10.9}$$

which we express in terms of the fragmentation kernel in the form

$$\bar{b}(\mathbf{n}) = \frac{1}{N}\sum_{i=2}^{\infty}\sum_{j=1}^{i-1} n_i B_{i-j,j} = \frac{M-N}{N}\bar{B}(\mathbf{n}). \tag{10.10}$$

The probability to obtain a fragment of size j from a cluster with mass i is proportional to $B_{i-j,j}$. Since a pair of fragments can appear in $1+\delta_{i-j,j}$ permutations, this probability is

$$p(j|i) = \frac{(1+\delta_{i-j,j})B_{j,i-j}}{\sum_{j=1}^{i-1} B_{i-j,j}} = \frac{(1+\delta_{i-j,j})B_{i-j,j}}{b_i}. \tag{10.11}$$

It is common in the literature to describe fragmentation via two separate functions, the fragmentation rate b_i and the distribution of fragments $p(j|i)$. Both quantities are contained in the fragmentation kernel, a single function that completely defines the process.

10.1.2 Parent–Offspring Relationships

A fragmentation event within a distribution of the parent ensemble $\mu C(M, N-1)$ produces a new distribution in the offspring ensemble $\mu C(M, N)$. We follow the convention in Chap. 8 and use primed variables to notate the parent and unprimed to notate the offspring. The parent–offspring relationship is schematically represented by the transition

$$\mathbf{n}' \xrightarrow{i-j,j} \mathbf{n}, \tag{10.12}$$

which further denotes that the event that produces the offspring is the fragmentation of cluster mass i into fragments $i-j$ and j. This identifies $\mathbf{n}'$ as the $(i-j, j)$-parent of $\mathbf{n}$. The fragmentation event reduces the number of n'_i by 1 and increases the number of n'_{i-j} and n'_j by 1, unless $i-j=j$ in which case the number is increased by 2:

$$\begin{aligned} n_i &= n'_i - 1, \\ n_{i-j} &= n'_{i-j} + 1 + \delta_{i-j,j}, \\ n_j &= n'_j + 1 + \delta_{i-j,j}. \end{aligned} \tag{10.13}$$

For all other cluster masses that do not participate in the fragmentation event the number of clusters is the same in both parent and offspring distributions. These equations can be solved for n'_i, n'_{i-j}, and n'_j to obtain the $(i-j, j)$-parent of distribution $\mathbf{n}$. Letting $i = 2, \cdots \infty$, $j = 1, \cdots i/2$, we obtain the complete set of parents of the distribution.

The parent–offspring relationships are demonstrated in Fig. 10.1 for $M = 6$. The process starts with a single cluster that contains all 6 monomers (bottom). Arrows represent parent–offspring transitions and distributions are shown pictorially by the clusters they contain. The fragmentation process moves upward until the state consists of all monomers. This graph is identical to that for aggregation in Fig. 8.1

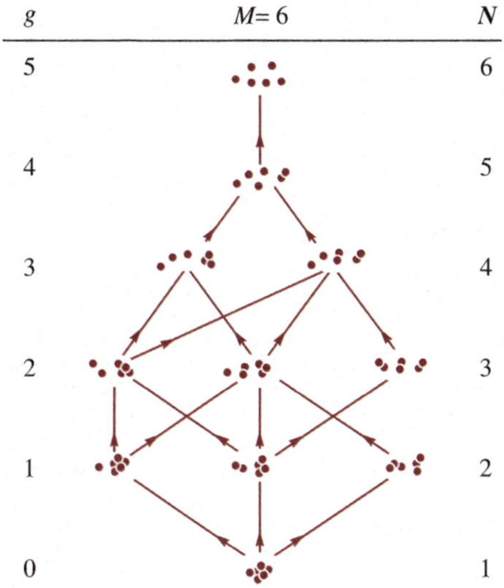

Fig. 10.1 Fragmentation graph for $M = 6$. Fragmentation begins at the bottom with a single cluster of mass M and ends at the top with a state of monomers. The fragmentation graph is identical to the aggregation graph with all arrows reversed. Generations are defined as $g = N - 1$, with $g = 0$ referring to the initial state and $g = M - 1$ to the final state

except that the direction of all arrows reversed. The symmetry between the graphs for binary aggregation and binary fragmentation is complete: a parent–offspring relationship in aggregation is an offspring-parent relationship in fragmentation and vice versa. Accordingly, there is complete correspondence between the parent and offspring ensembles in fragmentation, just as in aggregation: the set of all parents in the offspring ensemble is identical to the complete parent ensemble.

10.1.3 Transition Probabilities

Since permutations in the order of fragments produce the same offspring, the probability for the parent–offspring transition in Eq. (10.12) is

$$P_{\mathbf{n}'\to\mathbf{n}} = \frac{2}{1+\delta_{i-j,j}}\mathcal{P}_{i-j,j|\mathbf{n}'} = C(\mathbf{n}')\frac{2n_i' B_{i-j,j}}{1+\delta_{i-j,j}}, \tag{10.14}$$

with $C(\mathbf{n}')$ from Eq. (10.5) with N replaced by $N - 1$[2]:

$$C(\mathbf{n}') = \frac{1}{(M-N+1)\bar{B}(\mathbf{n}')}. \tag{10.15}$$

The transition probability is normalized to unity over all unordered pairs in $\mathbf{n}$:

[2]Equation (10.5) as written is for the offspring ensemble $\mu C(M, N)$. In the parent ensemble $\mu C(M, N-1)$, N must be replaced by $N - 1$.

$$\sum_{i=2}^{\infty}\sum_{j=1}^{i/2} P_{\mathbf{n}'\to\mathbf{n}} = 1. \tag{10.16}$$

It is understood in all of these expressions that $\mathbf{n}'$ is the "$(i - j, j)$-parent of $\mathbf{n}$." We note now some useful identities involving summations over the fragmentation domain:

$$S \equiv \sum_{i=2}^{\infty}\sum_{j=1}^{i/2} \frac{2n_i B_{i-j,j}}{1+\delta_{i-j,j}} \tag{10.17}$$

$$= \sum_{i=2}^{\infty}\sum_{j=1}^{i-1} n_i B_{i-j,j} \tag{10.18}$$

$$= \sum_{i=2}^{\infty}\sum_{j=1}^{i} \frac{2n_{i+j} B_{i,j}}{1+\delta_{i,j}} \tag{10.19}$$

$$= \sum_{i=2}^{\infty}\sum_{j=i}^{\infty} \frac{2n_{i+j} B_{i,j}}{1+\delta_{i,j}} \tag{10.20}$$

$$= \sum_{i=1}^{\infty}\sum_{j=1}^{\infty} n_{i+j} B_{i,j}. \tag{10.21}$$

These amount to a change of variables that can be interpreted as follows: The first two equations refer to the reaction

$$(i) \longrightarrow (i - j) + (j),$$

with (10.17) going over all ordered pairs and (10.18) over all unordered pairs; Eqs. (10.19), (10.20), and (10.21) refer to the equivalent reaction

$$(i + j) \longrightarrow (i) + (j)$$

and perform the summation in different domains, over the lower half, upper half, or over the entire $i \times j$ plane, respectively. Using Eqs. (10.17) and (10.18) it is easy to show that Eq. (10.16) is satisfied with $C(\mathbf{n}')$ from Eq. (10.15).

10.2 The Partition Function of Fragmentation

Parent distribution $\mathbf{n}'$ with microcanonical probability $P(\mathbf{n})$ transitions to offspring distribution $\mathbf{n}$ with transition probability $P(\mathbf{n} \to \mathbf{n}')$. The microcanonical probability of the offspring is given by the master equation,

$$P(\mathbf{n}) = \sum_{\mathbf{n}'} P(\mathbf{n}')P(\mathbf{n}' \to \mathbf{n}), \tag{10.22}$$

with the summation going over all parents of **n**. We express the probability of distribution in the form,

$$P(\mathbf{n}) = \mathbf{n}!\frac{W(\mathbf{n})}{\Omega_{M,N}}; \quad P(\mathbf{n}') = \mathbf{n}'!\frac{W(\mathbf{n}')}{\Omega_{M,N-1}},$$

and use Eq. (10.14) for the transition probability to write the master equation in the form

$$P(\mathbf{n}) = \frac{\mathbf{n}!W(\mathbf{n})}{\Omega_{M,N}} = \sum_{i=2}^{\infty}\sum_{j=1}^{i/2} \frac{\mathbf{n}'!W(\mathbf{n}')}{\Omega_{M,N-1}} \frac{2}{1+\delta_{i-j,j}} \frac{n_i' B_{i-j,j}}{(M-N+1)\bar{B}(\mathbf{n}')}. \tag{10.23}$$

The double summation on the right-hand side goes over all $(i-j, j)$ parents of distribution **n**. Using the parent–offspring relationships, Eq. (10.13), the ratio of the factorials can be expressed as

$$\frac{\mathbf{n}'!}{\mathbf{n}!} = \frac{n_{i-j}(n_j - \delta_{i-j,j})}{Nn_i'}. \tag{10.24}$$

We substitute this result and Eq. (10.18) and after some straightforward manipulation we obtain the following result for the evolution of the partition function from the parent ensemble to the ensemble of the offspring:

$$\frac{\Omega_{M,N-1}}{\Omega_{M,N}} = \frac{N-1}{(M-N+1)\langle B\rangle_{M,N-1}} \times \left\{ \langle B\rangle_{M,N-1} \sum_{i=2}^{\infty}\sum_{j=1}^{i-1} \frac{n_{i-j}(n_j - \delta_{i-j,j})}{N(N-1)} \frac{B_{i-j,j}}{\bar{B}(\mathbf{n}')} \frac{W(\mathbf{n}')}{W(\mathbf{n})} \right\}. \tag{10.25}$$

The double summation goes over the elements of the offspring distribution; each term refers to a parent (fragment pairs are treated as ordered) and involves the selection functional and mean kernel of that parent. Overall the summation goes over all parents of **n**. The quantity in the curly braces involves distribution **n** and its parents, and applies to all distributions in the microcanonical ensemble $\mu C(M, N)$. Since the left-hand side is independent of **n** and a function of M and N only, we must have

$$\langle B\rangle_{M,N-1}\sum_{i=2}^{\infty}\sum_{j=1}^{i-1}\frac{n_{i-j}(n_j-\delta_{i-j,j})}{N(N-1)}\frac{B_{i-j,j}}{\bar{B}(\mathbf{n}')}\frac{W(\mathbf{n}')}{W(\mathbf{n})}=a_{M,N+1}, \tag{10.26}$$

where $a_{M,N+1}$ is a function of M and N. As with Eq. (8.23) in aggregation, the selection of $a_{M,N}$ has no effect on the microcanonical probabilities. This indeterminacy is resolved by requiring *thermodynamic consistency*: $a_{M,N}$ must be such that the unbiased selection functional that is obtained by recursive solution of Eq. (10.26) is $W(\mathbf{n})=1$. The value that satisfies this condition is

$$a_{M,N}=1. \tag{10.27}$$

We will prove this claim in Sect. 10.3. Equation (10.25) then produces two separate recursions, one of the partition function,

$$\frac{\Omega_{M,N-1}}{\Omega_{M,N}}=\frac{N-1}{(M-N+1)\,\langle B\rangle_{M,N-1}}, \tag{10.28}$$

and one for the selection functional,

$$\langle B\rangle_{M,N-1}\sum_{i=2}^{\infty}\sum_{j=1}^{i-1}\frac{n_{i-j}(n_j-\delta_{i-j,j})}{N(N-1)}\frac{B_{i-j,j}}{\bar{B}(\mathbf{n}')}\frac{W(\mathbf{n}')}{W(\mathbf{n})}=1. \tag{10.29}$$

Starting with $\Omega_{M,1}=1$, the recursion for the partition function is solved to produce

$$\Omega_{M,N}=\binom{M-1}{N-1}\prod_{L=1}^{N-1}\langle B\rangle_{M,L}\,. \tag{10.30}$$

The recursion for the selection functional cannot be solved for the general case, but we may be expressed it more clearly as

$$W(\mathbf{n})=\langle B\rangle_{M,N-1}\sum_{i=2}^{\infty}\sum_{j=1}^{i-1}\frac{n_{i-j}(n_j-\delta_{i-j,j})}{N(N-1)}\frac{B_{i-j,j}}{\bar{B}(\mathbf{n}')}W(\mathbf{n}'). \tag{10.31}$$

In this form $W(\mathbf{n})$ may be calculated if the selection functionals of all parents are known. In principle, this can be done iteratively starting from the single-cluster state with $W(\mathbf{n}_0)=1$.

We obtain the canonical parameters β and q from the partition function using their finite difference forms:

$$\beta = \log \frac{\Omega_{M+1,N}}{\Omega_{M,N}} = \log \left(\frac{M}{M-N+1} \right) + \sum_{L=1}^{N-1} \log \frac{\langle B \rangle_{M+1,L}}{\langle B \rangle_{M,L}}, \tag{10.32}$$

$$q = \frac{\Omega_{M,N+1}}{\Omega_{M,N}} = \left(\frac{M-N}{N} \right) \langle B \rangle_{M,N}. \tag{10.33}$$

Equations (10.30)–(10.33) summarize the results of the theory for the generic binary fragmentation process with kernel $B_{i,j}$.

10.3 An Exact Solution: Unbiased Ensemble

The case $B_{ij} = 1$ is special because it assigns equal probability to all ordered pairs of fragments (*random fragmentation*). Since a cluster with mass i produces $i-1$ ordered pairs, the probability of fragmentation is proportional to $i-1$, essentially proportional to the cluster mass for large i. In this sense this model is physically realistic since large particles are generally found to break easier than smaller ones. Its theoretical importance is that it links to the unbiased ensemble and resolves the indeterminacy of $a_{M,N}$.

For the constant fragmentation kernel we have the trivial identities

$$B_{i,j} = \bar{B}(\mathbf{n}) = \langle B \rangle_{M,N} = 1$$

for all $i, j > 1$, all $\mathbf{n}$ and all $M \geq N > 1$. The partition function then in Eq. (10.30) becomes

$$\Omega_{M,N} = \binom{M-1}{N-1}, \tag{10.34}$$

which we recall is the volume of the microcanonical ensemble. From Eq. (10.31) we obtain

$$\sum_{i=2}^{\infty} \sum_{j=1}^{i-1} \frac{n_{i-j}(n_j - \delta_{i-j,j})}{N(N-1)} \frac{W(\mathbf{n}')}{W(\mathbf{n})} = 1. \tag{10.35}$$

This applies to all distributions $\mathbf{n}$ of the $\mu C(M, N)$ ensemble. Given the identity

$$\sum_{i=2}^{\infty}\sum_{j=1}^{i-1}\frac{n_{i-j}(n_j-\delta_{i-j,j})}{N(N-1)}=1,$$

also true for all $\mathbf{n}$ in the ensemble, we must have $W(\mathbf{n}) = W(\mathbf{n}')$. The selection bias is the same for all distributions for all M, N, therefore,

$$W(\mathbf{n}) = 1; \quad w_i = 1; \qquad \text{for all } \mathbf{n}. \tag{10.36}$$

Therefore we have obtained the unbiased example.

Equations (10.34) and (10.36) establish that fragmentation with $B_{i,j} = 1$ produces the unbiased ensemble. All results of the unbiased ensemble for the mean distribution and the MPD apply and will not be repeated here.

Note 10.2 (On $a_{M,N}$)

If we leave $a_{M,N}$ in Eq. (10.31) unspecified, the partition function becomes

$$\Omega_{M,N} = \binom{M-1}{N-1}\prod_{L=1}^{N-1}\frac{\langle B\rangle_{M,L}}{a_{M,L}}$$

and for the constant kernel $B_{i,j} = 1$ we obtain

$$\Omega_{M,N} = \binom{M-1}{N-1}\Big/\prod_{L=1}^{N-1} a_{M,L}\,.$$

The corresponding result for the selection bias is

$$\sum_{i=2}^{\infty}\sum_{j=1}^{i-1}\frac{n_{i-j}(n_j-\delta_{i-j,j})}{N(N-1)}\frac{W(\mathbf{n}')}{W(\mathbf{n})}=a_{M,N-1}, \tag{10.37}$$

and from this we conclude

$$\frac{W(\mathbf{n}')}{W(\mathbf{n})} = a_{M,N-1},$$

which states that all distributions within the same generation have the same selection bias. We must have then

$$W(\mathbf{n}') = W(\mathbf{n}) = a_{M,N-1} = 1,$$

which fixes the value of $a_{M,N}$.

★

10.4 Mean Distribution

We will now derive the equation that governs the evolution of the mean distribution $\langle n_k \rangle$ from one generation to the next. We begin with the master equation in Eq. (10.22) which we write in the form

$$P(\mathbf{n}) = \sum_{\mathbf{n}'} P(\mathbf{n}')P(\mathbf{n}' \to \mathbf{n}) = \sum_{i=2}^{\infty}\sum_{j=1}^{i-1} \mathcal{P}'_{i-j,j} P(\mathbf{n}') \tag{10.38}$$

with

$$\mathcal{P}'_{i-j,j} = \frac{n'_i B_{i-j,j}}{(M-N+1)\bar{B}(\mathbf{n}')} \tag{10.39}$$

subject to normalization

$$\sum_{i=1}^{\infty}\sum_{j=1}^{i-1} \mathcal{P}'_{i-j,j} = 1. \tag{10.40}$$

Equation (10.39) is obtained from (10.14) with N replaced by $N-1$.[3] When cluster mass i in parent distribution $\mathbf{n}'$ breaks into masses $i-j$ and j, the offspring distribution is

$$n_k = n'_k - \delta_{k,i} + \delta_{k,i-j} + \delta_{k,j}, \tag{10.41}$$

which is a concise form of the parent–offspring relationships in Eq. (10.13). We multiply Eq. (10.41) by $P(\mathbf{n})$ in Eq. (10.38) and sum both sides over all $\mathbf{n}$. On the left-hand side we obtain the ensemble average $\langle n_k \rangle$ in the offspring ensemble. On the right-hand side we have a summation over all parents of the offspring ensemble. This is equivalent to a summation over the entire parent ensemble, i.e., an ensemble average over the parent ensemble:

$$\begin{aligned}\langle n_k \rangle &= \sum_{\mathbf{n}}\sum_{i=2}^{\infty}\sum_{j=1}^{i-1} \mathcal{P}'_{i-j,j}(n'_k - \delta_{k,i} + \delta_{k,i-j} + \delta_{k,j})P(\mathbf{n}') \\ &= \sum_{\mathbf{n}'}\sum_{i=2}^{\infty}\sum_{j=1}^{i-1} \mathcal{P}'_{i-j,j}(n'_k - \delta_{k,i} + \delta_{k,i-j} + \delta_{k,j})P(\mathbf{n}')\end{aligned}$$

[3]Recall that Eq. (10.14) is written in the offspring ensemble $\mu C(M, N)$ while Eq. (10.39) is written in the parent ensemble $\mu C(M, N-1)$.

$$= \left\langle \sum_{i=2}^{\infty} \sum_{j=1}^{i-1} \mathcal{P}'_{i-j,j} (n'_k - \delta_{k,i} + \delta_{k,i-j} + \delta_{k,j}) \right\rangle. \tag{10.42}$$

The right-hand side is a summation of ensemble averages over the parent distribution and can be calculated individually. The result is

$$\langle n_k \rangle = \langle n'_k \rangle + \left\langle \frac{-\sum_{j=1}^{k-1} n'_k B_{k-j,j} + 2\sum_{j=k+1}^{\infty} n'_j B_{j-k,k}}{(M-N+1)\bar{B}(\mathbf{n}')} \right\rangle. \tag{10.43}$$

This equation governs the change in the mean number of k-mers when the ensemble transitions from one generation to the next and is exact for all discrete finite M and N. The negative term on the right-hand side accounts for the depletion of k-mers due to fragmentation and the positive term for production of k-mers via fragmentation of larger clusters. As written this equation applies to $k > 1$. For monomers the depletion term is not present.

In the thermodynamic limit ($M > N \gg 1$) the ensemble reduces to the MPD and Eq. (10.44) becomes

$$\frac{\delta \tilde{n}_k}{\delta N} = \frac{1}{(M-N)\tilde{B}} \left(-\sum_{j=1}^{k-1} \tilde{n}_k B_{k-j,j} + 2 \sum_{j=k+1}^{\infty} \tilde{n}_j B_{j-k,k} \right), \tag{10.44}$$

where $\delta\tilde{n}_k$ is the change in the number of k-mers during the transition $(M, N-1) \to (M, N)$, $\delta N = 1$, and $\tilde{B} = \bar{B}$ is the mean kernel in the MPD. Equation (10.44) is the classical fragmentation equation in population balance modeling. As in the case of the Smoluchowski equation in aggregation, this equation is obtained in the limit that the mean distribution and the most probable distribution converge.

10.5 Homogeneity in the Thermodynamic Limit

The partition function in Eq. (10.30) is exact for any discrete finite system (M, N). To develop expressions in the thermodynamic limit we must first re-examine the notion of this limit in the context of fragmentation. We have defined the thermodynamic limit by the conditions

$$M, N \to \infty; \quad M/N = \bar{x} = \text{fixed}.$$

In this limit the size of the system increases while the mean cluster size is fixed and the normalized distribution $f(x) = \tilde{n}_i/N$ achieves an intensive limit that is independent of M and N individually and depends only on their ratio. In this limit the ensemble is homogeneous and thermodynamics applies. This picture applies on the clustering side of the aggregation/fragmentation graph. On the fragmentation side the situation is different. In fragmentation we start with a mass M and progressively increase the number of clusters by producing fragments. In the first fragmentation event the number of fragments is $N = 2$ and the mean cluster size decreases from M to $M/2$. If we increase the mass to $M' = \lambda M$ and start again, once again the first fragmentation step produces $N = 2$ fragments and the mean cluster size decreases from M' to $M'/2$. In the process of increasing the size of the system we increase M while keeping N fixed. If we rescale the mass of the clusters by the total mass, then the initial mass is always 1 and the first fragmentation step always produces a mean cluster size of $1/2$. In this rescaled distribution the system behaves homogeneously: as M increases indefinitely at fixed N the rescaled distribution achieves an intensive limit that depends on N but not on M. We will now put these ideas in mathematical form.

We will be treating the cluster distribution as a continuous function of cluster and use $F(x)$, $x \in (0, \infty)$ to represent the extensive MPD $\tilde{n}_i$, $i = 1, 2\cdots$, and $f(x) = F(x)/N$ for the intensive distribution $\tilde{n}_i/N$. We define the rescaled size

$$z = x/M \tag{10.45}$$

whose corresponding distribution $\phi(z)$ is

$$\phi(z) = Mf(zM). \tag{10.46}$$

We can easily confirm that $\phi(z)$ is normalized to unit area and that its mean is

$$\bar{z} = \int \phi(z)dz = \frac{\bar{x}}{M} = \frac{1}{N}, \tag{10.47}$$

where $\bar{x} = M/N$ is the mean of $f(x)$. This simple relationship, $\bar{z} = 1/N$, allows us to use $\bar{z}$ and N interchangeably. We define the thermodynamic limit in fragmentation by the conditions

$$M \to \infty; \quad N = \text{fixed}. \tag{10.48}$$

In this limit the rescaled distribution $\phi(z)$ becomes independent of M: it reaches an intensive limit that is independent of the size of the system and depends only on N (or $\bar{z}$). The partition function of $f(x)$ is

$$\log \omega(\bar{x}) = \frac{\log \Omega_{M,N}}{N}, \tag{10.49}$$

with $\Omega_{M,N}$ given by Eq. (10.30). Let $\omega^*(\bar{z})$ be the partition function of $\phi(z)$. We will establish the relationship between ω and ω^*.

First we obtain the relationship between the entropies of $f(x)$ and $\phi(z)$. We apply the entropy functional on f and use Eq. (10.46) to express the result in terms of $\phi(z)$[4]:

$$S[f] = -\int f(x) \log f(x) dx = -\int \phi(z) \log \frac{\phi(z)}{M} dz$$
$$= -\int \phi(z) \log \phi(z) dz + \log M,$$

and finally

$$S[f] = S[\phi] + \log M. \tag{10.50}$$

The partition function of f is

$$\log \omega(\bar{x}) = S[f] + \log W[f], \tag{10.51}$$

where W is the selection functional of f. Using Eq. (10.50) this becomes

$$\log \omega(\bar{x}) = S[f] + \log W[f] = S[\phi] + \log[M] + \log W[\phi/M]. \tag{10.52}$$

Distribution f maximizes the microcanonical functional $S[\cdots]+\log W[\cdots]$ among all normalized distributions with mean $\bar{x}$. Accordingly ϕ maximizes the right-hand side of Eq. (10.52) at fixed M among all normalized distributions with mean $\bar{z} = \bar{x}/M$: if it did not and instead some other $\phi'(z)$ produced an even higher value on the right-hand side, then $f'(x) = \phi(x/M)/M$ would produce a higher microcanonical weight that f; but this is not possible because f was stipulated to maximize this weight. Since M is constant in Eq. (10.52), the term $\log M$ is irrelevant in the maximization, i.e., ϕ maximizes the quantity

$$S[\phi] + \log W[\phi/M] \tag{10.53}$$

for all M among all distributions with fixed mean $\bar{z}$. Yet at fixed $\bar{z}$, ϕ is independent of M. The only way that this can be consistent with Eq. (10.53) is if M separates out of $\log W[\phi/M]$ in the form

$$\log W[\phi/M] = \log W[\phi] + A(M), \tag{10.54}$$

where $A(M)$ is some function of M. We conclude that ϕ maximizes the functional

$$S[\phi] + \log W[\phi]. \tag{10.55}$$

[4]Notice that $f(x)dx = \phi(z)dz$ as an immediate consequence of Eq. (10.46).

It also maximizes its own microcanonical functional,

$$S[\phi] + \log W^*[\phi], \tag{10.56}$$

where W^* is the selection bias that produces ϕ as its MPD. But this means that $\log W$ and $\log W^*$ are different at most be a constant, and since constant factors have no effect on the selection functional, we must have $W^* = W$ and

$$\log \omega^*(\bar{z}) = S[\phi] + \log W[\phi]. \tag{10.57}$$

Incorporating these results into Eq. (10.52) we have

$$\log \omega(\bar{x}) = S[\phi] + \log W[\phi] + \log M + A(M) \tag{10.58}$$

and finally,

$$\log \omega(\bar{x}) = \log \omega^*(\bar{z}) + \log M + A(M). \tag{10.59}$$

The result establishes the relationship between the partition function of f and that of ϕ and also requires the unknown function $A(M)$. If the partition function $\omega(M/N)$ is known function of M and N, then we may obtain ω^* and $A(M)$ as follows: separate the partition function into two terms, one that contains only M and one that contains only N. The log of ω^* is the part that depends on N (with $\bar{z} = 1/N$) and $A(M)$ is the part that depends on M minus $\log M$. If this separation is not possible, then distribution $\phi(z)$ does not have an intensive limit and the system does not exhibit homogeneity.

10.5.1 Application to $B_{i,j} = 1$

To demonstrate the derivation of ω^* we consider the case of the constant kernel, $B_{i,j} = 1$ for which the partition function reduces to that of the unbiased ensemble. If N is large[5] the intensive partition function takes the form

$$\frac{\log \Omega_{M,N}}{N} = -\frac{M}{N} \log\left(1 - \frac{N}{M}\right) + \log\left(\frac{M}{N} - 1\right). \tag{10.60}$$

[5]The thermodynamic limit requires M to be large and N fixed. This condition ensures homogeneous behavior of ϕ at any N, even small N of the order 1. If we allow N to be large, but still much smaller than M, then we can treat $\log \Omega$ as a continuous function of M and N to write $\log \Omega = M\beta + N \log q$.

In the thermodynamic limit $M \to \infty$ at fixed N. Applying this condition the above equation becomes

$$\log \omega(\bar{x}) = \frac{\log \mathcal{V}_{M,N}}{N} = 1 + \log \frac{M}{N} = \underbrace{1 + \log \frac{1}{N}}_{\log \omega^*} + \log M. \tag{10.61}$$

Comparing this result with Eq. (10.59) we identify the partition function ω^* as

$$\log \omega^*(\bar{z}) = 1 + \log \bar{z}, \tag{10.62}$$

and the unknown function is $A(M) = 0$. The result for A in this case can be confirmed independently: in the unbiased ensemble $W[h] = 1$ for any function h; then $W[M\phi] = W[\phi]$ and it follows from Eq. (10.54) that $A(M) = 0$.

This analysis implies that $\phi(z)$ in the unbiased ensemble indeed converges to an intensive form that is independent of M. We can demonstrate this directly. The mean distribution of the unbiased ensemble is[6]

$$\frac{\langle n_{i;M,N} \rangle}{N} = \binom{M-i}{N-1} \Big/ \binom{M}{N}, \quad i = 1, 2 \cdots M - N + 1.$$

As an example, with $N = 4$ we obtain

$$\frac{\langle n_{i;M,N} \rangle}{N} = \frac{3(M-i-2)(M-i-1)}{(M-3)(M-2)(M-1)}.$$

The scaled distribution $\phi(z; \bar{z})$, $\bar{z} = 1/4$, is

$$\phi(z; \bar{z} = 1/4) = \frac{3M(M(-z) + M - 2)(M(-z) + M - 1)}{(M-3)(M-2)(M-1)}$$

and with $M \to \infty$ we obtain

$$\phi(z; \bar{z} = 1/4) = 3z^2 - 6z + 3, \tag{10.63}$$

which is indeed independent of M. The same is true for all N. Table 10.1 shows $\phi(z; \bar{z})$ for $N = 2, 3, 4$, and 5.

[6] Here we work with the *mean* rather than the most probable distribution. For large M and N the two are equal to each other.

Table 10.1 First few elements of $\phi(z; \bar{z})$

N	$\langle n_{i;M,N} \rangle / N$	$\bar{z} = 1/N$	$\phi(z; \bar{z})$
2	$\dfrac{1}{M-1}$	1/2	1
3	$\dfrac{2(M-1-i)}{(M-2)(M-1)}$	1/3	$2 - 2z$
4	$\dfrac{3(M-i-2)(M-i-1)}{(M-3)(M-2)(M-1)}$	1/4	$3z^2 - 6z + 3$
5	$\dfrac{4(M-i-3)(M-i-2)(M-i-1)}{(M-4)(M-3)(M-2)(M-1)}$	1/5	$-4z^3 + 12z^2 - 12z + 4$

10.6 Shattering: A Phase Transition

Stability requires the partition function to be a concave function of its argument. If stability is violated, the system splits into multiple phases. To study phase splitting in fragmentation we must adopt some specific model for the kernel. As with gelation, a general family of kernels that are capable of exhibiting phase splitting are of the power-law type. First we recall Eq. (10.10)

$$\bar{b}(\mathbf{n}) = \frac{1}{N} \sum_{i=2}^{\infty} \sum_{j=1}^{i-1} n_i B_{i-j,j} = \frac{M-N}{N} \bar{B}(\mathbf{n}). \qquad [10.10]$$

For $M \gg N$, a condition that is satisfied in the thermodynamic limit, we have

$$\bar{b}(\mathbf{n}) \sim \left(\frac{M}{N}\right) \bar{B}(\mathbf{n}). \qquad (10.64)$$

For $B_{i,j} = 1$ we obtain $\bar{b} = M/N$. Suppose now that b_i has power-law dependence on i,

$$b_i = i^{\alpha}.$$

Then we have $\bar{b}(\mathbf{n}) \sim (M/N)^{\alpha}$ and

$$\bar{B}(\mathbf{n}) \sim \left(\frac{M}{N}\right)^{\alpha-1}. \qquad (10.65)$$

With $\alpha = 1$ this reverts to the case of the constant kernel, $B_{i,j} = 1$. We will now allow α to vary and study the stability of the system.

To the extent that the scaling between $\bar{B}$ and the mean cluster mass M/N holds,[7] all distributions of the cluster ensemble have the same mean kernel, therefore the same scaling applies to the ensemble average kernel,

$$\langle B\rangle_{M,N} \sim \left(\frac{M}{N}\right)^{\alpha-1}. \tag{10.66}$$

We can now calculate the partition function and its derivatives. Using Eq. (10.30) and (10.66) the partition function is

$$\Omega_{M,N} = \binom{M-1}{N-1}\left(\frac{M^{N-1}}{(N-1)!}\right)^{\alpha-1}. \tag{10.67}$$

The corresponding canonical variables are

$$\beta = \beta_0 + \alpha\left(\frac{N}{M}\right) = \log\left(\frac{M}{M-N}\right) + \alpha\left(\frac{N}{M}\right), \tag{10.68}$$

$$q = q_0\left(\frac{M}{N}\right)^{\alpha-1} = \left(\frac{M-N}{N}\right)\left(\frac{M}{N}\right)^{\alpha-1}. \tag{10.69}$$

where β_0 and q_0 are the corresponding parameters of the unbiased ensemble.

Now that we have the partition function of extensive distribution $\tilde{n}_i$, we will obtain the partition function of the scaled distribution $\phi(z)$ using the method discussed in Sect. 10.5. We write the partition function in the form

$$\begin{aligned}\log\omega(\bar{x}) &= \frac{\log\Omega_{M,N}}{N} = \frac{M}{N}\beta + \log q\\ &= -\frac{M}{N}\log\left(1-\frac{N}{M}\right) + \log\left(\frac{M}{N}-1\right) + \alpha + (\alpha-1)\log\frac{M}{N},\end{aligned}$$

and apply the condition $M \gg N$. The result is

$$\log\omega(\bar{x}) = \alpha\left(1 + \log\frac{1}{N} + \log M\right). \tag{10.70}$$

Comparing this result with Eq. (10.59) we make the identifications,

$$\log\omega^*(\bar{z}) = \alpha(1 + \log\bar{z}), \tag{10.71}$$

[7] This scaling presumes that the cluster distribution is concentrated in a relatively narrow range of cluster masses.

and $A(M) = (\alpha - 1)\log M$.[8] The corresponding canonical parameters are

$$\beta^* = \frac{d\log\omega^*}{d\bar{z}} = \frac{\alpha}{\bar{z}}, \tag{10.72}$$

$$\log q^* = \log\omega^* - \beta^*\bar{z} = \alpha\left(1 - \frac{1}{\bar{z}} + \log\bar{z}\right). \tag{10.73}$$

Equations (10.71), (10.72), and (10.73) summarize the thermodynamics of power-law kernels.

10.6.1 Stability of the Power-Law Kernel

We may study the stability of the power-law by application of the stability criteria but the easiest approach is to look directly at the curvature of the partition function. The log of $\bar{z}$ is concave function of $\bar{z}$ and is multiplied by α in Eq. (10.71). For $\alpha \geq 0$, the partition function is concave, thus stable. For $\alpha < 0$ it is convex at all $\bar{z}$, the system is therefore unstable and produces a mixture of two phases at *all* $\bar{z}$. What are these phases? It will be easier to examine this question by simulation. Figure 10.2 juxtaposes the scaled distribution $\phi(z)$ for $\alpha = 1/2$ (stable but near the border of instability) and $\alpha = -1$ (well in the unstable region) using $M = 50$, 100, and 200. When α is positive, large particles break up preferentially over smaller ones. Even though the process begins with a giant mass at $z = 1$, large particles are depleted immediately to produce a population that consists mostly of small particles. The distribution detaches from $z = 1$ and as fragmentation advances, the entire distribution shifts to smaller sizes to the left. Homogeneity is obeyed and is indicated by the collapse of all distributions on the same curve as M increases at constant N. Negative values of the exponent α promote the breakup of smaller particles over larger ones. This has two consequences: it generates a very large amount of very small particles, while permitting the existence of very large particles of order 1. The result is a distribution that stretches over the entire available space of cluster sizes. The fact that the rate of fragmentation increases with decreasing size results in a very high current of mass from the larger to the smaller sizes. This current hits the low end of the distribution as a wave and accumulates there, since monomers cannot break any further. In the rescaled coordinate z the size of the monomer is $z = 1/M$ and approaches zero as M increases. The accumulation of mass at the monomer appears as a separation of a sharp peak, a delta function at $z = 1/M$ that moves closer to zero as M increases. With the exception of this sharp peak, the rest of the distribution obeys homogeneity, as indicated by the collapse onto a single curve for all M, in this case a power-law distribution.

[8] Function $A(M)$ is of no particular interest once $\log\omega^*(\bar{z})$ has been determined.

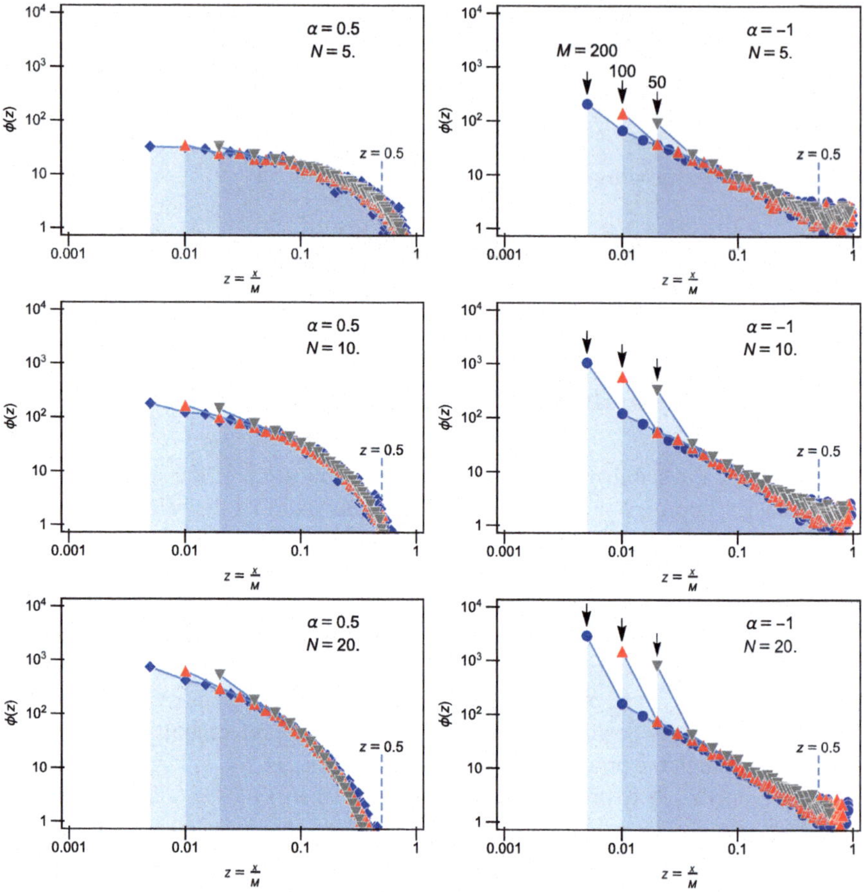

Fig. 10.2 Monte Carlo results of the scaled distribution $\phi(z)$ for $a = 1/2$ (left row) and $\alpha = -1$ (right row). The symbols correspond to $M = 50$ (*triangledown*), $M = 100$ (*triangle*), $M = 200$ (*opencircle*). Fragmentation advances from top ($N = 5$) to bottom ($N = 20$). The case $\alpha = -1$ is unstable. The instability is manifested through the appearance of a delta function at the smallest possible mass, $z = 1/M$ (arrows). This component of the distribution does not satisfy the homogeneity of the rest of the distribution and represents shattering, a phase transition that produces clusters of infinitesimal mass. The dashed line at $z = 0.5$ marks the sol/gel boundary

The distribution at $z = 0$ represents a new phase. When homogeneity cannot be maintained over the entire population under a single distribution, the system responds by splitting up into two distributions, each satisfying homogeneity on its own. One phase is the distribution of finite sizes, the other phase consists of particles of zero mass.[9] The formation of a phase of zero mass by fragmentation

[9]The physical impossibility of zero mass is a consequence of the passage to continuous space. In discrete finite systems the smallest mass in scaled coordinates is finite and equal to $1/M$.

has been called *shattering* in the literature and its connection to phase transitions has been discussed in qualitative terms.[10] We have seen here that shattering is a true phase transition whose emergence can be predicted by formal stability analysis of the partition function.

10.6.2 Shattering and Gelation

There are strong analogies between shattering and gelation, a point that was first made by McGrady and Ziff (1987), who also coined the term "shattering." In both cases a population of finite size clusters coexists with a delta function at an extreme size. In gelation, the delta function is located at infinity, in shattering at zero. This antisymmetric relationship between gelation and shattering is due to the different condition of homogeneity in each case. In aggregation the cluster distribution $f(x)$ attains homogeneous behavior in the limit

$$M \to \infty, \; N \to \infty, \; M/N = \text{fixed},$$

whereas in the case of fragmentation homogeneity is achieved by the scaled distribution $Mf(x/M)$ in the limit

$$M/N \to \infty, \; N = \text{fixed}.$$

Nonetheless, shattering and gelation are fundamentally and mathematically the *same* process. In Chap. 9 we obtained the canonical parameter q of the power-law aggregation kernel with exponent ν (Eq. 9.63), which we may express as

$$q = q = \theta(1-\theta)^{\nu-1}.$$

Compare this with the canonical q of the fragmentation power-law kernel from Eq. (10.68),

$$q = \left(\frac{M-N}{N}\right)\left(\frac{M}{N}\right)^{\alpha-1} = \theta(1-\theta)^{1-\alpha}.$$

The two expressions are identical if we set

$$\nu - 1 = 1 - \alpha.$$

[10] See McGrady and Ziff (1987), Singh and Hassan (1996), Ziff (1992), Ernst and Szamel (1993). These works study shattering by examining the kinetic equation for the mean distribution (our Eq. (10.44)) and the conditions under which it fails to preserve mass, a breakdown that identifies the presence of shattering.

This condition establishes opposite motion between the two exponents: increasing ν has the same effect on q as decreasing α, and vice versa. With $a = 0$, $\nu = 1$, both processes produce the unbiased solution. Instability arises when a very large current of mass emerges towards one of the two end zones of cluster sizes: the monomer (shattering), or the giant cluster. Recall that the region of the giant cluster is defined by the condition

$$\frac{M - N + 1}{2} < i < M - N + 1.$$

This region is relevant not only in gelation but in shattering as well. In stable fragmentation ($\alpha > 0$) the process begins with one giant cluster but the distribution moves quickly out of the gel region to become fully contained in the sol range. The sol range is defined by the condition

$$1 \le i < \frac{M - N + 1}{2},$$

which in terms of the rescaled size becomes

$$0 \le z < \frac{M - N + 1}{2M} \to 0.5.$$

Returning to Fig. 10.2, we confirm that in the case $\alpha = 0.5$ the distribution becomes confined in the region $z < 0.5$ except for the very early stages. For $\alpha = -1$ this region remains populated at all times. The system at $\alpha = 0.5$, though stable, is close to the border of instability and we notice weak features of shattering, such as presence of clusters in the gel region at early stages ($N < 20$) and a hint of a shattering phase at the low end of the distribution at $N = 20$ and higher. These features however do not persist: the gel region is quickly depleted as fragmentation advances, and the shattering phase dissolves as M increases.

The sustained presence of clusters in the gel region is directly linked to the presence of the shattering phase. The gel region may only contain a single cluster and this cluster is very likely to exit the region if it breaks up in two fairly large fragments. The gel region can remain populated at all times only if the probability of fragmentation of cluster mass $z \approx 0.5$ is much smaller compared to all smaller sizes. For power-law kernels this means negative exponent α. In light of this discussion, shattering, just like gelation, is associated with the presence of clusters in the gel region. In terms of the raw cluster size x, the gel region is relegated to $x = \infty$, but in terms of the scaled variable z it is brought into full view at the expense of rendering monomers invisible by relegating them to $z = 0$, where they carry no mass.

Appendix: Derivations

Equation (10.24)

Consider parent and offspring distributions for the fragmentation reaction

$$(i+j) \to (i) + (j).$$

Only the cluster masses involved in this reaction have different number of clusters in the parent and offspring distributions. Then we have

$$\frac{\mathbf{n}'!}{\mathbf{n}!} = \frac{(N-1)!}{N!}\frac{(n_i!)(n_j!)(n_{i+j}!)}{(n_i'!)(n_j'!)(n_{i+j}'!)} = \frac{1}{N}\frac{(n_i!)(n_j!)(n_{i+j}'-1)!}{(n_i-1!)(n_j'-1)!(n_{i+j}'!)} = \frac{1}{N}\frac{n_i n_j}{n_{i+j}'}$$

where we have used the parent–offspring relationships with $i \neq j$. Repeating with $i = j, n_i > 1$:

$$\frac{\mathbf{n}'!}{\mathbf{n}!} = \frac{1}{N}\frac{(n_i!)(n_{2i})!}{(n_i'!)(n_{2i}')!} = \frac{1}{N}\frac{(n_i!)(n_{2i}'-1)!}{(n_i-2)!(n_{2i}')!} = \frac{1}{N}\frac{n_i(n_i-1)}{n_{2i}'}$$

and we notice that we no longer need to specify $n_i > 1$ since $n_i = 1$ makes this ratio 0. The results can be summarized into a single expression as follows:

$$\frac{\mathbf{n}'!}{\mathbf{n}!} = \frac{1}{N}\frac{n_i(n_j-\delta_{ij})}{n_{i+j}'} \tag{10.74}$$

which is Eq. (10.24) in the text.

Equation (10.44)

We start with Eq. (10.42)

$$\langle n_k \rangle = \left\langle \sum_{i=2}^{\infty}\sum_{j=1}^{i-1} \mathcal{P}_{i-j,j}'(n_k' - \delta_{k,i} + \delta_{k,i-j} + \delta_{k,j}) \right\rangle$$

$$= \left\langle \sum_{i=2}^{\infty}\sum_{j=1}^{i-1} P_{i-j,j}' n_k' \right\rangle - \left\langle \sum_{i=2}^{\infty}\sum_{j=1}^{i-1} P_{i-j,j}' \delta_{i,k} \right\rangle$$

$$+ \left\langle \sum_{i=2}^{\infty}\sum_{j=1}^{i-1} P_{i-j,j}' \delta_{k,i-j} \right\rangle + \left\langle \sum_{i=2}^{\infty}\sum_{j=1}^{i-1} P_{i-j,j}' \delta_{k,j} \right\rangle \qquad [10.42]$$

and calculate each term on the right-hand side separately. For the first term we have

$$\left\langle \sum_{i=2}^{\infty} \sum_{j=1}^{i-1} P'_{i-j,j} n'_k \right\rangle = \left\langle n'_k \sum_{i=2}^{\infty} \sum_{j=1}^{i-1} P'_{i-j,j} \right\rangle = \langle n'_k \rangle$$

with the last result obtained by the normalization condition for $\mathcal{P}'_{i-j,j}$. The second term gives

$$-\left\langle \sum_{i=2}^{\infty} \sum_{j=1}^{i-1} P'_{i-j,j} \delta_{i,k} \right\rangle = -\left\langle \sum_{j=1}^{k-1} P'_{k-j,j} \right\rangle = -\left\langle \sum_{j=1}^{k-1} \frac{n'_k B_{k-j,j}}{(M-N+1)\bar{B}(\mathbf{n}')} \right\rangle.$$

The third and fourth terms are identical due to symmetry between all ordered pairs of fragments. The two terms combined are

$$\left\langle \sum_{i=2}^{\infty} \sum_{j=1}^{i-1} P'_{i-j,j} (\delta_{k,i-j} + \delta_{k,j}) \right\rangle = 2\left\langle \sum_{i=2}^{\infty} \sum_{j=1}^{i-1} P'_{i-j,j} \delta_{k,j} \right\rangle = 2\left\langle \sum_{i=k+1}^{\infty} P'_{i-k,k} \right\rangle$$

$$= 2\left\langle \sum_{i=k+1}^{\infty} \frac{n_i B_{i-k,k}}{(M-N+1)\bar{B}(\mathbf{n}')} \right\rangle$$

Incorporating these results into Eq. (10.42) we obtain

$$\langle n_k \rangle = \langle n'_k \rangle - \left\langle \sum_{j=1}^{k-1} \frac{n'_k B_{k-j,j}}{(M-N+1)\bar{B}\mathbf{n}'} \right\rangle + 2\left\langle \sum_{i=k+1}^{\infty} \frac{n_i B_{i-k,k}}{(M-N+1)\bar{B}(\mathbf{n}')} \right\rangle.$$

which leads directly to Eq. (10.44) in the main text.

List of Symbols

Acronyms

MPD	Most probable distribution
ThL	Thermodynamic limit

Symbols

$a_{M,N}$	Intensive function of M, N, used in derivations of the partition function in aggregation and fragmentation
$B_{i,j}$	Fragmentation kernel
b_i	Fragmentation rate (total fragmentation probability of size i)
$\bar{B}(\mathbf{n})$	Mean fragmentation kernel in distribution $\mathbf{n}$
$\bar{b}(\mathbf{n})$	Total fragmentation rate in distribution $\mathbf{n}$
$f(x)$	Most probable distribution in the continuous domain
f_i	Discretized form of most probable distribution
$h(x)$	Generic normalized probability distribution
$H(x)$	Generic extensive distribution
H_i	Discretized form of generic distribution
J	Generic functional
J_k	Functional that gives the k order moment of $h(x)$
$K_{i,j}$	Aggregation kernel
$\langle K \rangle_{M,N}$	Ensemble average aggregation kernel in (M, N) ensemble
$\bar{K}(\mathbf{n})$	Mean aggregation kernel in distribution $\mathbf{n}$
M	Mass of distribution
m_i	Cluster mass
$\mathbf{m}$	Configuration of cluster masses
$\mathbf{n}$	Cluster distribution in discrete domain
$\mathbf{n}!$	Multinomial coefficient of $\mathbf{n} = (n_1, n_2, \cdots)$
$\tilde{\mathbf{n}}$	Most probable distribution in discrete domain
$\langle \mathbf{n} \rangle$	Ensemble average distribution of clusters
$\mathbf{n}'$	Parent distribution of $\mathbf{n}$ in clustering and fragmentation

T. Matsoukas, *Generalized Statistical Thermodynamics*,
Understanding Complex Systems, https://doi.org/10.1007/978-3-030-04149-6

N	Number of clusters
p_i	Discrete probability distribution
q	Canonical parameter of most probable distribution
S	Entropy functional
$\mathcal{V}$	Total number of configurations in the ensemble (ensemble volume)
W	Selection functional
$w_{i\|\mathbf{n}}$	Cluster bias of cluster size i in distribution $\mathbf{n}$
$w(x; h)$	Cluster bias (continuous domain) of size x in distribution $h(x)$
$w(x; f)$	Cluster bias (continuous domain) of size x in the most probable distribution
$w(x; h)$	Cluster bias (continuous domain) of size x in distribution $h(x)$
$w(x, \bar{x})$	Equivalent notation to $w(x; f)$
w_i or $w(x)$	Linear cluster bias, discrete or continuous
$\tilde{w}_i$ or $\tilde{w}(x)$	Linear cluster bias evaluated at the most probable distribution, discrete or linear, respectively
x	Continuous size variable or generic independent variable
$\bar{x}$	Mean value of distribution
z	Rescaled size in fragmentation

Greek

β	Canonical parameter (generalized inverse temperature) of most probable distribution
θ	Progress coordinate in aggregation graph, $\theta = 1 - 1/\bar{x}$
ϱ	Thermodynamic functional
ρ_f	Linearized thermodynamic functional
Ω	Extensive partition function
ω	Intensive partition function

References

D.J. Aldous, Deterministic and stochastic models for coalescence (aggregation and coagulation): a review of the mean-field theory for probabilists. Bernoulli **5**(1), 3–48 (1999)

N. Berestycki, J. Pitman, Gibbs distributions for random partitions generated by a fragmentation process. J. Stat. Phys. **127**(2), 381–418 (2007). https://doi.org/10.1007/s10955-006-9261-1

M. Bóna, *A Walk Through Combinatorics - An Introduction to Enumeration and Graph Theory*, 2nd edn. (World Scientific, Singapore, 2006)

T.M. Cover, J.A. Thomas, *Elements of Information Theory*, 2nd edn. (Wiley, Hoboken, 2006)

C. Darwin, R. Fowler, LXXI. On the partition of energy.—Part II. Statistical principles and thermodynamics. Lond. Edinb. Dublin Philos. Mag. J. Sci. **44**(263), 823–842 (1922). https://doi.org/10.1080/14786441208562558

R.L. Drake, A mathematical survey of the coagulation equation, in *Topics in Current Aerosol Research*, vol. 3, ed. by C.M. Hidy, J.R. Brock (Pergamon Press, New York, 1972), pp. 204–379

R. Durrett, B. Granovsky, S. Gueron, The equilibrium behavior of reversible coagulation-fragmentation processes. J. Theor. Probab. **12**(2), 447–474 (1999). https://doi.org/10.1023/A:1021682212351

M.H. Ernst, G. Szamel, Fragmentation kinetics. J. Phys. A Math. Gen. **26**(22), 6085 (1993). URL http://stacks.iop.org/0305-4470/26/i=22/a=011

M.H. Ernst, R.M. Ziff, E.M. Hendriks, Coagulation processes with a phase transition. J. Colloid Interface Sci. **97**(1), 266–277 (1984). ISSN 0021-9797. https://doi.org/10.1016/0021-9797(84)90292-3. URL http://www.sciencedirect.com/science/article/B6WHR-4CX6Y6K-FJ/2/395b9786c48f68bc7a08c0946f79e89e

P.J. Flory, Molecular size distribution in three dimensional polymers. II. trifunctional branching units. J. Am. Chem. Soc. **63**(11), 3091–3096 (1941). https://doi.org/10.1021/ja01856a062. URL http://pubs.acs.org/doi/abs/10.1021/ja01856a062

I.M. Gelfand, S.V. Fromin, *Calculus of Variations* (Dover, Mineola, NY, 2000). (Reprint of the 1963 edition)

J.W. Gibbs, *Elementary Principles in Statistical Mechanics* (University Press, John Wilson and Son, Cambridge, USA, 1902). (Available online from https://archive.org/details/elementaryprinci00gibbrich)

E.M. Hendriks, J.L. Spouge, M. Eibl, M. Schreckenberg, Exact solutions for random coagulation processes. Zeitschrift für Physik B Condensed Matter **58**(3), 219–227 (1985). https://doi.org/10.1007/BF01309254. URL http://dx.doi.org/10.1007/BF01309254

T.L. Hill, *Statistical Mechanics Principles and Selected Applications* (Dover, Mineola, NY, 1987). (Reprint of the 1956 edition)

T. Matsoukas, *Generalized Statistical Thermodynamics*,
Understanding Complex Systems, https://doi.org/10.1007/978-3-030-04149-6

E. Jaynes, Prior probabilities. IEEE Trans. Syst. Sci. Cybern. **4**(3), 227–241 (1968). ISSN 0536-1567. https://doi.org/10.1109/TSSC.1968.300117
E.T. Jaynes, Information theory and statistical mechanics. Phys. Rev. **106**(4), 620–630 (1957). https://doi.org/10.1103/PhysRev.106.620
E.T. Jaynes, Gibbs vs Boltzmann entropies. Am. J. Phys. **33**(5), 391–398 (1965). https://doi.org/10.1119/1.1971557. URL http://link.aip.org/link/?AJP/33/391/1
E.T. Jaynes, *Papers on Probability, Statistics and Statistical Physics* (Kluwer Academic Publishers, 1983)
J.N. Kapur, *Maximum Entropy Methods in Science and Engineering* (Wiley Eastern Limited, Brisbane, 1989)
F.P. Kelly, *Reversibility and Stochastic Networks* (Cambridge, 2011). (Reprint of the 1979 edition by Wiley)
F. Leyvraz, Scaling theory and exactly solved models in the kinetics of irreversible aggregation. Physics Reports **383**(2–3), 95–212 (2003). https://doi.org/10.1016/S0370-1573(03)00241-2. URL http://dx.doi.org/10.1016/S0370-1573(03)00241-2
A.A. Lushnikov, Evolution of coagulating systems. J. Colloid Interface Sci. **45**(3), 549–556 (1973). URL http://www.sciencedirect.com/science/article/pii/0021979773901719
A.A. Lushnikov, Evolution of coagulating systems. ii. asymptotic size distributions and analytical properties of generating functions. J. Colloid Interface Sci. **48**(3), 400–409 (1974). URL http://www.sciencedirect.com/science/article/B6WHR-4CV7YC5-1H/2/ec1b7fbc574d183f73c0a3ba64a567ea
A.A. Lushnikov, Evolution of coagulating systems : III. coagulating mixtures. J. Colloid Interface Sci. **54**(1), 94–101 (1976). URL http://www.sciencedirect.com/science/article/B6WHR-4CV92N7-KB/2/10caea82d7ada3c813013e71748d27a3
A.A. Lushnikov, Coagulation in finite systems. J. Colloid Interface Sci. **65**(2), 276–285 (1978). URL http://dx.doi.org/10.1016/0021-9797(78)90158-3
A.A. Lushnikov, From sol to gel exactly. Phys. Rev. Lett. **93**, 198302 (2004). https://doi.org/10.1103/PhysRevLett.93.198302. URL http://link.aps.org/doi/10.1103/PhysRevLett.93.198302
A.A. Lushnikov, Exact kinetics of the sol-gel transition. Phys. Rev. E **71**, 046129 (2005). https://doi.org/10.1103/PhysRevE.71.046129
A.A. Lushnikov, Gelation in coagulating systems. Phys. D Nonlinear Phenomena **222**(1–2), 37–53 (2006a). https://doi.org/http://dx.doi.org/10.1016/j.physd.2006.08.002. URL http://www.sciencedirect.com/science/article/pii/S016727890600306X
A.A. Lushnikov, Sol-gel transition in a source-enhanced coagulating system. Phys. Rev. E **74**, 011103 (2006b). https://doi.org/10.1103/PhysRevE.74.011103. URL http://link.aps.org/doi/10.1103/PhysRevE.74.011103
A.A. Lushnikov, Exact kinetics of a coagulating system with the kernel k = 1. J. Phys. A Math. Theor. **44**(33), 335001 (2011). URL http://stacks.iop.org/1751-8121/44/i=33/a=335001
A.A. Lushnikov, Supersingular mass distributions in gelling systems. Phys. Rev. E **86**, 051139 (2012). https://doi.org/10.1103/PhysRevE.86.051139. URL http://link.aps.org/doi/10.1103/PhysRevE.86.051139
A.A. Lushnikov, Postcritical behavior of a gelling system. Phys. Rev. E **88**, 052120 (2013). https://doi.org/10.1103/PhysRevE.88.052120. URL http://link.aps.org/doi/10.1103/PhysRevE.88.052120
A. Marcus, Stochastic coalescence. Technometrics **10**(1), 133–143 (1968). URL http://www.jstor.org/stable/1266230
T. Matsoukas, Y. Lin, Fokker-Planck equation for particle growth by monomer attachment. Phys. Rev. E **74**(3), 031122 (2006). URL http://link.aps.org/abstract/PRE/v74/e031122
J.E. Mayer, M.G. Mayer, *Statistical Mechanics* (Wiley, New York, 1940)
E.D. McGrady, R.M. Ziff, "Shattering" transition in fragmentation. Phys. Rev. Lett. **58**(9), 892–895 (1987). URL http://journals.aps.org/prl/abstract/10.1103/PhysRevLett.58.892
I. Müller, *A History of Thermodynamics: The Doctrine of Energy and Entropy* (Springer, Berlin, Heidelberg, 2007). https://doi.org/10.1007/978-3-540-46227-9

J. Pitman, *Combinatorial Stochastic Processes*, vol. 1875 (Springer, Berlin, 2006)
Y.A. Rozanov, *Probability Theory: A concise course* (Dover Pubications, New York, NY, 1977)
E. Schrödinger, *Statistical Thermodynamics* (Dover Publications, 1989). (Reprint of the 2nd edition originally published by Cambridge University Press, 1952, under the subtitle A Course Seminar Lectures Delivered in January-March 1944 at the School of Theoretical Physics, Dublin Institute for Advanced Studies)
C.E. Shannon, A mathematical theory of communication. Bell Syst. Tech. J. **27**(3), 379–423 (1948)
P. Singh, M.K. Hassan, Kinetics of multidimensional fragmentation. Phys. Rev. E **53**, 3134–3144 (1996). https://doi.org/10.1103/PhysRevE.53.3134. URL http://link.aps.org/doi/10.1103/PhysRevE.53.3134
M. Smith, T. Matsoukas, Constant-number Monte Carlo simulation of population balances. Chem. Eng. Sci. **53**(9), 1777–1786 (1998). ISSN 0009-2509. https://doi.org/10.1016/S0009-2509(98)00045-1. URL http://www.sciencedirect.com/science/article/pii/S0009250998000451
J. Spencer, The giant component: The golden anniversary. Not. AMS **57**(6), 720–724 (2010)
J.L. Spouge, Solutions and critical times for the monodisperse coagulation equation when $a_{ij} = A + B(i + j) + Cij$. J. Phys. A Math. Gen. **16**(4), 767–774 (1983a). URL http://www.iop.org/EJ/abstract/0305-4470/16/4/014
J.L. Spouge, The size distribution for the $a_g rb_{f-g}$ model of polymerization. J. Stat. Phys. **31**, 363–378 (1983b). ISSN 0022-4715. https://doi.org/10.1007/BF01011587. URL http://dx.doi.org/10.1007/BF01011587
J.L. Spouge, Equilibrium polymer size distributions. Macromolecules **16**(1), 121–127 (1983c). URL http://pubs.acs.org/doi/abs/10.1021/ma00235a024
J.L. Spouge, Asymmetric bonding of identical units: A general $a_g rb_{f-g}$ polymer model. Macromolecules **16**(5), 831–838 (1983d). URL http://pubs.acs.org/doi/abs/10.1021/ma00239a021
J.L. Spouge, Monte Carlo results for random coagulation. J. Colloid Interface Sci. **107**(1), 38–43 (1985a). URL http://www.sciencedirect.com/science/article/B6WHR-4CX6YFY-HD/2/e615b0fedacbd185beaa2dc9a0be60b7
J.L. Spouge, Analytic solutions to Smoluchowski's coagulation equation: a combinatorial interpretation. J. Phys. A Math. Gen. **18**(15), 3063 (1985b). URL http://stacks.iop.org/0305-4470/18/i=15/a=028
W.H. Stockmayer, Theory of molecular size distribution and gel formation in branched-chain polymers. J. Chem. Phys. **11**(2), 45–55 (1943). https://doi.org/10.1063/1.1723803. URL http://link.aip.org/link/?JCP/11/45/1
H. Touchette, The large deviation approach to statistical mechanics. Phys. Rep. **478**(1), 1–69 (2009). ISSN 0370-1573. https://doi.org/10.1016/j.physrep.2009.05.002. URL http://www.sciencedirect.com/science/article/pii/S0370157309001410
M. Tribus, E.C. McIrvine, Energy and information. Scientific American **225**(3), 179–190 (1971). ISSN 00368733, 19467087. URL http://www.jstor.org/stable/24923125
R.M. Ziff, An explicit solution to a discrete fragmentation model. J. Phys. A Math. Gen. **25**(9), 2569 (1992). URL http://stacks.iop.org/0305-4470/25/i=9/a=027
R.M. Ziff, G. Stell, Kinetics of polymer gelation. J. Chem. Phys. **73**(7), 3492–3499 (1980). https://doi.org/10.1063/1.440502. URL http://link.aip.org/link/?JCP/73/3492/1
R.M. Ziff, E.M. Hendriks, M.H. Ernst, Critical properties for gelation: A kinetic approach. Phys. Rev. Lett. **49**, 593–595 (1982). https://doi.org/10.1103/PhysRevLett.49.593. URL http://link.aps.org/doi/10.1103/PhysRevLett.49.593
R.M. Ziff, M.H. Ernst, E.M. Hendriks, Kinetics of gelation and universality. J. Phys. A Math. Gen. **16**(10), 2293–2320 (1983). URL http://stacks.iop.org/0305-4470/16/2293

Index

A

Aggregation kernel
  cluster population at time t, 272
  colloidal spheres of radius r, 259
  continuous domain, 283–287
  diffusion coefficient D_i, 259
  in distribution n, 245
  merging probability, 244
  Spouge calculation, 279

B

Bayesian inference, 197
Biased sampling, 199–201
Bicomponent ensemble
  bicomponent distribution, 165–166
  cluster and configuration, 164–165
  color-blind size distribution, 166–167
  distinct species, 163
  mean cluster size, 164
  microcanonical probability
    bicomponent entropy, 172–174
    color-blind clusters, 171
    multiplicity of distribution, 171
    permutations, 170
  MPD, 176–179
  nonrandom mixing
    cluster function, 188
    composition-dependent bias, 188
    composition-independent bias, 189
    entropy of mixing, 191–192
    homogeneity condition, 188
    mean cluster size, 189
    microcanonical weight, 189
    phase splitting, 194–195
    scaling, 192–193
  probability distribution, 168–170
  random mixing
    cluster function, 183
    cluster size MPD, 184
    color-blind distribution, 186
    entropy of, 186–188
    familiar canonical MPD, 184
    Lagrange multipliers, 185
    microcanonical weight, 184
  selection functional, 175–176
  sieve-cut distribution, 167–168
  sieve-cut ensemble
    binomial distribution, 181
    Lagrange multipliers, 180
    one-component entropy, 181–183
  total mass, 163
Bicomponent entropy, 172–174
Bilinear kernel, 312
Binary aggregation, *see* Irreversible clustering
Binary clustering graph, 242–243
Binary exchange reaction
  equilibrium constant, 39, 41
  exchange reactions, 42–43
  Markov process, 40
  metropolis method, 40
  transition probability, 40
Binary fragmentation
  average fragmentation kernel, 327
  distribution of fragments, 328
  double summation, 327
  fragmentation kernel, 325–326
  mean distribution, 335–336

T. Matsoukas, *Generalized Statistical Thermodynamics*,
Understanding Complex Systems, https://doi.org/10.1007/978-3-030-04149-6

Binary fragmentation (*cont.*)
mean rate of, 327
ordered fragmentation events, 326
parent–offspring relationship, 328–329, 349–350
partition function
canonical parameters, 333
microcanonical probability, 330–332
thermodynamic consistency, 332
shattering
canonical variables, 342–343
gelation, 345–346
power-law, 341, 343–345
scaled distribution, 342
thermodynamic limit
aggregation/fragmentation graph, 337
application to $B_{i,j} = 1$, 339–341
cluster distribution, 337
defined, 336
entropy functional, 338
maximization, 338–339
transition probabilities, 329–330
unbiased ensemble, 333–334
Brownian aggregation, 259

C
Canonical configuration
microcanonical ensemble μC, 54
microcanonical slices, 53
multiplicity of, 53
N–N' clusters, 52
reconstruction, 55–57
total mass M', 52
Canonical ensemble
average cluster mass, 52
canonical slices, 49
cluster mass, 50
configuration (*see* Canonical configuration)
microcanonical table, 49–50
microcanonical volume, 51
multiplicity of cluster, 51
probability, 58–61
same cluster masses, 51
$V_{M,N}$ elements, 49
Canonical fluctuations, 79–80
Canonical probability, 58–61, 202–203
canonical MPD, 77–79
fluctuations of, 79–80
homogeneous sample, 76
larger microcanonical ensemble, 74
microcanonical ensemble, 75
microcanonical partition function, 75
N clusters, 76
partition function, 76–77
Canonical sampling, 202–203, 219
Classical entropy
attributes of, 2
contact with physics, 3–4
differential of entropy, 3
equilibrium state, 2
Euler's theorem, 2–3
Cluster ensemble, xi–xiv
binary exchange reaction
equilibrium constant, 39, 41
exchange reactions, 42–43
Markov process, 40
metropolis method, 40
transition probability, 40
canonical
average cluster mass, 52
canonical slices, 49
cluster mass, 50
configuration (*see* Canonical configuration)
microcanonical table, 49–50
microcanonical volume, 51
multiplicity of cluster, 51
probability, 58–61
same cluster masses, 51
$V_{M,N}$ elements, 49
definition, 23
entropy, 26–30, 38–39
linear, 43–46
M'-canonical, 23–26, 61–64
multinomial distribution
Euler equation, 35
probability of distribution n, 36
weight factor W(n), 37
multiplicity, 26–30
selection functional
cluster bias, 32
cluster function, 32
cluster weight, 32
Euler form, 32
Gibbs-Duhem equation, 32
linear selection functional, 33–35
microcanonical partition function, 31
microcanonical weight, 31
probability of configuration, 32–33
selection bias, 31
unbiased functional, 46–48
Conditional probabilities, 7–9
Continuous distribution, 198–199
Continuous domain
aggregation kernel, 283–287

β and log q, 276–277
Gibbs-Duhem equation, 276
homogeneity condition, 275
mean size, 275
Sterling approximation, 283–284
ThL, 277–278
unbiased ensemble, 283–287
Continuous limit
definition, 99
exponential distribution, 119–120
Gaussian distribution, 121–122
inverse problem, 116–119
limits of integration, 100
nonlinear ensembles
cluster function, 111
"entropic" selection functional, 110
entropic *vs.* unbiased ensemble, 112
Laplace-transform pair, 113–114
microcanonical probability of distribution, 110
non-uniqueness, 114–115
number of monomers, fluctuation in, 112
standard Lagrange maximization, 111
power-law cluster bias
Laplace transform, 107, 108
Legendre transformation, 108
microcanonical partition function, 109
microcanonical probability, 107
Monte Carlo sampling, 107
properties in, 100–102
unbiased ensemble, 102–106
uniform distribution, 123
Weibull distribution, 122–123

D
Degree of homogeneity, 68
Discrete probability distribution, 264–265

E
Entropy
classical
attributes of, 2
contact with physics, 3–4
differential of entropy, 3
equilibrium state, 2
Euler's theorem, 2–3
continuous distribution, 205
discrete approximation, 206–208
Gibbs entropy
equilibrium distribution, 5–7
Lagrange multipliers, 5
microstate, 4–5
Newtonian mechanics, 4
partitioning, 5
Stirling approximation, 5
Jaynes entropy
maximum uncertainty, 11
Newtonian mechanics, 12
normalization condition, 11–12
priori probabilities, 13
unknown distribution, 12
mathematical calculus, 15
curvature, 16
entropy equation, 20
homogeneity, 16–17
Legendre transformation, 18–19
second law, 19
stability criteria, 21
thermodynamic potentials, 20–21
maximum entropy distribution, 13–15
multinomial coefficient, 206–207
partition function, 208
Renyi entropy, 15
Shannon entropy, uncertainty functional
additivity, 7–9
continuity, 7
monotonicity, 7, 9
partitioning, 9–10
reversibility, 10–11
Entropy functional, viii
Entropy inequality, 28–30
Euler's theorem, 27, 72, 91
Exponential distribution, 119–120

F
Fragmentation, *see* Binary fragmentation

G
Gaussian distribution, 121–122
Gaussian function, 67
Gelation, *see* Kinetic gelation
Generalized canonical distribution, x, xi
Generalized cluster ensemble
canonical family, 95–97
cluster bias, 90
entropy of distribution, 90
extensive variables, 88

Generalized cluster ensemble (*cont.*)
generalized canonical ensemble, 93–95
Lagrange maximization generalized MPD, 91–93
microcanonical ensemble, 89
multidimensional vector, 89
multinomial coefficient, 90
single extensive attribute, 88
total number of members, 89
Generalized microcanonical cluster ensemble, 96
Generalized statistical thermodynamics, vii, viii
Generalized thermodynamics
biased sampling, 199–201
calculus of variations, continuous functionals (*see* Variational calculus)
canonical relationships, 219–220
canonical sampling, 202–203, 219
curvature of
concave function, 208–210
convex function, 211
stability analysis, cluster ensemble, 210–212
entropy
continuous distribution, 205
discrete approximation, 206–207
multinomial coefficient, 207–208
partition function, 208
linearized selection functional
Euler equation, 213
inverse problem, 214–217
microcanonical probability functional, 213–214
MEP
bicomponent distribution, 222
examples, 222
Gaussian distribution, 222
improper prior, 223
indeterminacy, 221
inference *vs.* deduction, 225–227
integral constraints, 222
invariant measure, 223–225
proper prior, 223–225
uninformed prior, 223–225
microcanonical sampling, 203–205, 219
random sampling
continuous distribution, 198–199
frequency distribution, 198
Gibbs inequality, 199
Kullback-Leibler divergence, 198
normalization, 197–198
probability space of distributions, 199
sampling h_0, 199
Stirling's formula, 198
selection functional W
binary clustering, 227
canonical probabilities, 228
irreversible process, 227
Markov chain, 227–228
parent-offering transition, 229
partition function, 228
stochastic process, 227
thermodynamic consistency, 229
transition probabilities, 228
statistical mechanics, 220–221
variational principle, 218
Giant cluster, 125, 129–132
Gibbs-Duhem equation, 21, 32, 73, 101
Gibbs entropy
equilibrium distribution, 5–7
Lagrange multipliers, 5
microstate, 4–5
Newtonian mechanics, 4
partitioning, 5
Stirling approximation, 5
Gibbs function, 20–21
Gibbs inequality, 199
Grand canonical, 64

H
Homogeneous–concave inequality, 30
Homogeneous properties, 68

I
i^2 model
linear and cluster function, 151
scaling limit, 160–161
sol-gel equilibrium, 151, 152
sol-gel point (*see* Sol-gel point)
tie line calculation, 152–155
Inference, 11
Intensive microcanonical partition function, 101
Invariant measure, 223
continuous entropy functional, 224
homogeneity, 224
Jaynes's continuous entropy, 224
microcanonical functional, 224
Shannon's discrete entropy, 224
three related functionals, 225
Inverse problem
canonical functions, 214–215
exponential distribution, 217
indeterminacy, 215–216

inequality, 215
negative Kullback-Leibler divergence, 215
Irreversible clustering
aggregation kernel, 241
binary clustering graph, 242–243
canonical parameters, 253–254
continuous domain
aggregation kernel, 283–287
β and log q, 276–277
Gibbs-Duhem equation, 276
homogeneity condition, 275
mean size, 275
Stirling approximation, 283–284
ThL, 276–277
unbiased ensemble, 283–287
double summation, 281–282
linear ensemble
average kernel, 256
in closed form, 254
constant kernel, 257–259
Kronecker delta, 256
log W, 255–256
mean kernel, 256–257
scaling limit, 266–268
sum kernel, 263–265
thermodynamic consistency, 260–262
mathematical literature, 243, 278–280
mean distribution, 268–270
parent–offspring relationship
cluster masses, 246–247
monomers, 246
phase space of distributions, 248–250
propagation of probability, 250–251
summation over parents, 251, 280–281
systematic calculation, 247–249
partition function, 252–253, 280
Smoluchowski theory
cluster distribution, 271–272
constant kernel, 273
distribution of clusters, 270–271
mass conservation, 271
number concentration, 272
parent ensemble, 282–283
particle concentration, 271
sum kernel, 274
time-free form, 272
transition probabilities, 244–246

J
Jaynes entropy
maximum uncertainty, 11
Newtonian mechanics, 12
normalization condition, 11–12
priori probabilities, 13
unknown distribution, 12

K
K-canonical ensemble, 95
K-dimensional vector, 92
Kinetic gelation
infinite series, 321–323
linear ensemble
cluster bias, 304
cluster size, 305
gel point, 306–307
isothermal condition, 304
nonlinear nature, 307
normalizations, 304–305
quasilinear systems, 307
temperature of, 305–306
locus of gel points, 321
Monte Carlo simulation
cluster distribution, 299–302
constant-volume, 299
mean aggregation kernel, 302–303
mean distribution, 302
power-law kernels
canonical parameters, 314
cluster functions, 314
gel fraction, 314–315
Monte Carlo simulation, 315–318
parameters, 319–321
quasilinear, 313
stability, 314
product kernel
partition function, 290–291
phase diagram, 295–298
quasilinear ensemble, 290, 292–293
recursion for, 291–292
stability condition, 293–295
Smoluchowski equation, 319
Stockmayer's model
aggregation kernel, 310
aggregation problem, 311
convergence, 310
equilibrium constant, 309
Flory's model, 308
irreversible aggregation, 310
irreversible reaction, 309
microcanonical weight, 308–309
monomers and functional sites, 308–309
product kernel, 312
statistical mechanics, 311–312
thermodynamic connection, 311
Kullback-Leibler divergence, 198–199

L

Lagrange multipliers, 177
Laplace transform, 101
Legendre transformation, 93, 95
Linear ensembles
  average kernel, 256
  in closed form, 254
  cluster bias, 304
  cluster size, 305
  constant kernel, 257–259
  continuous domain, 102
  evolution of distribution, 147–148
  gel point, 143, 306–307
  isothermal condition, 304
  Kronecker delta, 256
  log W, 255–256
  mean cluster sizes, 136, 137, 144
  mean kernel, 256–257
  mean size, 144
  nonlinear nature, 307
  normalizations, 304–305
  partition function and cluster bias, 43–46
  phase space, 71
  power-law cluster function, 136
  quasilinear systems, 307
  scaling limit, 140–143, 266–268
  scaling theory *vs.* Monte Carlo simulation, 146–147
  selection functional, 71
  sum kernel, 263–265
  temperature of, 305–306
  thermodynamic consistency, 260–262
  ThL, 70
  tie line, 138–140
Lushnikov's approach, 278–279

M

Marcus-Lushnikov process, 278
Mass, 88, 163
Mathematical calculus, 15
  curvature, 16
  entropy equation, 20
  homogeneity, 16–17
  Legendre transformation, 18–19
  second law, 19
  stability criteria, 21
  thermodynamic potentials, 20–21
Maximum entropy distribution, viii, ix
Maximum entropy principle (MEP)
  bicomponent distribution, 224
  examples, 224
  Gaussian distribution, 224
  improper prior, 223
  indeterminacy, 221
  inference *vs.* deduction, 225–227
  integral constraints, 222
  invariant measure, 223–225
  proper prior, 223
  uninformed prior, 223
Maximum uncertainty, 11
M'-canonical ensemble, 61–64
M-canonical probability
  bicomponent entropy, 172–174
  color-blind clusters, 171
  mass *vs.* fluctuations, 83–84
  M-canonical MPD, 85–86
  mean cluster size, 82
  microcanonical ensemble, 80
  microcanonical weights, 81
  multinomial interpretation, 81
  multiplicity of distribution, 171
  normalization constant, 81
  partition function, 84–85
  permutations, 170
Mean cluster distribution, 74
Microcanonical probability functional, 203–205
Microcanonical sampling, 203–205, 219
Microcanonical table, 26
Microcanonical volume, 25
Microcanonical weight, 66, 176
Monte Carlo simulation
  cluster distribution, 299–302
  constant-volume, 299
  mean aggregation kernel, 302–303
  mean distribution, 302
  power-law kernels, 315–318
Most probable distribution (MPD), 70
  continuous domain, 275–276
  continuous limit (*see* Continuous limit)
  inverse problem, 215–216
  Lagrange multipliers, 69
  linear ensemble
    average kernel, 256
    in closed form, 254
    constant kernel, 257–259
    Kronecker delta, 256
    log W, 255–256
    mean kernel, 256–257
    phase space, 71
    scaling limit, 266–268
    selection functional, 71
    sum kernel, 263–265
    thermodynamic consistency, 260–262
    ThL, 70
  microcanonical weight, 68

stability analysis, cluster ensemble, 210–212
Stirling approximation, 69
Multinomial coefficient, ix

N
Natural multiplicity, 24
Nonlinear ensembles
cluster function, 111
"entropic" selection functional, 110
entropic *vs.* unbiased ensemble, 112
Laplace-transform pair, 113–114
microcanonical probability of distribution, 110
non-uniqueness, 114–115
number of monomers, fluctuation in, 112
standard Lagrange maximization, 111
Nonrandom mixing, bicomponent ensemble
cluster function, 188
composition-dependent bias, 188
composition-independent bias, 189
entropy of mixing, 191–192
homogeneity condition, 188
mean cluster size, 189
microcanonical weight, 189
phase splitting, 194–195
scaling, 192–193

P
Phase transition
canonical variables, 342–343
equilibrium phase, 128–129
fluctuations, 147
gelation, 345–346
gel cluster, 148–149
giant cluster, 125, 129–132
i^2 model
linear and cluster function, 151
scaling limit, 160–161
sol-gel equilibrium, 151, 152
sol-gel point (*see* Sol-gel point)
tie line calculation, 152–155
linear ensemble
evolution of distribution, 147–148
gel point, 143
mean cluster, 136, 137, 144
mean size, 144
power-law cluster function, 136
scaling limit, 140–143
scaling theory *vs.* Monte Carlo simulation, 146–147
tie line, 138–140
order parameter, 148–151
partition function
composite ensemble, 126
functions of $\bar{x} = M/N$, 127
homogeneous copies, 126
intensive properties, 126
microcanonical weight, 125
polymer gelation, 125
power-law, 341, 343–345
scaled distribution, 342
shattering in fragmentation, 125
sol-gel equilibrium, 133–135
Physical model, x
Power-law cluster bias
Laplace transform, 107, 108
Legendre transformation, 108
microcanonical partition function, 109
microcanonical probability, 107
Monte Carlo sampling, 107
Power-law kernels
canonical parameters, 314
cluster functions, 314
gel fraction, 314–315
Monte Carlo simulation, 315–318
parameters, 319–321
quasilinear, 313
stability, 314
stability of, 343–345
thermodynamic limit, 341
Prior probability distribution, xi
Probability distribution, vii, viii, 168–170, 197, 243

R
Random mixing, bicomponent ensemble
cluster function, 183
cluster size MPD, 184
color-blind distribution, 186
entropy of, 186–188
familiar canonical MPD, 184
Lagrange multipliers, 185
microcanonical weight, 184
Random sampling
continuous distribution, 198–199
frequency distribution, 198
Gibbs inequality, 199
Kullback-Leibler divergence, 198
normalization, 197–198
probability space of distributions, 199
sampling h_0, 199
Stirling's formula, 198
Relative entropy, 198–199
Renyi entropy, 15

S
Shannon entropy
additivity, 7–9
continuity, 7
monotonicity, 7, 9
partitioning, 9–10
reversibility, 10–11
Shattering
canonical variables, 342–343
gelation, 345–346
power-law
stability of, 343–345
thermodynamic limit, 341
scaled distribution, 342
Sieve-cut ensemble
binomial distribution, 181
Lagrange multipliers, 180
one-component entropy, 181–183
Simply MaxEnt, *see* Maximum entropy principle
Smoluchowski coagulation equation
cluster distribution, 271–272
constant kernel, 273
distribution of clusters, 270–271
mass conservation, 271
mathematical literature, 243, 278–280
number concentration, 272
parent ensemble, 282–283
particle concentration, 271
sum kernel, 263, 274
time-free form, 272
Sol-gel equilibrium, 133–135, 151, 152
Sol-gel point
maximum cluster size, 157
mean cluster, 155
minimum gel size, 159
minimum possible gel cluster, 159
Monte Carlo simulation, 158
stability conditions, 157
Statistical mechanics, 220–221
Stirling's asymptotic formula, 103
Stockmayer's model
aggregation kernel, 310
aggregation problem, 311
convergence, 310
equilibrium constant, 309
Flory's model, 308
irreversible aggregation, 310
irreversible reaction, 309
microcanonical weight, 308–309
monomers and functional sites, 308–309
product kernel, 312
statistical mechanics, 311–312
thermodynamic connection, 311
Stockmayer's treatment of polymerization, 279
Stokes-Einstein equation, 259

T
Thermodynamic consistency
linear ensemble, 260–262
partition function, 252–253
selection functional W, 229
Thermodynamic entropy, 220–221
Thermodynamic limit (ThL), 65–68
canonical probability
canonical MPD, 77–79
canonical partition function, 76–77
fluctuations of, 79–80
homogeneous sample, 76
larger microcanonical ensemble, 74
microcanonical ensemble, 75
microcanonical partition function, 75
N clusters, 76
continuous domain, 275–276
convergence of ensembles, 86–87
definition, 65
generalized cluster ensemble
canonical family, 95–97
cluster bias, 90
entropy of distribution, 90
extensive variables, 88
generalized canonical ensemble, 93–95
Lagrange maximization generalized MPD, 91–93
microcanonical ensemble, 89
multidimensional vector, 89
multinomial coefficient, 90
single extensive attribute, 88
total number of members, 88
homogeneity in
aggregation/fragmentation graph, 337
application to $B_{i,j} = 1$, 339–341
cluster distribution, 337
defined, 336
entropy functional, 338
maximization, 338–339
M-canonical probability
mass *vs.* fluctuations, 83–84
M-canonical MPD, 85–86
mean cluster size, 82
microcanonical ensemble, 80
microcanonical weights, 81
multinomial interpretation, 81
normalization constant, 81
partition function, 84–85
microcanonical equations, 72–74
microcanonical surface, 87–88

MPD, 70
Lagrange multipliers, 69
linearized ensemble, 70–71
microcanonical weight, 68
Stirling approximation, 69
phase space, 248–250
power-law kernels, 319–321
quasistatic process, 87–88
Tie line, 138–140
Transition probabilities, 329–330

U
Unbiased ensemble, 46–49, 105, 178–179
Uniform distribution, 123

V
Variational calculus
entropy functional, 236–238
Euler's theorem for homogeneous functions, 234
examples, 230–231
functional derivative, 233
Gibbs-Duhem equation, 234–235
maximization, 238–239
types of functionals, 230
variation of functionals, 231–232

W
Weibull distribution, 122–123

Printed by Printforce, the Netherlands